Detection of SARS-CoV-2 Neutralizing Antibodies and Vaccine Development

Detection of SARS-CoV-2 Neutralizing Antibodies and Vaccine Development

Guest Editor

Jianhui Nie

MDPI

Basel • Beijing • Wuhan • Barcelona • Belgrade • Novi Sad • Cluj • Manchester

Guest Editor
Jianhui Nie
Division of HIV/AIDS and
Sex-Transmitted
Virus Vaccines
National Institutes for Food
and Drug Control
Beijing
China

Editorial Office
MDPI AG
Grosspeteranlage 5
4052 Basel, Switzerland

This is a reprint of the Special Issue, published open access by the journal *Vaccines* (ISSN 2076-393X), freely accessible at: https://www.mdpi.com/journal/vaccines/special_issues/2JK78332PC.

For citation purposes, cite each article independently as indicated on the article page online and as indicated below:

Lastname, A.A.; Lastname, B.B. Article Title. *Journal Name* **Year**, *Volume Number*, Page Range.

ISBN 978-3-7258-6534-5 (Hbk)
ISBN 978-3-7258-6535-2 (PDF)
https://doi.org/10.3390/books978-3-7258-6535-2

Contents

vaccines

MDPI

Editorial

SARS-CoV-2-Neutralizing Antibodies

Yawen Liu and Jianhui Nie *

Division of HIV/AIDS and Sex-Transmitted Virus Vaccines, National Institutes for Food and Drug Control (NIFDC), State Key Laboratory of Drug Regulatory Science, NHC Key Laboratory of Research on Quality and Standardization of Biotech Products, NMPA Key Laboratory for Quality Research and Evaluation of Biological Products, Beijing 102629, China; yawenliu13@163.com

* Correspondence: niejianhui@nifdc.org.cn

The COVID-19 pandemic, triggered by the SARS-CoV-2 virus, has profoundly and permanently affected many aspects of the world. The COVID-19 vaccination is the most critical strategy to induce a protective immune response and may be the only way to prevent the spread of infection and progression to severe disease and death. The types of COVID-19 vaccines mainly include inactivated vaccines, mRNA vaccines, adenovirus vector vaccines, and protein subunit vaccines [1]. These vaccines are highly effective at preventing severe illness, hospitalization, and death. Neutralizing antibody levels are important indicators of vaccine effectiveness [2]. This Special Issue features the latest scientific and technological advancements in the field of SARS-CoV-2 neutralization assay and vaccine development. It outlines new techniques for neutralization assays and highlights the critical importance of neutralizing antibody evaluation for vaccine and antiviral development.

Vaccine-induced protection depends on a broad humoral immune response and a strong cellular immune response. These responses work together to effectively prevent infection caused by pathogens. The COVID-19 vaccines, which encode the spike (S) protein of SARS-CoV-2, can broadly induce anti-spike antibodies and neutralizing antibodies (nAbs) [3,4]. The humoral immune response plays a key role in antiviral defense. Specifically, the production of specific antibodies in response to SARS-CoV-2 antigens controls viral replication through neutralization [5].

NAbs play a vital role in the prevention and control of COVID-19 infection, providing immune protection to individuals by blocking the virus from entering host cells, reducing viral transmission, enhancing immune clearance, preventing reinfection, and influencing the course of the disease. The following are some common methods for detecting neutralizing antibodies against COVID-19: (1) The first method is virus neutralization tests (VNT), including the classical virus neutralization test (cVNT), and plaque reduction neutralization test (PRNT) [6–8]. (2) The second is the surrogate virus neutralization test (sVNT), a protein-based binding assay for detecting neutralizing antibodies (nAbs), offering a simpler and faster alternative. The PremaLabs Diagnostics COVID-19 rapid nAbs test enables large-scale testing. The evaluation of this test kit is based on the use of the international standard for anti-SARS-CoV-2 immunoglobulin (NIBSC 20/136) [9]. In addition, the commercial sVNT microarray quantifies the distinct characteristics of variant-specific neutralizing antibodies for different SARS-CoV-2 variants and Omicron subvariants [10]. The main target antigen of inactivated and recombinant COVID-19 vaccines is the spike (S) protein, which is typically quantified using methods such as ELISA [11,12]. The ELISA method based on Isotope Dilution Tandem Mass Spectrometry (IDMS) can accurately quantify the S protein content of the prototype, Delta, and Omicron strains, enabling quality control for variant-based vaccines and multivalent vaccines against variants. (3) The third method is the Pseudovirus Neutralization Test (pVNT), (4) the fourth is Lateral Flow Assays (LFA), and (5) the fifth is Chemiluminescent Immunoassays (CLIA). The pseudovirus-based neutralization assay quantifies the neutralizing activity of human serum through some reporter genes. This method can also be used to develop antiviral drug molecules and to assess vaccine efficacy [13].

The ongoing evolution and mutation of SARS-CoV-2 underscore the necessity for more potent vaccines or novel COVID-19 vaccination strategies.

Identifying the precise binding epitope of potent neutralizing antibodies is crucial for rational vaccine design and the development of antibody-based therapies. By comparing the differences in sera neutralizability among individuals with acquired immunity, infection-induced humoral immunity appears to establish a more mature and finely selected antibody repertoire. The application of peptide microarray technology to analyze B cell epitopes may facilitate the development and implementation of peptide-based vaccines [14–16].

It was proven that heterologous vaccination induces more effective immunogenicity and a longer-lasting antibody response [17]. Based on clinical research results, compared to the homologous vaccination regimen, a booster dose of the mRNA vaccine BNT162b2 after receiving the viral vector vaccine ChAdOx1 induces higher levels of binding and neutralizing antibodies.

Neutralizing antibody testing plays a critical role in research on drug treatments for COVID-19, and is of significant importance in drug development, efficacy evaluation, and monitoring immune responses. In order to effectively respond to the ongoing mutations of the SARS-CoV-2 virus, some novel molecules (molnupiravir, baricitinib, sotrovimab, novel Mabs, etc.) [18] and repurposed drugs (dexamethasone, naproxen, remdesivir, hydroxychloroquine, etc.) have been approved for use [19]. Due to the lack of sufficient clinical data, the efficacy and safety of these novel molecules and repurposed drugs have not been determined.

Adjuvants are defined as various components that enhance the immunogenicity of vaccines when administered in conjunction with vaccine antigens. They include synthetic small-molecule compounds, complex natural extracts, and particulate materials. Clinical trials have shown that they help enhance the intensity, breadth, and durability of the immune response [20,21]. Xu and colleagues have compared the immune effects of inactivated vaccines with different adjuvants in mice. They found that the vaccine group with adjuvants induced higher levels of neutralizing antibodies compared to the group without adjuvants. Furthermore, the levels of neutralizing antibodies and binding antibodies in the adjuvant group showed significant differences, suggesting that adjuvants may also have different mechanisms of action and immunological characteristics [22].

The relationship between vaccine evaluation methods and vaccine development is dynamic and reciprocal, with the processes including preclinical evaluation, clinical trials, accelerated evaluation methods, post-marketing surveillance, immunological evaluation, variant surveillance, and adaptation. Evaluation methods provide critical data that inform and shape the development process, ensuring that vaccines are safe, effective, and able to meet public health needs.

Author Contributions: Conceptualization, J.N. and Y.L.; writing—original draft preparation, Y.L.; writing—review and editing, J.N. and Y.L.; supervision, critical comments, and suggestions, and manuscript revision, J.N. and Y.L.; funding acquisition, J.N. All authors have read and agreed to the published version of the manuscript.

Funding: This work was supported by the General Program of the National Natural Science Foundation of China (grant nos. 82172244) and the National Key R and D Program of China (grant nos. 2021YFC2302500).

Conflicts of Interest: The authors declare no conflicts of interest.

References

1. Chi, W.Y.; Li, Y.D.; Huang, H.C.; Chan, T.E.H.; Chow, S.Y.; Su, J.H.; Ferrall, L.; Hung, C.F.; Wu, T.C. COVID-19 vaccine update: Vaccine effectiveness, SARS-CoV-2 variants, boosters, adverse effects, and immune correlates of protection. *J. Biomed. Sci.* **2022**, *29*, 82. [CrossRef] [PubMed]

2. Cheng, H.L.; Lim, S.M.; Jia, H.; Chen, M.W.; Ng, S.Y.; Gao, X.; Somani, J.; Sengupta, S.; Tay, D.M.Y.; Chua, P.W.L.; et al. Rapid Evaluation of Vaccine Booster Effectiveness against SARS-CoV-2 Variants. *Microbiol. Spectr.* **2022**, *10*, e0225722. [CrossRef] [PubMed]
3. Nam, M.; Cha, J.H.; Kim, S.-W.; Kim, S.B.; Lee, K.-B.; Chung, Y.-S.; Yun, S.G.; Nam, M.-H.; Lee, C.K.; Cho, Y. Performance Evaluation of Three Antibody Binding Assays, a Neutralizing Antibody Assay, and an Interferon-Gamma Release Assay for SARS-CoV-2 According to Vaccine Type in Vaccinated Group. *Diagnostics* **2023**, *13*, 3688. [CrossRef] [PubMed]
4. Edara, V.V.; Norwood, C.; Floyd, K.; Lai, L.; Davis-Gardner, M.E.; Hudson, W.H.; Mantus, G.; Nyhoff, L.E.; Adelman, M.W.; Fineman, R.; et al. Infection- and vaccine-induced antibody binding and neutralization of the B.1.351 SARS-CoV-2 variant. *Cell Host Microbe* **2021**, *29*, 516–521.E3. [CrossRef] [PubMed]
5. Zheng, J.; Deng, Y.; Zhao, Z.; Mao, B.; Lu, M.; Lin, Y.; Huang, A. Characterization of SARS-CoV-2-specific humoral immunity and its potential applications and therapeutic prospects. *Cell Mol. Immunol.* **2022**, *19*, 150–157. [CrossRef]
6. Medits, I.; Springer, D.N.; Graninger, M.; Camp, J.V.; Holtl, E.; Aberle, S.W.; Traugott, M.T.; Hoepler, W.; Deutsch, J.; Lammel, O.; et al. Different Neutralization Profiles After Primary SARS-CoV-2 Omicron BA.1 and BA.2 Infections. *Front. Immunol.* **2022**, *13*, 946318. [CrossRef]
7. Rossler, A.; Knabl, L.; von Laer, D.; Kimpel, J. Neutralization Profile after Recovery from SARS-CoV-2 Omicron Infection. *N. Engl. J. Med.* **2022**, *386*, 1764–1766. [CrossRef]
8. Rossler, A.; Netzl, A.; Knabl, L.; Bante, D.; Wilks, S.H.; Borena, W.; von Laer, D.; Smith, D.J.; Kimpel, J. Characterizing SARS-CoV-2 neutralization profiles after bivalent boosting using antigenic cartography. *Nat. Commun.* **2023**, *14*, 5224. [CrossRef]
9. McLean, G.R.; Zhang, Y.; Ndoyi, R.; Martin, A.; Winer, J. Rapid Quantification of SARS-CoV-2 Neutralising Antibodies Using Time-Resolved Fluorescence Immunoassay. *Vaccines* **2022**, *10*, 2149. [CrossRef]
10. Lynch, K.L.; Zhou, S.; Kaul, R.; Walker, R.; Wu, A.H. Evaluation of Neutralizing Antibodies against SARS-CoV-2 Variants after Infection and Vaccination Using a Multiplexed Surrogate Virus Neutralization Test. *Clin. Chem.* **2022**, *68*, 702–712. [CrossRef]
11. von Rhein, C.; Scholz, T.; Henss, L.; Kronstein-Wiedemann, R.; Schwarz, T.; Rodionov, R.N.; Corman, V.M.; Tonn, T.; Schnierle, B.S. Comparison of potency assays to assess SARS-CoV-2 neutralizing antibody capacity in COVID-19 convalescent plasma. *J. Virol. Methods* **2021**, *288*, 114031. [CrossRef] [PubMed]
12. Kohmer, N.; Rühl, C.; Ciesek, S.; Rabenau, H.F. Utility of Different Surrogate Enzyme-Linked Immunosorbent Assays (sELISAs) for Detection of SARS-CoV-2 Neutralizing Antibodies. *J. Clin. Med.* **2021**, *10*, 2128. [CrossRef] [PubMed]
13. Izac, J.R.; Kwee, E.J.; Tian, L.; Elsheikh, E.; Gaigalas, A.K.; Elliott, J.T.; Wang, L. Development of a Cell-Based SARS-CoV-2 Pseudovirus Neutralization Assay Using Imaging and Flow Cytometry Analysis. *Int. J. Mol. Sci.* **2023**, *24*, 12332. [CrossRef] [PubMed]
14. Nitahara, Y.; Nakagama, Y.; Kaku, N.; Candray, K.; Michimuko, Y.; Tshibangu-Kabamba, E.; Kaneko, A.; Yamamoto, H.; Mizobata, Y.; Kakeya, H.; et al. High-Resolution Linear Epitope Mapping of the Receptor Binding Domain of SARS-CoV-2 Spike Protein in COVID-19 mRNA Vaccine Recipients. *Microbiol. Spectr.* **2021**, *9*, e0096521. [CrossRef] [PubMed]
15. Frische, A.; Krogfelt, K.A.; Fomsgaard, A.; Lassauniere, R. Antigen-Heterologous Vaccination Regimen Triggers Alternate Antibody Targeting in SARS-CoV-2-DNA-Vaccinated Mice. *Vaccines* **2024**, *12*, 218. [CrossRef]
16. Pflumm, D.; Seidel, A.; Klein, F.; Gross, R.; Krutzke, L.; Kochanek, S.; Kroschel, J.; Munch, J.; Stifter, K.; Schirmbeck, R. Heterologous DNA-prime/protein-boost immunization with a monomeric SARS-CoV-2 spike antigen redundantizes the trimeric receptor-binding domain structure to induce neutralizing antibodies in old mice. *Front. Immunol.* **2023**, *14*, 1231274. [CrossRef]
17. Bae, S.; Ko, J.H.; Choi, J.Y.; Park, W.J.; Lim, S.Y.; Ahn, J.Y.; Song, K.H.; Lee, K.H.; Song, Y.G.; Chan Kim, Y.; et al. Heterologous ChAdOx1 and Bnt162b2 vaccination induces strong neutralizing antibody responses against SARS-CoV-2 including delta variant with tolerable reactogenicity. *Clin. Microbiol. Infect.* **2022**, *28*, 1390.e1–1390.e7. [CrossRef]
18. Hwang, Y.C.; Lu, R.M.; Su, S.C.; Chiang, P.Y.; Ko, S.H.; Ke, F.Y.; Liang, K.H.; Hsieh, T.Y.; Wu, H.C. Monoclonal antibodies for COVID-19 therapy and SARS-CoV-2 detection. *J. Biomed. Sci.* **2022**, *29*, 1. [CrossRef]
19. Meng, Q.F.; Tai, W.; Tian, M.; Zhuang, X.; Pan, Y.; Lai, J.; Xu, Y.; Xu, Z.; Li, M.; Zhao, G.; et al. Inhalation delivery of dexamethasone with iSEND nanoparticles attenuates the COVID-19 cytokine storm in mice and nonhuman primates. *Sci. Adv.* **2023**, *9*, eadg3277. [CrossRef]
20. Dehghan, M.; Askari, H.; Tohidfar, M.; Siadat, S.O.R.; Fatemi, F. Improvement of RBD-FC Immunogenicity by Using Alum-Sodium Alginate Adjuvant Against SARS-COV-2. *Influenza Other Respir. Viruses* **2024**, *18*, e70018. [CrossRef]
21. Pandey, B.; Wang, Z.; Jimenez, A.; Bhatia, E.; Jain, R.; Beach, A.; Maniar, D.; Hosten, J.; O'Farrell, L.; Vantucci, C.; et al. A Dual-Adjuvanted Parenteral-Intranasal Subunit Nanovaccine generates Robust Systemic and Mucosal Immunity Against SARS-CoV-2 in Mice. *Adv. Sci.* **2024**, e2402792. [CrossRef] [PubMed]
22. Xu, K.; Li, J.; Lu, X.; Ge, X.; Wang, K.; Wang, J.; Qiao, Z.; Quan, Y.; Li, C. The Immunogenicity of CpG, MF59-like, and Alum Adjuvant Delta Strain Inactivated SARS-CoV-2 Vaccines in Mice. *Vaccines* **2024**, *12*, 60. [CrossRef] [PubMed]

vaccines

Article

High-Content Imaging-Based Assay for SARS-CoV-2-Neutralizing Antibodies

Vinícius Pinto Costa Rocha [1,2,*], Bruna Aparecida Souza Machado [1], Helenita Costa Quadros [2], Antônio Márcio Santana Fernandes [1], Bianca Sampaio Dotto Fiuza [1], Cássio Santana Meira [1,2], Vitória Torres Barbosa da Silva [1], Afrânio Ferreira Evangelista [1], Larissa Moraes dos Santos Fonseca [1,2], Roberto José da Silva Badaró [1] and Milena Botelho Pereira Soares [1,2,*]

1 Institute of Health Technology, National Industrial Learning Service—Integrated Manufacturing and Technology Campus, SENAI CIMATEC, Salvador 41650-010, Bahia, Brazil; brunam@fieb.org.br (B.A.S.M.); antonio.fernandes@fbter.org.br (A.M.S.F.); bianca.fiuza@fbter.org.br (B.S.D.F.); cassio.meira@fieb.org.br (C.S.M.); vitorres2015@outlook.com (V.T.B.d.S.); afranio.evangelista@fieb.org.br (A.F.E.); larissa.fonseca@fieb.org.br (L.M.d.S.F.); badaro@fieb.org.br (R.J.d.S.B.)

2 Laboratory of Tissue Engineering and Immunopharmacology, Oswaldo Cruz Foundation, Gonçalo Moniz Institute—Fiocruz, Salvador 40296-710, Bahia, Brazil; helenita.quadros@fiocruz.br

* Correspondence: vinicius.rocha@fieb.org.br (V.P.C.R.); milena@bahia.fiocruz.br (M.B.P.S.)

Abstract: The COVID-19 pandemic and the consequent emergence of new SARS-CoV-2 variants of concern necessitates the determination of populational serum potency against the virus. Here, we standardized and validated an imaging-based method to quantify neutralizing antibodies against lentiviral particles expressing the spike glycoprotein (pseudovirus). This method was found to efficiently quantify viral titers based on ZsGreen-positive cells and detect changes in human serum neutralization capacity induced by vaccination with up to two doses of CoronaVac, Comirnaty, or Covishield vaccines. The imaging-based protocol was also used to quantify serum potency against pseudoviruses expressing spikes from Delta, Omicron BA.1.1.529, and BA.4/5. Our results revealed increases in serum potency after one and two doses of the vaccines evaluated and demonstrated that Delta and Omicron variants escape from antibody neutralization. The method presented herein represents a valuable tool for the screening of antibodies and small molecules capable of blocking viral entry and could be used to evaluate humoral immunity developed by different populations and for vaccine development.

Keywords: SARS-CoV-2; antibodies; neutralization; high-content screening; variants

1. Introduction

To date, 50 vaccines have been approved, authorized, licensed, or granted emergency-use authorization in at least one country [1]. Vaccination is the most effective way to avoid the development of severe acute respiratory syndrome, as it was estimated that 14.4 million deaths in 185 countries were prevented during the first year of vaccination [2]. Several SARS-CoV-2 variants have emerged since the declaration of the coronavirus disease (COVID-19) pandemic. The RNA-based viral genome facilitates mutation acquisition over time, and some mutations may make the virus more transmissible. More transmissible viral strains will overtake the previous one by infecting people, regardless of vaccination status, since intramuscular administration does not induce effective mucosal immunity capable of blocking viral spread through the upper airways [3–6].

SARS-CoV-2 has undergone great diversification, with more than 1500 PANGO lineages identified during the pandemic [7,8]. Those presenting increased transmissibility, virulence, and immune evasion are classified by the World Health Organization as variants of concern (VoC). Five VoCs have been reported, in addition to their subvariants, at the time of this study [9]. Since the emergence of the Omicron variant (B.1.1.529), which harbors an unprecedented number of mutations, the monitoring of neutralizing antibodies has become even more necessary to determine a given population's neutralizing capacity, mainly four to six months after vaccine administration, when antibody levels tend to decrease.

Viral particles carrying a reporter gene sequence and expressing the spike are important tools in terms of evaluating the neutralizing capacity of antibodies [10–12]. Muruato and colleagues (2020), for example, used a reverse genetic system to produce replicating recombinant SARS-CoV-2 expressing the mNeonGreen reporter gene [10,13]. When tested with human convalescent plasma, their assay was comparable to the gold standard plaque reduction neutralization test (PRNT) and is being used in a phase I/II clinical trial [10,14]. Other groups developed similar assays using the lentiviral approach, which has the advantage of being manipulated at BSL-2 laboratories [15–21]. Based on the importance of evaluating the neutralizing capacity of human serum and developing new vaccines, when considering the variation among different populations, we standardized and validated an imaging-based high-content assay, which has been shown to be safe, high-throughput compatible, and able to determine the neutralizing potency of plasma in two days, which is half of the time required by the PRNT [22]. Our assay is based on commercially available plasmids used to produce lentiviral non-replicating particles expressing spike glycoproteins, which can transduce ACE2-overexpressing human embryonic kidney cells (HEK293T). Infection results in the integration of the ZsGreen sequence into the cell genome. Automated image analysis employs a predefined algorithm that can detect fluorescent cells. This assay was shown to detect neutralizing antibodies in previously analyzed human samples regarding the production of total IgG and neutralizing antibodies following the administration of the first and second doses of three clinically used vaccines. Screening can be expanded to detect monoclonal antibodies and small molecules that are able to block viral infection and could also be used for vaccine development.

2. Materials and Methods

2.1. Plasmids and Reagents

The plasmid encoding Spike Wuhan-Hu-1 was obtained from BEI Resources, as well as all packaging vectors for lentiviral production (NIAID, NIH: SARS-Related Coronavirus 2, Wuhan-Hu-1 Spike-Pseudotyped Lentiviral Kit V2, NR-53816). Delta B.1.617.2 (plasmid #179905), Omicron B.1.1.529 (plasmid #179907), and Omicron BA.4/5 (plasmid #186031) plasmids were purchased from Addgene (Watertown, MA, USA) [23,24]. All other general laboratory chemicals and solvents used were of analytical or HPLC grade. The plasmids pHAGE-CMV-Luc2-IRES-ZsGreen-W, HDM-Hgpm2, HDM-Tat1b, and pRC-CMV-Rev1b were obtained from BEI Resources (NIAID, NIH: SARS-Related Coronavirus 2, Wuhan-Hu-1 Spike-Pseudotyped Lentiviral Kit V2, NR-53816).

2.2. Ethics Statement

Thirty volunteers kindly participated in a cohort study aimed at analyzing humoral response prior to and after receiving two doses of SARS-CoV-2 vaccines (CoronaVac, Comirnaty, or Covishield) at the beginning of the vaccination campaign in Brazil between February and August 2021. The study was conducted in accordance with the Declaration of Helsinki and approved by the Ethics Committee of National Industrial Learning Service–Integrated Manufacturing and Technology Campus (SENAI-CIMATEC) (protocol code 4.334.505; date of approval: 2 September 2021). The samples were collected by convenience, following the provision of informed consent, from employees of SENAI-CIMATEC. The studied population consisted of men and women aged 18 years or older. All volunteers received results from serological testing performed using an Enzyme-Linked Immunosorbent Assay (ELISA), and antibody neutralization was conducted using an ELISA Kit from Elabscience (Houston, TX, USA).

2.3. Determination of Anti-SARS-CoV-2 IgG Antibodies

The ELISA for total IgG against the Spike protein followed the previously published protocol [25]. Serum samples from all participants were collected before vaccination, 30 days after receiving the first dose, and 30 days after the second dose. Briefly, 96-well flat-bottom microplates were sensitized with 50 µL of 1 µg/mL *spike* D614G recombinant protein (kindly provided by Sean Grey from PAI Life Sciences, Seattle, WA, USA) diluted in phosphate-buffered saline (PBS, pH 7.2), and subsequently incubated at 4 °C overnight. After 16 h of incubation, the microplates were washed three times with a washing solution containing PBS plus 0.05% Tween 20 (Sigma Aldrich, San Luis, MO, USA) and then blocked with a PBS solution containing 3% skim milk for 1 h. Fifty microliters of serum samples at an initial dilution of 1:10 was dispensed into the wells, followed by a two-fold serial dilution in PBS with 1% skim milk, ranging from 1:10 to 1:1280.

The detection of neutralizing antibodies was performed using the SARS-CoV-2 Neutralization Antibody ELISA Kit from Elabscience (Houston, TX, USA), following the manufacturer's instructions. This kit, based on the competition ELISA technique, aims to detect the presence of neutralizing antibodies against SARS-CoV-2. All testing was performed using serum dilutions ranging from 1:40 to 1:2560.

2.4. Cell Cultures

Human Embryonic Kidney (HEK) 293T cell line and HEK 293T overexpressing ACE2 (293T-ACE2) were cultivated in Dulbecco's Modified Eagle Medium (DMEM, Thermo Fisher, Waltham, MA, USA) supplemented with 10% FBS (Vitrocell, Campinas, São Paulo, Brazil), 50 U/mL of penicillin, 50 µg/mL of streptomycin (PenStrep; Thermo Fisher), and 100 mM of HEPES (Thermo Fisher). The cells were incubated at 37 °C and 5% CO_2 in T25 or T75 vented cap flasks, with cell maintenance performed every 3 to 4 days by cell trypsinization (LGC Biotecnologia, Cotia, São Paulo, Brazil). The 293T ACE2 cells were obtained through BEI Resources (NIAID, NIH: Human Embryonic Kidney Cells Expressing Human Angiotensin-Converting Enzyme 2, HEK-293T-hACE2 Cell Line, NR-52511).

2.5. Generation of Pseudotyped Lentiviral Particles

Pseudotyped lentiviral particles expressing the *spike* Wuhan-Hu-1 protein (GenBank: NC_045512), Delta B.1.617.2, Omicron B.1.1.529, and BA.4/5 were generated via the co-transfection of the HEK 293T cell line, according to a previously described protocol [26]. For this, HEK 293T cells were plated on 6-well plates at 5×10^5 in 2 mL of supplemented DMEM and then incubated at 37 °C under 5% CO_2 for 24 h. Next, the cells were transfected using Lipofectamine 3000 (Thermo Fisher) and 2 µg of a DNA mix containing the following lentiviral constructs to produce the pseudotyped virus expressing the Wuhan-Hu-1 spike surface glycoprotein: a total of 1 µg of pHAGE-CMV-Luc2-IRES-ZsGreen-W, 0.22 µg of HDM-Hgpm2, 0.22 µg of HDM-Tat1b, 0.22 µg of pRC-CMV-Rev1b, and 0.34 µg of HDM-IDTSpikeΔCter. For pHAGE-CMV-Luc2-IRES-ZsGreen-W, a map and the sequence are provided in Figures S1 and S2, respectively. To produce Delta, Omicron B.1.1.529, or BA.4/5 pseudotyped particles, the plasmid encoding Wuhan-Hu-1 spike glycoprotein was replaced by 0.34 µg of pTwist-SARS-CoV-2 Δ18 B.1.617.2v1, pTwist-SARS-CoV-2 Δ18 B.1.1.529, or pCAGGS SARS-CoV-2 BA.4/5 Spike, respectively [23,24]. After 18 h of transfection, the media was replaced in each well, and the plates were incubated for an additional 48 h. Sixty hours after transfection, the supernatants containing the pseudoviruses were collected and filtered using a 0.45 µm PES membrane filter. Pseudovirus preparations were titrated and used in neutralization assays. Viral particles with no entry protein ("bald" pseudoviruses) were used as negative controls in the titration assays, while a pseudotyped particle expressing the vesicular stomatitis virus G (VSV G) protein was used as a positive control.

2.6. Assay Precision

Repeatability (intra-assay) was determined by one researcher through the performance of three independent experiments (three viral batches) in triplicate. Reproducibility (inter-assay) was determined by two researchers who performed three independent experiments with five replicates [27]. The coefficient of variation (CV) was calculated by the ratio between the standard deviation and the pNT_{50} mean values. The CV up to 15% was accepted for both precision parameters [28].

2.7. Cell Transduction, Image-Based Analysis, and Titration

Using a 96-well flat-bottom microplate, the produced pseudovirus was 2-fold serially diluted in supplemented DMEM at a final volume of 100 μL per dilution, which resulted in seven concentrations, including the non-diluted stock solution. The 293T-ACE2 cells were added to a 96-well plate at a density of 1.25×10^4 in 50 μL of supplemented DMEM and then incubated at 37 °C under 5% CO_2 for 48 h. Cell nuclei were labeled using Hoechst 33342 (Abcam, Cambridge, UK) at 20 μM and, after 15 min of incubation at 37 °C, the percentage of 293T-ACE2 cells expressing ZsGreen was quantified on a CellInsight CX7 LZR platform (Thermo Fisher) and analyzed using Cellomics Scan software (version 6.6.2).

The algorithm used to quantify ZsGreen-positive cells was based on the Cellomics Scan Target Activation protocol, resulting in 25 analyzed fields per well (the entire well). To identify the primary objects using the Hoechst 33342 channel, the algorithm applied a mask in the nuclear region. Primary object validation and selection were performed in accordance with size and fluorescence intensity. Objects validated using the Hoechst 33342 channel were then used to determine the number of identified cells per well. To identify objects using the ZsGreen channel, a mask was applied to the cytoplasm region, excluding the nucleus. Object validation and selection were performed in accordance with fluorescence intensity by performing comparisons among 293T-ACE2 plus pseudovirus groups, 293T-ACE2 plus the "bald" pseudovirus, and 293T-ACE2 cells only. Reference levels were established based on negative controls, with average fluorescence intensity over the reference used to determine the percentage of ZsGreen-positive cells. Pseudovirus titers were determined using the Poisson formula, which revealed the number of viral particles per milliliter in accordance with the following equation: $-\ln(1 - P/100) \times$ (number of cells per well)/(volume of pseudovirus added per well in milliliters), where P is the percentage of $ZsGreen^+$ cells [26].

2.8. Neutralization Assay

To validate this new method, the same samples from 30 subjects were used after being collected at three different times: before vaccination, 30 days after the first immunization, and 30 days after the second immunization. Using a 96-well microplate, 5 μL of serum from each subject was added to wells A1 through A12, followed by the addition of 95 μL of supplemented DMEM medium, corresponding to a 1:20 serum dilution. From well B1 to G12, 50 μL of supplemented DMEM was added in the absence of serum. The samples were serially diluted by taking 50 μL from the sample diluted at 1:20 and passing it to the next well below in the same column. Twelve wells (H1 to H12) were left without serum, and 50 μL of supplemented DMEM was added. Fifty microliters of the pseudotyped particles was added to each well already containing 50 μL of diluted serum, corresponding to a final dilution curve ranging from 1:40 to 1:2560. The wells corresponding to the negative and positive controls for neutralization received 50 μL of spike-expressing pseudoviruses (H1 to H6) or control bald pseudoviruses (H7 to H12), respectively. The plates were then incubated at 37 °C under 5% CO_2 for 1 h to allow the antibodies to interact with the spike. Next, 293T-ACE2 cells were added at a density of 1.25×10^4 in 50 μL of supplemented DMEM, followed by an additional 48 h of incubation at 37 °C under 5% CO_2. Hoechst 33342 staining and image analysis were performed as described above. The serum titration curve was used to determine the half-maximum effective concentration of each sample against the pseudoviruses (pseudovirus neutralization—pNT_{50}). This parameter corresponds to the serum dilution capable of neutralizing 50% of the pseudoviruses and is represented by the reciprocal value ($1/pNT_{50}$), i.e., higher values indicate greater sample potency.

2.9. Statistical Analysis

Statistical analyses were performed using GraphPad Prism 8.0 (Boston, MA, USA). Data normality was evaluated by using the Shapiro–Wilk test. Statistical differences among the three groups were evaluated using one-way ANOVA or Friedman test, with results considered significant when $p < 0.05$. Statistical differences between the two groups were evaluated via the Mann–Whitney test and considered significant when $p < 0.05$. pNT_{50} was determined via nonlinear regression by the Log_{10} from serum dilutions and the percentage of reduction in infection.

3. Results

3.1. Standardization of a High-Content Imaging-Based Neutralization Assay

To standardize the high-content imaging-based neutralization assay using 293T-ACE2 cells, we first produced lentiviral particles expressing the spike glycoprotein from the Wuhan-Hu-1 variant (GenBank: NC_045512). Viable viral particles transduced the target cells and integrated the ZsGreen sequence in the host genome, generating a green fluorescent transgenic cell line. Using the Target Activation protocol in Cellomics software, the percentage of ZsGreen-positive cells ($ZsGreen^+$) was automatically quantified (Figure 1a). No fluorescent cells were identified via transduction with control pseudoviruses not expressing the spike protein (bald virus, Figure 1b), as only cells transduced by spike-expressing pseudoviruses exhibited fluorescence (Figure 1c). Cell nuclei were visualized using the Hoechst 33324 channel, while $ZsGreen^+$ cells were identified by using the algorithm's region of interest in the cytoplasm. Average ZsGreen intensity was used to separate populations exhibiting negativity or positivity for ZsGreen (Figure 1d–g). Using a viral dilution curve, we identified significantly more $ZsGreen^+$ objects in the groups infected with undiluted viral spike (S) stock or 1:10 VSV-G-expressing pseudoviruses, as well as among the dilutions, characterizing a titer detection relationship. HEK293T cells were only able to be infected by VSV-G-expressing pseudovirus, confirming the need for ACE2 in the model involving spike-expressing particles (Figure 2a,b). Linear regression curves are shown in Figure S3A–D. The reduction in $ZsGreen^+$ cells in relation to viral titers is evidenced in Figure 2c–g. Moreover, we established correlations between not only the percentage of $ZsGreen^+$ cells and the number of pseudotyped viral particles per milliliter but also between average ZsGreen fluorescence intensity and the number of viral particles/mL (Figure 2h,i). We observed a linear trend between both parameters and viral titers. A Pearson's correlation coefficient (r) value of 0.9961 was found in relation to the percentage of $ZsGreen^+$ cells and 0.9865 regarding ZsGreen average fluorescence intensity (Figure S4A,B). As the mean ZsGreen intensity was used to select $ZsGreen^+$ cells, the cut-off was calculated as the mean of ZsGreen intensity from negative controls plus $3\times$ the standard deviation. The cut-off value was calculated at 39 (a.u.); however, it is important to highlight that we determined the cut-off for each analyzed plate based on the positive control for neutralization (the lowest ZsGreen percentage) represented by infection with pseudoviruses without entry proteins. The assay precision was determined by intra- and inter-assay variability analysis using two WHO standard samples. Fifty microliters of fresh pseudoviruses were used, corresponding to a multiplicity of infection (MOI) of approximately 0.6. The CV measured in the intra-assay research was 14% and 11.9% for 21/388 and 21/340, respectively. Two researchers independently tested the WHO standards to determine the inter-assay reproducibility. The CV calculated between the researchers was 19% for 21/338 and 3.2% for 21/340, demonstrating less than 15% of variability (Figure 3a,b), except for the inter-assay regarding sample 21/388. Nonlinear regression curves are shown in Figure 3c,d.

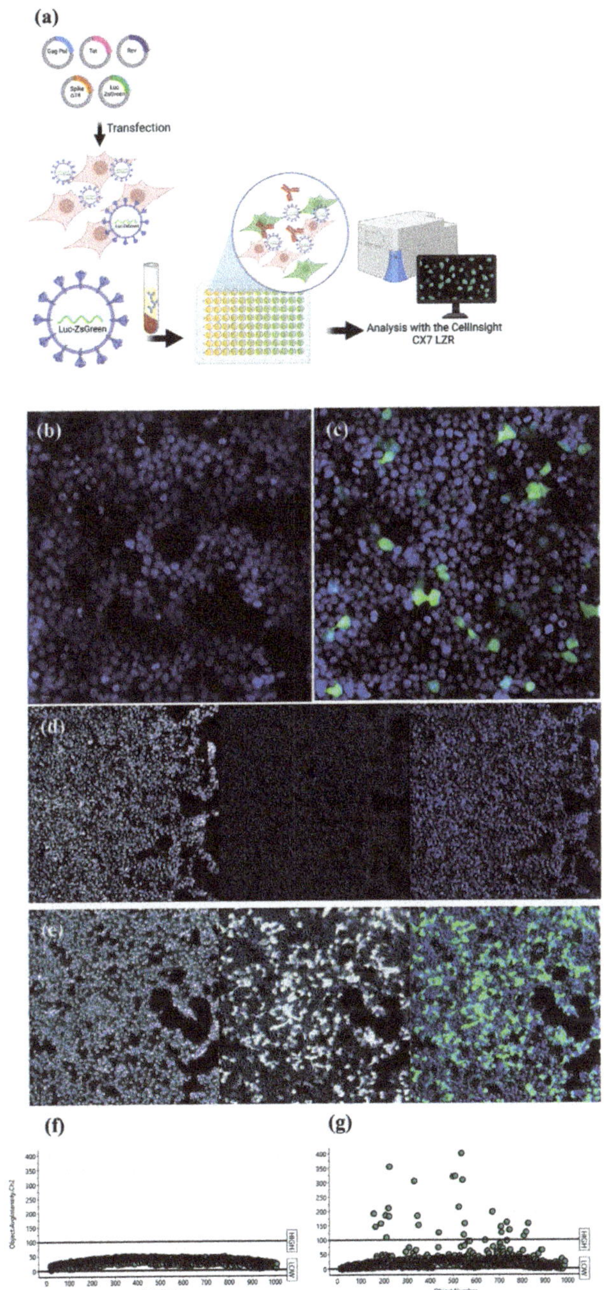

Figure 1. Automated imaging-based analysis workflow. (**a**) HEK-293T cells were transfected with plasmids to generate pseudoviruses. Viral particles were collected from supernatants and used in neutralization assays using HEK-293T-ACE2 cells together with serum from volunteers. The quantification of ZsGreen^{+} cells was determined automatically via Cellomics software using a CX7LZR high-content system. (**b**) Pseudoviruses without an entry spike protein (bald virus) did not generate fluorescent cells, (**c**) while spike expression promotes viral entry into ACE2-overexpressing

cells. (**d**,**e**). The Target Activation protocol from Cellomics was applied to detect cell nuclei and determine the region of interest (ROI) to detect the ZsGreen fluorescence. Representative imaging analysis of the average fluorescence intensity used to separate the cellular population of (**f**) $ZsGreen^-$ and (**g**) $ZsGreen^+$ cells. The cut-off established as 100 was only for representative purposes. The cut-off is finely adjusted in each assay. Nuclei was stained in blue with Hoechst and ZsGreen fluorescence is shown in green.

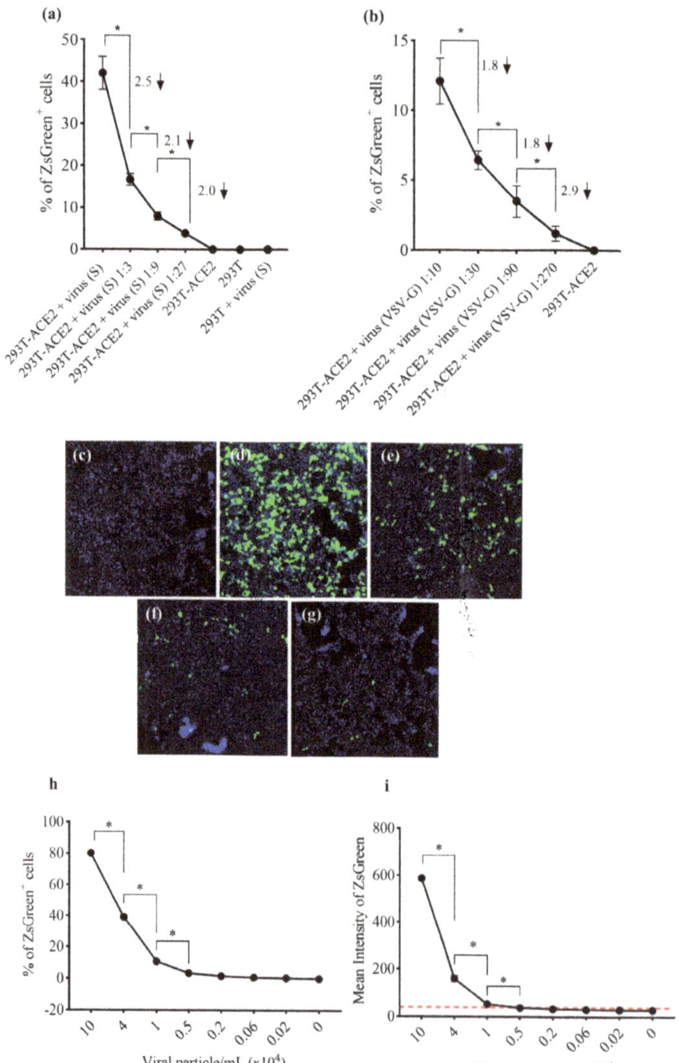

Figure 2. Detection of viral titers via automated imaging analysis protocol: the percentage of $ZsGreen^+$ cells and average fluorescence intensity were measured by using the image-based analysis protocol. (**a**) Pseudoviruses expressing the Wuhan-Hu-1 spike glycoprotein are functional, as only HEK293T-ACE2 cells were transduced. Numbers above the graphic line and the arrows denote the reduction fold calculated by the ratio between viral amounts. (**b**) VSV g-expressing particles were used as a transduction control since both cell lines express the receptor for this ligand. Numbers above

the graphic line denote the reduction fold calculated via the ratio between viral amounts. Fluorescence images of (**c**) bald pseudoviruses, (**d**) undiluted Wuhan-Hu-1 spike-expressing pseudoviruses, (**e**) Wuhan-Hu-1 pseudoviruses diluted 3-fold, (**f**) diluted 9-fold, and (**g**) diluted 27-fold, as illustrated by decreasing numbers of $ZsGreen^+$ cells. The protocol was shown to both (**h**) detect decreasing levels of Wuhan-Hu-1 pseudovirus titers based on the percentage of $ZsGreen^+$ and (**i**) on the average fluorescence intensity. The red line represents the calculated cut-off value (39 a.u.). Data are shown as mean ± SD. * One-way ANOVA, followed by Holm–Sidak's multiple comparisons test, $p < 0.05$, calculated by using GraphPad Prism, version 8.0.

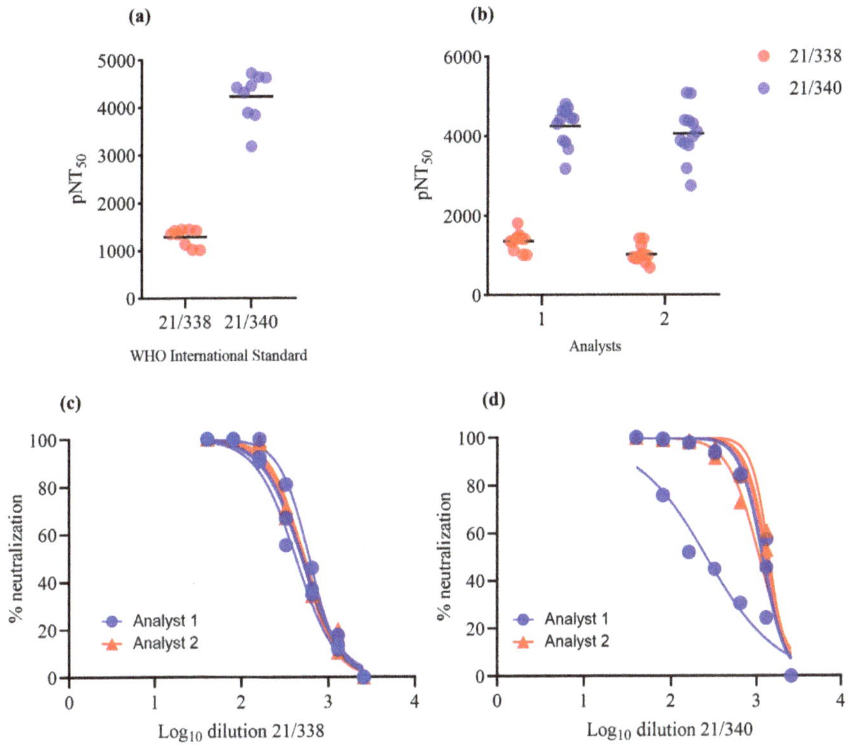

Figure 3. Precision tests of the automated imaging analysis. Pseudoviruses expressing the Wuhan-Hu-1 spike glycoprotein were included with WHO 21/338 and 21/340 standard serum samples. The pNT50 was calculated, as well as the CV in the same (intra-assay) or independent experiments (inter-assay). (**a**) Intra-assay: 3 independent experiments were performed by one researcher. Data were pooled to represent the variation among the samples. (**b**) Inter-assay analysis: 2 researchers performed 3 independent assays. Data were pooled to represent the variation among the samples. (**c**,**d**) Nonlinear curves regarding the precision assays.

3.2. Imaging-Based Analysis to Detect the Neutralizing Capacity of Volunteers after Two Vaccination Doses

To validate the imaging-based protocol, serum samples from cohorts of volunteers vaccinated with two doses of three different vaccines were used. Of the 30 participants, 15 (50%) were immunized with the Covishield vaccine, representing the largest study group, followed by 8 (26.7%) participants vaccinated with Comirnaty, and 7 (23.3%) with CoronaVac (Figure 4a). Regarding participant sex, females predominated (n = 22, 75.9%). The predominant race in the three subgroups was white (n = 14, 48.3%), followed by black (n = 7, 24.1%). Only two (6.9%) individuals tested positive for COVID-19 via RT-PCR prior to vaccination. At the time of vaccination, these volunteers were already negative for COVID-19, as determined by RT-PCR. Firstly, the samples were evaluated using a neutralization assay kit using recombinant HRP-conjugated receptor-binding domain (RBD) of the Wuhan-Hu-1 spike glycoprotein. The percentage of neutralizing antibodies was determined in all samples at a 1:40 dilution. As shown in Figure 4, immunization with CoronaVac induced 19.83% and 60.63% neutralization after the first and second doses, respectively. Immunization with Comirnaty and Covishield induced 20.36% and 34.56% neutralization after the first dose, with 84.61% and 71.38% following the second dose, respectively (Figure 4b–d). Serum potency (presented as pNT_{50}) was determined via nonlinear regression. Samples from volunteers vaccinated once with CoronaVac presented pNT_{50} values at the detection limit of 40, as those prior to immunization and after the first dose. Only after receiving the booster did pNT_{50} increase to 53.61. Samples from volunteers vaccinated with Comirnaty presented a slight increase (pNT_{50} = 44.96) after the first dose and an approximately six-fold increase after the second dose (pNT_{50} = 224.88). Volunteers vaccinated with Covishield presented respective pNT_{50} values of 40, 69.46, and 151.40 before immunization, after the first, and following the second dose (Figure 4e–g). In addition, total IgG titers demonstrated an increase in antibody production after the second dose (Figure S5).

These samples were submitted to the imaging-based protocol to determine the potency of neutralizing antibodies. Pseudoviruses at MOI 0.6 were used for each spike variant, resulting in 34.4 $\pm$ 7.2% of ZsGreen^{+} cells for wild-type virus, 26.2 $\pm$ 13.8% for Delta, 43.6 $\pm$ 6.5% for Omicron B.1.1.529, and 9.8 $\pm$ 1% for Omicron BA4/5. This viral amount was enough to produce at least a 30-fold increase in the ZsGeen mean fluorescence intensity, which can distinguish transduced and non-transduced cells. The samples were tested at a 1:40 dilution to determine the percentage of neutralization against the Wuhan-Hu-1 spike glycoprotein pseudovirus. Samples from volunteers prior to vaccination presented a maximum level of 32% pseudovirus neutralization. The first dose of Covishield significantly increased neutralization. A similar profile was not observed after receiving the first dose of CoronaVac and Comirnaty. Following the second dose, Comirnaty and Covishield significantly increased neutralization capacity compared to before vaccination (Figure 5a–c). Serum potency was determined via nonlinear regression and is represented as pNT_{50}. Samples from volunteers vaccinated with CoronaVac presented pNT_{50} at the detection limit of 40, with enhanced serum potency of 213 only following the second dose. However, based on paired analysis, the second dose of CoronaVac did not significantly increase the neutralization capacity of the sera. Samples from naïve volunteers before receiving Comirnaty presented a potency at the detection limit, which increased to 265 after the first dose. In these individuals, pNT_{50} was significantly higher than before vaccination after receiving a second dose of Comirnaty. A similar profile was found in samples of volunteers vaccinated with Covishield. In this case, pNT_{50} significantly increased from 40 to 231 after the first dose and remained at 203 after the second dose (Figure 6a–c). Nonlinear regression curves are shown in Figure S6A–C. These data are consistent with the neutralizing antibody measurements obtained using ELISA despite slight increases in pNT_{50} mean values detected by the imaging-based assay (Figure S7).

The imaging-based protocol was also used to analyze the neutralizing ability of the samples against the main VoCs (at the time of this research), including Delta (B.1.617.2), and Omicron B.1.1.529 and B.A4/5. Our results show that vaccination with CoronaVac did not induce antibodies capable of neutralizing these VoCs (Figure 6d,g,j). The mRNA and vector-based vaccines were found to significantly increase the levels of serum-neutralizing potency against the Delta VoC only following the second dose (Figure 6e,f). These vaccines did not significantly increase serum potency against Omicron B.1.1.529 (Figure 6h,i) and BA.4/5 (Figure 6k,l). Nonlinear regression curves are shown in Figure S6D–L.

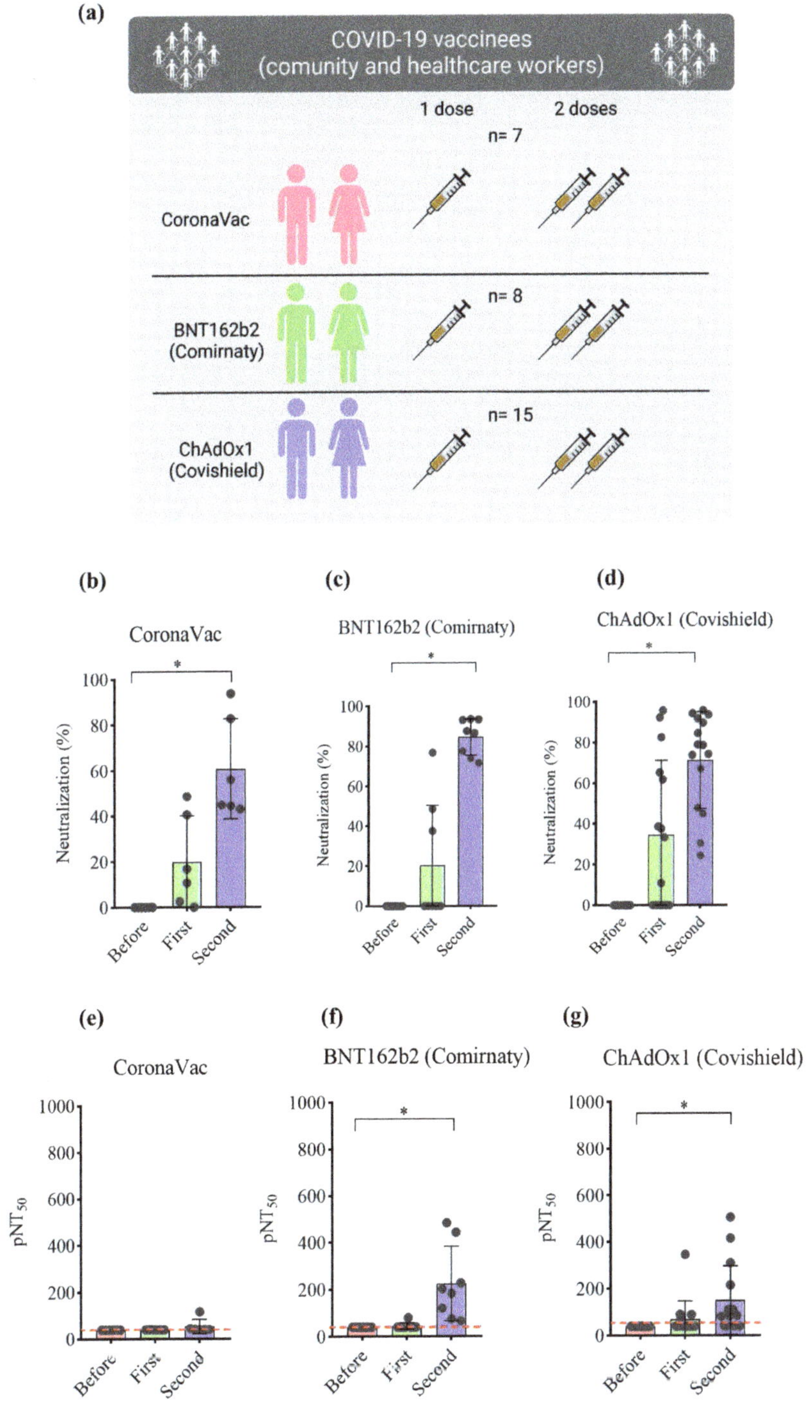

Figure 4. Percentage of neutralization and pNT_{50} values in volunteers immunized with different vaccines. (**a**) Schematic representation of the cohorts of volunteers tested. Blood from healthy volunteers

was collected prior to and 30 days after the first and second doses of three different vaccines against SARS-CoV-2. Serum was collected and stored at −80 °C until use. Samples were tested with neutralization assay kit using recombinant HRP-conjugated RBD of the Wuhan-Hu-1 spike glycoprotein. (**b–d**) Percentage of neutralization of serum from volunteers immunized with CoronaVac (n = 7), Comirnaty (n = 8), or Covishield (n = 15), respectively, at 1:40 dilution. (**e–g**) pNT_{50} of volunteers immunized with CoronaVac (n = 7), Comirnaty (n = 8), or Covishield (n = 15), respectively. Values considered statistically significant when $p < 0.05$ (*) via the Friedman test, followed by Dunn's multiple comparisons test, calculated by using GraphPad Prism, version 8.0. Data are shown as mean ± SD. Circles indicate the volunteers, and red line indicate the lowest pNT_{50} value equal to 40.

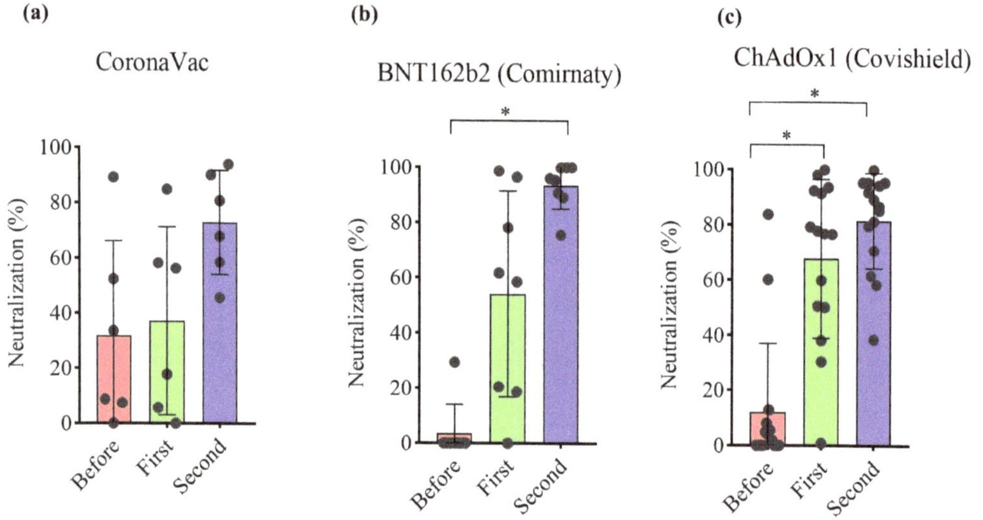

Figure 5. Imaging-based analysis detects the neutralizing capacity of volunteers after vaccination and following boost with (**a**) CoronaVac, (**b**) Comirnaty, or (**c**) Covishield against Wuhan-Hu-1 spike glycoprotein pseudovirus. Samples were submitted to the neutralization image-based assay for method validation. The neutralization capacity of serum specimens was determined by the transduction of HEK293T-ACE2 cells using Wuhan-Hu-1 spike-expressing pseudovirus. Values were considered statistically significant when $p < 0.05$ (*) examined using Friedman test, followed by Dunn's multiple comparisons test, calculated by GraphPad Prism, version 8.0. Data are shown as mean ± SD. Circles indicate the volunteers.

We observed that the ability of the vaccines to induce neutralizing antibodies against VoCs became reduced following the evolution of the variants. Following the second dose of CoronaVac, the calculated pNT_{50} showed a three-fold and a five-fold lower value, respectively, when the serum samples were subjected to neutralization assays using Delta and omicron VoC spikes, relative to the pNT_{50} calculated using pseudoviruses expressing Wuhan-Hu-1 spike (Figure 7a). The same profile was observed regarding the wane of serum potency following one and two doses of Comirnaty or Covishield (Figure 7b,c). These data demonstrated the antibody neutralization escape of VoCs detected using our imaging-based protocol. It is noted that some volunteers presented high-potency serum against Delta after the second dose of Comirnaty and Omicron B.1.1.529 after the first and second doses of Covishield.

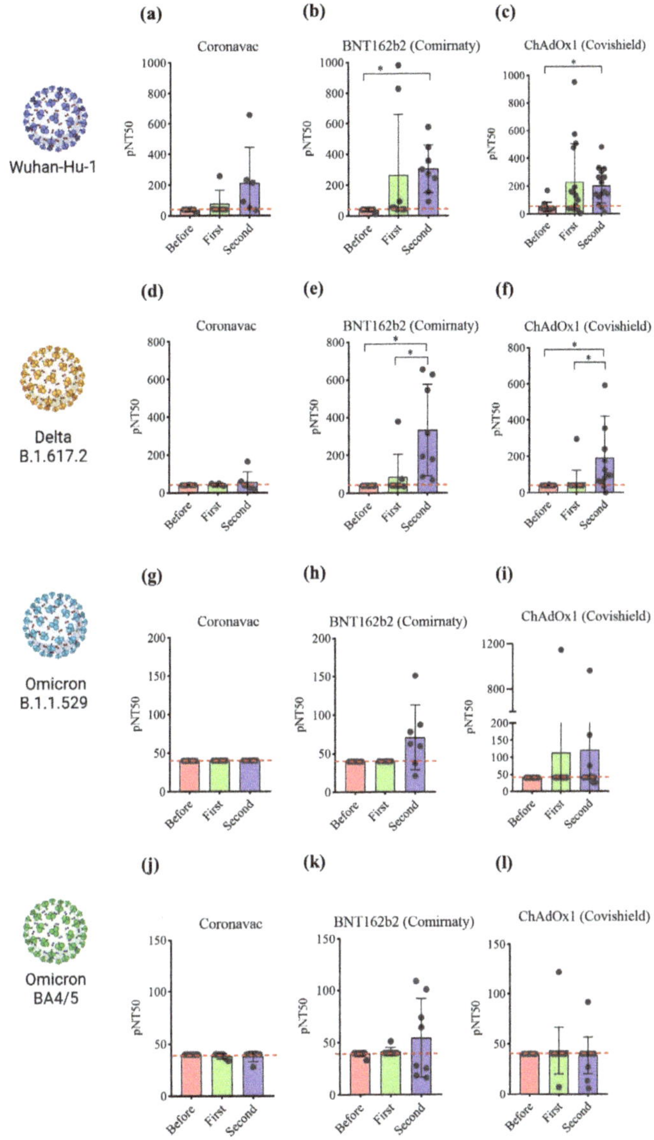

Figure 6. Imaging-based analysis characterizes cohorts based on the vaccine administered and number of doses. Validated samples were diluted and submitted to the neutralization image-based assay. Serum potency (pNT_{50}) was determined via nonlinear regression as a function of titers log_{10} and the inhibition of pseudovirus infection. The pseudoviruses produced expressed (**a–c**) Wuhan-Hu-1, (**d–f**) Delta B.1.617.2, (**g–i**) Omicron B.1.1.529, or (**j–l**) Omicron B.A4/5 spike proteins. Values considered statistically significant when $p < 0.05$ (*) by using Friedman test, followed by Dunn's multiple comparisons test, calculated by using GraphPad Prism, version 8.0. Data are shown as mean ± SEM. Circles indicate the volunteers, and red line indicate the lowest pNT_{50} value equal to 40.

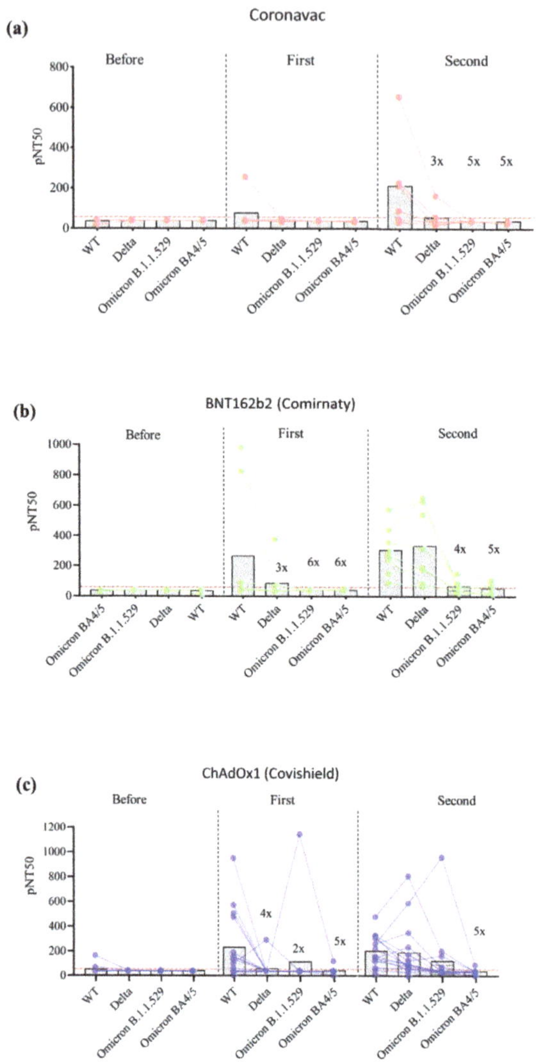

Figure 7. Imaging-based analysis can detect VoC neutralization escape. pNT_{50} values were used to compare serum potency with respect to different VoCs. Graphs show VoC neutralization escape after two doses of (**a**) Coronavac, (**b**) Comirnaty, or (**c**) Covishield. The numbers above the bars (such as 3×, 4×, 5×, and 6×) are the ratio between wild-type and VoCs pNT_{50} and denote the resistance of variants to the serum tested.

4. Discussion

In this work, we standardized a method of automated imaging analysis to perform a neutralization assay using pseudovirus. The pseudotyped viral particles produced were functional and capable of transducing ACE2-overexpressing cells. The protocol for automated imaging analysis can detect different levels of viral titers during the infection assay presenting accepted levels of variability among the experiments. Moreover, the protocol can also detect changes in serum titers induced by vaccination and measure the potency of neutralizing antibodies. Therefore, this protocol represents a useful tool for screening neutralizing antibodies and small molecules to prevent SARS-CoV-2 VoC infection, as well as for vaccine development [29,30]. Our method presented similar results on samples previously characterized by using ELISA with respect to the neutralization of Wuhan-Hu-1 RBD. Moreover, our results were near to those reported in previous studies [23,31]. It is noteworthy that slight differences in the pNT_{50} neutralization results were measured by using ELISA and our imaging-based method. It can be explained by the higher sensitivity of the imaging-based assay with respect to neutralizing antibodies targeting the S1 spike domain outside the RBD [32]. As the commercial kit employs RBD, no neutralizing antibodies outside this region are not detected.

Different groups standardized methods like ours, which were also demonstrated to produce results in agreement with the PRNT and identify neutralizing escape when the samples were tested against VoCs [15–21,23,31]. PRNT requires a BSL-3, rendering it non-safety compatible since it requires live viruses, is laborious, time-consuming, and is not well adapted to high-throughput studies [22,33]. Reverse genetics was used to develop SARS-CoV-2 particles carrying a reporter gene; however, it still needs BLS-3 infrastructure to handle this transgenic virus and is incompatible with most laboratories around the world [10]. It reflects the importance of this type of neutralization assay presented here. The use of pseudovirus presents the disadvantage of being restricted to the study of viral entry mechanisms since it is non-replicating. However, pseudoviruses are generally stable compared to wild-type viruses, which need to be propagated in animal cells. Each interaction with the host cells may generate point mutations, meaning that the viral batch tested is not homogeneous. With the use of plasmids, it is more controlled, and viral particles contain a unique clone of spike glycoprotein. Furthermore, the use of plasmids eases the assays with viral variants. As demonstrated in this work, vectors for expressing mutated spikes are interchanged to produce new pseudoviruses.

The continuous emergence of VoCs highlights the importance of regularly monitoring the population's neutralizing capacity, as well as evaluating vaccine efficacy against newly evolving viral strains. BQ.1.1 and XBB.1, for example, have demonstrated high resistance against vaccine-induced neutralizing antibodies. The neutralizing antibody titer against BQ.1.1 and XBB.1 was shown to decrease by 53- and 127-fold, respectively, compared to the Wuhan-Hu-1 strain after monovalent boost (two doses followed by monovalent booster) and by 80- and 232-fold after bivalent boost (two doses followed by bivalent booster), respectively [34].

In this work, we used the imaging analysis to evaluate pseudovirus neutralization; however, a pseudovirus based on pHAGE-CMV-Luc2-IRES-ZsGreen-W vector also integrates luciferase reporter gene in the host cell, making it possible to validate neutralization by using other methods, such as a bioluminescence plate reader [26,27]. Therefore, neutralizing protocols like ours should be widespread in different countries to monitor and characterize the immune response by neutralizing antibodies in a specific population.

It is noteworthy that even though the number of subjects evaluated herein was small, it was not our intention to describe biological patterns in the volunteers. Rather, we intended to use the samples to evaluate if our imaging-based method could provide consistent results through the quantification of serum potency. We also showed that some volunteers could efficiently neutralize Delta or Omicron B.1.1.529. The reason for this high neutralizing capacity was not investigated and is unknown. This can be a consequence of VoC infection, or they have naturally high responsiveness regarding the production of neutralizing antibodies against Delta or Omicron B.1.1.529. For the evaluation of the results produced by using the imaging-based protocol, the cohorts used were of sufficient size to generate data that were close to those reported in previous publications [23].

5. Conclusions

In sum, the method presented in this work was shown to efficiently detect changes in neutralizing antibody titers induced by vaccination. This pseudovirus imaging-based assay represents a valuable tool in the fight against COVID-19, which is presently focused on the vigilance of vaccine-induced neutralizing antibodies against newly emerging variants and subvariants, the development of new COVID-19 vaccines, as well as the screening of compounds and antibodies designed to prevent or reduce viral infection. Furthermore, this method can be used to monitor the waning of protective neutralizing antibodies acquired by infection or vaccination.

Supplementary Materials: The following supporting information can be downloaded at https://www.mdpi.com/article/10.3390/vaccines12030236/s1, Figure S1: pHAGE-CMV-Luc2-IRES-ZsGreen-W vector map. Figure S2: pHAGE-CMV-Luc2-IRES-ZsGreen-W vector sequence. Figure S3: Linear regression related to Figure 2. Figure S4: Correlation curves. Figure S5: IgG titers measured in serum samples via indirect ELISA using spike protein as antigen. Figure S6: Nonlinear regression curves related to Figure 6. Figure S7: Comparison between pNT_{50} values obtained from imaging-based neutralization analysis and by commercial ELISA kit.

Author Contributions: V.P.C.R.: conceptualization, assay execution, data curation, formal analysis, visualization, writing and final review; B.A.S.M.: funding aquisition; final review, conceptualization and project administration; H.C.Q., A.M.S.F. and B.S.D.F.: assay execution and data curation (H.C.Q. and A.M.S.F. contributed equally to the work); C.S.M.: assay execution; data curation and writing; V.T.B.d.S.: assay execution; A.F.E.: conceptualization; methodology; L.M.d.S.F.: resources; L.M.d.S.F. and R.J.d.S.B.: resources; writing—review and editing and M.B.P.S.: conceptualization; investigation, data curation, visualization, final review. All authors have read and agreed to the published version of the manuscript.

Funding: This study was supported by the National Research and Development Council (Grant Number: 424911/2021-3) and the Ministry of Science, Technology, and Innovation of the Federal Government.

Institutional Review Board Statement: This study was conducted in accordance with the Declaration of Helsinki and approved by the Ethics Committee of SENAI CIMATEC (protocol code 4.334.505; date of approval: 02/09/2021). Informed consent was obtained from all subjects involved in the study.

Informed Consent Statement: Informed consent was obtained from all subjects involved in the study.

Data Availability Statement: Created data is reported in the manuscript.

Acknowledgments: We would like to thank SENAI CIMATEC for infrastructure support, including the resources provided by the National System of the Nanotechnology Laboratories (SisNANO/MCTI/Brazil) Plataforma Aberta CIMATEC Nano, which greatly contributed to the successful completion of this research. The authors would like to thank Andris K. Walter for the critical analysis, English language revision, and manuscript copyediting assistance.

Conflicts of Interest: The authors declare no conflicts of interest.

References

1. Vaccines—COVID-19 Vaccine Tracker. Available online: https://covid19.trackvaccines.org/vaccines/approved/#vaccine-list (accessed on 26 December 2023).
2. Watson, O.J.; Barnsley, G.; Toor, J.; Hogan, A.B.; Winskill, P.; Ghani, A.C. Global Impact of the First Year of COVID-19 Vaccination: A Mathematical Modelling Study. *Lancet Infect. Dis.* **2022**, *22*, 1293–1302. [CrossRef]
3. Acharya, C.B.; Schrom, J.; Mitchell, A.M.; Coil, D.A.; Marquez, C.; Rojas, S.; Wang, C.Y.; Liu, J.; Pilarowski, G.; Solis, L.; et al. Viral Load among Vaccinated and Unvaccinated, Asymptomatic and Symptomatic Persons Infected with the SARS-CoV-2 Delta Variant. *Open Forum Infect. Dis.* **2022**, *9*, ofac135. [CrossRef]
4. Riemersma, K.K.; Haddock, L.A.; Wilson, N.A.; Minor, N.; Eickhoff, J.; Grogan, B.E.; Kita-Yarbro, A.; Halfmann, P.J.; Segaloff, H.E.; Kocharian, A.; et al. Shedding of Infectious SARS-CoV-2 despite Vaccination. *PLoS Pathog.* **2022**, *18*, e1010876. [CrossRef] [PubMed]
5. Alu, A.; Chen, L.; Lei, H.; Wei, Y.; Tian, X.; Wei, X. Intranasal COVID-19 Vaccines: From Bench to Bed. *EBioMedicine* **2022**, *76*, 103841. [CrossRef]
6. Chavda, V.P.; Vora, L.K.; Pandya, A.K.; Patravale, V.B. Intranasal Vaccines for SARS-CoV-2: From Challenges to Potential in COVID-19 Management. *Drug Discov. Today* **2021**, *26*, 2619–2636. [CrossRef]
7. Rambaut, A.; Holmes, E.C.; O'Toole, Á.; Hill, V.; McCrone, J.T.; Ruis, C.; du Plessis, L.; Pybus, O.G. A Dynamic Nomenclature Proposal for SARS-CoV-2 Lineages to Assist Genomic Epidemiology. *Nat. Microbiol.* **2020**, *5*, 1403–1407. [CrossRef]
8. Cov-Lineages. Available online: https://cov-lineages.org/lineage_list.html (accessed on 18 January 2023).

9. Tracking SARS-CoV-2 Variants. Available online: https://www.who.int/activities/tracking-SARS-CoV-2-variants (accessed on 18 January 2023).
10. Muruato, A.E.; Fontes-Garfias, C.R.; Ren, P.; Garcia-Blanco, M.A.; Menachery, V.D.; Xie, X.; Shi, P.Y. A High-Throughput Neutralizing Antibody Assay for COVID-19 Diagnosis and Vaccine Evaluation. *Nat. Commun.* **2020**, *11*, 4059. [CrossRef]
11. Nie, J.; Li, Q.; Wu, J.; Zhao, C.; Hao, H.; Liu, H.; Zhang, L.; Nie, L.; Qin, H.; Wang, M.; et al. Establishment and Validation of a Pseudovirus Neutralization Assay for SARS-CoV-2. *Emerg. Microbes Infect.* **2020**, *9*, 680–686. [CrossRef]
12. Xie, X.; Muruato, A.E.; Zhang, X.; Lokugamage, K.G.; Fontes-Garfias, C.R.; Zou, J.; Liu, J.; Ren, P.; Balakrishnan, M.; Cihlar, T.; et al. A Nanoluciferase SARS-CoV-2 for Rapid Neutralization Testing and Screening of Anti-Infective Drugs for COVID-19. *Nat. Commun.* **2020**, *11*, 5214. [CrossRef]
13. Xie, X.; Muruato, A.; Lokugamage, K.G.; Narayanan, K.; Zhang, X.; Zou, J.; Liu, J.; Schindewolf, C.; Bopp, N.E.; Aguilar, P.V.; et al. An Infectious CDNA Clone of SARS-CoV-2. *Cell Host Microbe* **2020**, *27*, 841–848.e3. [CrossRef]
14. Mulligan, M.J.; Lyke, K.E.; Kitchin, N.; Absalon, J.; Gurtman, A.; Lockhart, S.; Neuzil, K.; Raabe, V.; Bailey, R.; Swanson, K.A.; et al. Phase I/II Study of COVID-19 RNA Vaccine BNT162b1 in Adults. *Nature* **2020**, *586*, 589–593. [CrossRef]
15. Cruz-Cardenas, J.A.; Gutierrez, M.; López-Arredondo, A.; Castañeda-Delgado, J.E.; Rojas-Martinez, A.; Nakamura, Y.; Enciso-Moreno, J.A.; Palomares, L.A.; Brunck, M.E.G. A Pseudovirus-Based Platform to Measure Neutralizing Antibodies in Mexico Using SARS-CoV-2 as Proof-of-Concept. *Sci. Rep.* **2022**, *12*, 17966. [CrossRef]
16. Condor Capcha, J.M.; Lambert, G.; Dykxhoorn, D.M.; Salerno, A.G.; Hare, J.M.; Whitt, M.A.; Pahwa, S.; Jayaweera, D.T.; Shehadeh, L.A. Generation of SARS-CoV-2 Spike Pseudotyped Virus for Viral Entry and Neutralization Assays: A 1-Week Protocol. *Front. Cardiovasc. Med.* **2021**, *7*, 618651. [CrossRef]
17. Sun, H.; Xu, J.; Zhang, G.; Han, J.; Hao, M.; Chen, Z.; Fang, T.; Chi, X.; Yu, C. Developing Pseudovirus-Based Neutralization Assay against Omicron-Included SARS-CoV-2 Variants. *Viruses* **2022**, *14*, 1332. [CrossRef]
18. Izac, J.R.; Kwee, E.J.; Tian, L.; Elsheikh, E.; Gaigalas, A.K.; Elliott, J.T.; Wang, L. Development of a Cell-Based SARS-CoV-2 Pseudovirus Neutralization Assay Using Imaging and Flow Cytometry Analysis. *Int. J. Mol. Sci.* **2023**, *24*, 12332. [CrossRef]
19. Zhang, F.; Li, W.; Feng, J.; Ramos da Silva, S.; Ju, E.; Zhang, H.; Chang, Y.; Moore, P.S.; Guo, H.; Gao, S.J. SARS-CoV-2 Pseudovirus Infectivity and Expression of Viral Entry-Related Factors ACE2, TMPRSS2, Kim-1, and NRP-1 in Human Cells from the Respiratory, Urinary, Digestive, Reproductive, and Immune Systems. *J. Med. Virol.* **2021**, *93*, 6671–6685. [CrossRef]
20. Dadonaite, B.; Crawford, K.H.D.; Radford, C.E.; Farrell, A.G.; Yu, T.C.; Hannon, W.W.; Zhou, P.; Andrabi, R.; Burton, D.R.; Liu, L.; et al. A Pseudovirus System Enables Deep Mutational Scanning of the Full SARS-CoV-2 Spike. *Cell* **2023**, *186*, 1263–1278.e20. [CrossRef]
21. Donofrio, G.; Franceschi, V.; Macchi, F.; Russo, L.; Rocci, A.; Marchica, V.; Costa, F.; Giuliani, N.; Ferrari, C.; Missale, G. A Simplified SARS-CoV-2 Pseudovirus Neutralization Assay. *Vaccines* **2021**, *9*, 389. [CrossRef]
22. Mendoza, E.J.; Manguiat, K.; Wood, H.; Drebot, M. Two Detailed Plaque Assay Protocols for the Quantification of Infectious SARS-CoV-2. *Curr. Protoc. Microbiol.* **2020**, *57*, cpmc105. [CrossRef]
23. Garcia-Beltran, W.F.; St Denis, K.J.; Hoelzemer, A.; Lam, E.C.; Nitido, A.D.; Sheehan, M.L.; Berrios, C.; Ofoman, O.; Chang, C.C.; Hauser, B.M.; et al. MRNA-Based COVID-19 Vaccine Boosters Induce Neutralizing Immunity against SARS-CoV-2 Omicron Variant. *Cell* **2022**, *185*, 457–466.e4. [CrossRef]
24. Tauzin, A.; Chatterjee, D.; Dionne, K.; Gendron-Lepage, G.; Medjahed, H.; Bo, Y.; Perreault, J.; Goyette, G.; Gokool, L.; Arlotto, P.; et al. SARS-CoV-2 BA.4/5 Spike Recognition and Neutralization Elicited after the Third Dose of MRNA Vaccine. *medRxiv* **2022**. [CrossRef]
25. Meira, C.; Silva, D.; Santos, I.; Barreto, B.; Rocha, V.; Santos, E.; Dos Reis, B.; Evangelista, A.; Ribeiro Dos Santos, R.; Machado, B.; et al. Diagnostic Performance of Three ELISAs for Detection of Antibodies against SARS-CoV-2 in Human Samples. *Sci. World J.* **2022**, *2022*, 7754329. [CrossRef]
26. Crawford, K.H.D.; Eguia, R.; Dingens, A.S.; Loes, A.N.; Bloom, J.D.; Crawford, K. Protocol and Reagents for Pseudotyping Lentiviral Particles with SARS-CoV-2 Spike Protein for Neutralization Assays. *Viruses* **2020**, *15*, 513. [CrossRef]
27. Merluza, J.; Ung, J.; Makowski, K.; Robinson, A.; Manguiat, K.; Mueller, N.; Audet, J.; Chen, J.C.-Y.; Strong, J.E.; Wood, H.; et al. Validation and Establishment of the SARS-CoV-2 Lentivirus Surrogate Neutralization Assay as a Prescreening Tool for the Plaque Reduction Neutralization Test. *Microbiol. Spectr.* **2023**, *11*, e0378922. [CrossRef]
28. *Committee for Medicinal Products for Human Use ICH Guideline M10 on Bioanalytical Method Validation and Study Sample Analysis Step5*; European Medicines Agency: Amsterdam, The Netherlands, 2022.
29. Acharya, A.; Pandey, K.; Thurman, M.; Klug, E.; Trivedi, J.; Sharma, K.; Lorson, C.L.; Singh, K.; Byrareddy, S.N. Discovery and Evaluation of Entry Inhibitors for SARS-CoV-2 and Its Emerging Variants. *J. Virol.* **2021**, *95*, e0143721. [CrossRef]
30. He, Q.; Wu, L.; Xu, Z.; Wang, X.; Xie, Y.; Chai, Y.; Zheng, A.; Zhou, J.; Qiao, S.; Huang, M.; et al. An Updated Atlas of Antibody Evasion by SARS-CoV-2 Omicron Sub-Variants Including BQ.1.1 and XBB. *Cell Rep. Med.* **2023**, *4*, 100991. [CrossRef] [PubMed]
31. Garcia-Beltran, W.F.; Lam, E.C.; St Denis, K.; Nitido, A.D.; Garcia, Z.H.; Hauser, B.M.; Feldman, J.; Pavlovic, M.N.; Gregory, D.J.; Poznansky, M.C.; et al. Multiple SARS-CoV-2 Variants Escape Neutralization by Vaccine-Induced Humoral Immunity. *Cell* **2021**, *184*, 2372–2383.e9. [CrossRef]
32. Fang, Y.; Sun, P.; Xie, X.; Du, M.; Du, F.; Ye, J.; Kalveram, B.K.; Plante, J.A.; Plante, K.S.; Li, B.; et al. An Antibody That Neutralizes SARS-CoV-1 and SARS-CoV-2 by Binding to a Conserved Spike Epitope Outside the Receptor Binding Motif. *Sci. Immunol.* **2022**, *7*, 9962. [CrossRef] [PubMed]

33. Suthar, M.S.; Zimmerman, M.G.; Kauffman, R.C.; Mantus, G.; Linderman, S.L.; Hudson, W.H.; Vanderheiden, A.; Nyhoff, L.; Davis, C.W.; Adekunle, O.; et al. Rapid Generation of Neutralizing Antibody Responses in COVID-19 Patients. *Cell Rep. Med.* **2020**, *1*, 100040. [CrossRef] [PubMed]
34. Miller, J.; Hachmann, N.P.; Collier, A.Y.; Lasrado, N.; Mazurek, C.R.; Patio, R.C.; Powers, O.; Surve, N.; Theiler, J.; Korber, B.; et al. Substantial Neutralization Escape by SARS-CoV-2 Omicron Variants BQ.1.1 and XBB.1. *N. Engl. J. Med.* **2023**, *388*, 662–664. [CrossRef]

vaccines

MDPI

Article

Rapid Quantification of SARS-CoV-2 Neutralising Antibodies Using Time-Resolved Fluorescence Immunoassay

Gary R. McLean [1,2,*], Yueke Zhang [3], Rene Ndoyi [3], Adam Martin [3] and Julian Winer [4]

1 School of Human Sciences, London Metropolitan University, London N7 8DB, UK
2 National Heart and Lung Institute, Imperial College London, London W2 1PG, UK
3 PremaLabs Diagnostics UK Ltd., London W1J 6ER, UK
4 Winer Clinic London, London W1G 7DB, UK
* Correspondence: g.mclean@imperial.ac.uk

Abstract: The quantification of neutralising antibodies (NAb) for SARS-CoV-2 has become an important tool for monitoring protective immunity following infection or immunisation. In this study, we evaluated using World-Health-Organisation-standard immunoglobulin preparations, a novel point-of-care test that quantitates NAb by time-resolved fluorescent immunoassay. The assay provided robust data of binding antibody units (BAU) in 15 min that were well correlated with NAb values obtained by traditional in vitro neutralisation assay. The data also correlated well to spike-receptor-binding domain-binding antibodies over a broad range of plasma dilutions. The assay was extremely sensitive, able to detect positive samples after dilution 1:10,000 and over a wide range of BAU. Assay specificity was estimated at 96% using Pre-COVID-19 serum samples when applying a cut-off value of 47 BAU/mL, although readings of up to 100 BAU/mL could be considered borderline. This point-of-care diagnostic test is useful for rapid population screening and includes the use of capillary blood samples. Furthermore, it provides results for SARS-CoV-2 NAb in 15 min, which can inform immediate decisions regarding protective immunity levels and the need for continued COVID immunisations.

Keywords: SARS-CoV-2; neutralising antibodies; point-of-care diagnostic test

1. Introduction

The emergence and subsequent spread of severe acute respiratory syndrome coronavirus (SARS-CoV-2) from late 2019 was responsible for the worldwide COVID-19 pandemic [1]. This has resulted in the rapid development of a plethora of diagnostic tests for the virus [2] and specific antibodies [3], including the immunological response to vaccines [4] and neutralising antibodies (NAb) [5].

At this current stage of the COVID-19 pandemic, population-level knowledge of protective immunity is necessary to establish vaccine efficacy and the proportion of individuals at risk from infection or re-infection with SARS-CoV-2. Although it is desirable to determine both cell-mediated and humoral protective immunity, the latter is comparatively technically easier to perform. However, traditional assays for the detection of SARS-CoV-2 NAb, such as the plaque reduction neutralisation test (PRNT), require assays in which protective antibodies in samples are detected using live virus and the infection of mammalian cells [6]. These assays require several days to complete, skilled operators, and the use of BSL3 laboratories. Assays using pseudo-typed virus (pVNT) are effective and overcome the safety requirement [7] but do not address the time issues, nor are they high-throughput enough for mass testing. More recently, protein-based binding methods without infectious virus and cells have been developed to detect Nab, allowing for simplicity and speed over the time-consuming in vitro authentic virus-neutralising assays [8]. These tests, known as surrogate virus neutralisation tests (sVNT), detect NAb via chemiluminescence or enzymatic readouts of interference between the SARS-CoV-2 spike (S) receptor binding domain (RBD) and the entry receptor angiotensin-converting enzyme (ACE-2) [9]. The sVNT provide good agreement with PRNT and pVNT. In addition, sVNT are well suited to large-scale population testing, even though the PRNT and, to some extent, the pVNT remain the gold-standard assays. Numerous diagnostic tests have been applied throughout the pandemic and have offered benefits at various stages. It is clear now, with a largely immune population, that large-scale detection of NAb levels would be the most useful diagnostic method (for review, see [10]). A study evaluating a direct comparison of vaccine-induced NAb by PRNT and pVNT amongst performing laboratories revealed significant discrepancies that could be alleviated with the use of World Health Organisation (WHO)-standard immunoglobulins [11]. Thus, standardisation and simplification of NAb testing methods is critical for establishing immune correlates of protection, which is a key factor that can influence vaccine development. The large-scale use of sVNT is one approach which could lead to the adoption of NAb as a surrogate endpoint for vaccine efficacy that could streamline future COVID-19 vaccine efforts systematically on a global scale. One such sVNT, the PremaLabs Diagnostics SARS-CoV-2 NAb test kit, is a rapid in vitro immunochromatographic test (Time-Resolved Dry Fluorescence Immunoassay, TRFIA) for the qualitative detection of NAb to SARS-CoV-2 in human serum, plasma, or whole blood. The test kit serves as an aid in identifying individuals with adaptive immune responses to SARS-CoV-2, which are indicative of past infection or successful immunisation. The test cassette device is similar to a standard lateral flow device for detecting IgG to SARS-CoV-2 [12] and is made up of two precoated lines: test and control. The area at the test line is coated with monoclonal anti-RBD antibody. During the test, SARS-CoV-2-specific NAb in the sample interact with recombinant RBD conjugated to fluorescent particles, forming an antibody–antigen–fluorescent particle complex that migrates by capillary action on the membrane until it reaches the test line. The complex is bound by the capture antibodies on the test line to develop time-resolved fluorescence proportional to the amount of NAb in the sample. For a negative reaction, the recombinant RBD (without bound NAb) can bind to the free recombinant ACE-2 receptor proteins, blocking the binding site of capture antibody at the test line, resulting in low fluorescence (see Figure 1).

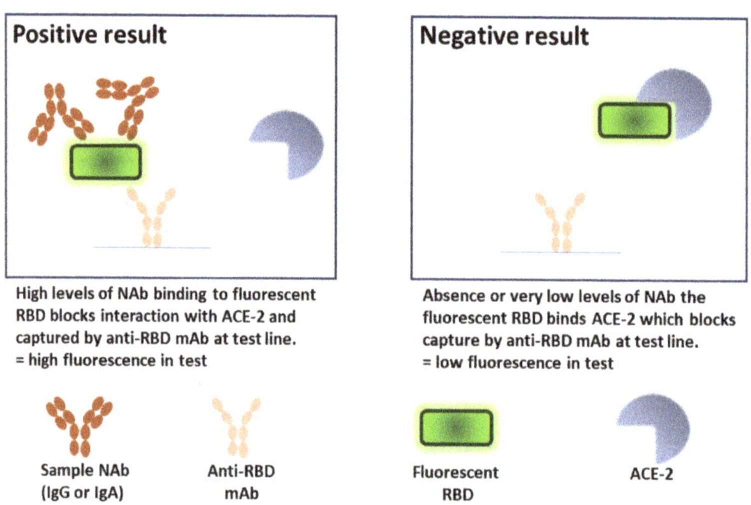

Figure 1. Schematic overview of PL-NAb test function. NAb in test sample bind to fluorescent SARS-CoV-2-RBD that is captured at the test line by specific mAb, causing a fluorescence signal proportional to level of NAb to be detected (**left panel**). In the absence of NAb within samples, the fluorescent RBD molecule is free to bind to ACE-2. This interaction blocks the capture at the test line, resulting in a low fluorescent signal (**right panel**). Graphics are not drawn to scale.

The aim of the current study was to evaluate the performance of the PremaLabs Diagnostics COVID-19 rapid NAb test (PL-NAb) to determine its sensitivity using SARS-CoV-2 reference immunoglobulins from the World Health Organisation (WHO) and specificity using Pre-COVID-19 serum samples that would be negative for NAb.

2. Materials and Methods

2.1. Samples and Standards

National Institute for Biological Standards and Control (NIBSC)-standard immunoglobulins were obtained from the Medicines and Healthcare products Regulatory Agency (MHRA). These reagents were the first WHO international reference panel for anti-SARS-CoV-2, developed in December 2020. The panel contained pooled plasma samples obtained from individuals that had recovered from COVID-19 and a negative control plasma obtained from healthy blood donors before 2019. The immunoglobulins panel, previously evaluated by a WHO international collaborative study [13], contained 5 individual components: NIBSC code 20/150 (high); 20/148 (mid); 20/144 (low anti-S, high anti-N); 20/140 (low); and 20/142 (negative).

Pre-COVID-19 serum samples from a prior study performed at Imperial College London [14] were used to define the range of negative values and determine assay specificity. Ethical approval had been obtained previously with informed consent.

2.2. Enzyme-Linked Immunosorbent Assay (ELISA)

Indirect ELISA was performed to determine the binding specificity of NIBSC standards to SARS-CoV-2 recombinant spike S1 and nucleocapsid (N) antigens (Cambridge Bioscience, Cambridge, UK). This analysis was designed to establish relative levels of S- and N-binding immunoglobulins contained within the NIBSC plasma preparations. We performed ELISA as described previously [15], with minor modifications. SARS-CoV-2 antigens were diluted to 1 μg/mL in PBS and coated into ELISA plate wells. Plasma was serially titrated 2-fold, beginning at 1:100 and ending at 1:51,200 using PBS containing between 20 and 5% milk powder (PBSTM). Bound antibodies were detected using anti-human IgG or anti-human IgA HRP conjugates (Southern Biotechnology Associates, Birmingham, AL, USA) diluted 1:1000 in PBSTM. Reactions were developed using TMB substrate (Thermo Fisher Scientific, Waltham, MA, USA) before obtaining optical density at 450 nm.

2.3. SARS-CoV-2-Neutralising Antibody Diagnostic Test (PL-NAb)

The method was followed in accordance with the manufacturer's instructions. Human plasma or serum samples were diluted where necessary in sterile PBS. An amount of 20 μL of sample (triplicate determinations) was added to 80 μL of the supplied sample diluent, mixed, and added to the test strip device. After 15 min incubation, dry fluorescence was measured using the LS-1100 (medi) or the LS-21100 (maxi) dry fluorescence analyser (Lansion Biotechnology Co., Ltd., Nanjing, China). Data were generated in BAU/mL with a detection range of 10–3000. Measurements >40 BAU/mL are considered a positive result based on prior analysis of 150 negative samples. According to the manufacturers, the positive test rate of SARS-CoV-2 NAb for 50 confirmed cases was (46/50) 92.00%. The negative coincidence rate of SARS-CoV-2 NAb for 150 negative samples was (143/150) 95.33%.

2.4. Data Graphics and Analysis

Data obtained were plotted using GraphPad Prism version 9.4.1 (GraphPad Software, LLC, San Diego, USA). Data were first evaluated for normal distribution to ascertain appropriate statistical tests and linear regression was assessed by Spearman's correlation and the r values stated.

3. Results

We evaluated the performance of the PremaLabs Diagnostics COVID-19 rapid NAb (PL-NAb) test (https://premalabsdiagnostics.com/covid-19-nab-test/ accessed on 23 September 2022) using the NIBSC standard immunoglobulins, which formed the first World Health Organisation (WHO) international reference panel for anti-SARS-CoV-2. The immunoglobulin panel contains pooled plasma samples obtained from individuals recovered from COVID-19 and a negative control plasma obtained from healthy blood donors before 2019. These had previously been evaluated by a WHO international collaborative study and harmonised against live virus neutralisation assays for NAb, anti-RBD IgG, anti-S1 IgG, anti-Spike IgG, and anti-N IgG (Table 1) [13]. The panel's five individual components were therefore assigned as high, mid, low anti-S and high anti-N, low, and negative for SARS-CoV-2 antibodies. Thus, individual components contained a range of values recorded as IU/mL for NAb (relative to the International Standard 20/136) and the standard BAU/mL for binding antibodies. Larger neutralising antibody values were contained in the high component (1473 IU/mL), with decreasing values seen as mid (210 IU/mL), low S high N (65 IU/mL), and low (44 IU/mL). The negative component was consistently very low or undetectable (Table 1). Values for binding antibodies to spike, receptor binding domain (RBD), spike S1 subunit (S1), and nucleocapsid (N) showed similar trends, with the high component containing the largest BAU/mL and decreasing values seen with the other plasma components (Table 1).

Table 1. Anti-SARS-CoV-2 antibody titres are calculated as the geometric mean of the potencies obtained from the collaborative study participants calibrated against the WHO International Standard for anti-SARS-CoV-2 immunoglobulin (NIBSC 20/136). The five components are shown at the top with corresponding NIBSC code in brackets. Neutralising antibody (Neut Ab), anti-receptor binding domain (anti-RBD), anti-spike 1 subunit (Anti-S1), anti-spike (Anti-S1/S2), anti-nucleocapsid (Anti-N) rows are listed. IU: International Units; BAU: binding antibody units; n.d.: not detected.

	High (20/150)	Mid (20/148)	Low S, High N (20/144)	Low (20/140)	Negative (20/142)
Neut Ab (IU/mL)	1473	210	65	44	1
Anti-RBD IgG (BAU/mL)	817	205	66	45	n.d.
Anti-S1 IgG (BAU/mL)	766	246	50	46	n.d.
Anti-Spike IgG (BAU/mL)	832	241	86	53	n.d.
Anti-N IgG (BAU/mL)	713	295	146	12	n.d.

Our aim was to use the NIBSC standards to determine the linearity, sensitivity, and specificity of the PremaLabs Diagnostic NAb test (referred to as PL-NAb). Serial 10-fold dilution of the four NIBSC standards (high, mid, low S high N, and low) that have binding activity was performed to determine linearity of the NAb content using PL-NAb. This analysis used the LS-1100 device and time-resolved fluorescence immunoassay (TRFIA) to determine antibody interference with SARS-CoV-2 S-RBD (original Wuhan strain accession number YP_009724390.1) binding to the entry receptor angiotensin-converting enzyme (ACE-2), which is represented as BAU/mL. Figure 2 shows that the four NIBSC standards show reproducible linearity of NAb detection over the titration range, with detection still apparent at plasma dilutions of 1:1000. Saturation of the high NIBSC sample was observed at dilutions lower than 1:10, which was not observed with other samples. At the lower end of detection, the high NIBSC standard still gave a positive detectable result (64 BAU/mL) when diluted 1:10,000, whereas other samples at this dilution were below the limit of detection (LOD) of 10 BAU/mL. Importantly, these PL-NAb data are consistent with data obtained using our ELISA binding assay, which was calibrated using the NIBSC standards (Table 2). Endpoint titration (first plasma dilution to become <0.2 OD450 nm) of the NIBSC standards and binding analyses by ELISA demonstrated that the high NIBSC standard contained the highest level of IgG and IgA to SARS-CoV-2 S and is followed sequentially with reducing levels of binding found with mid, low S high N, and low NIBSC standards. IgG- and IgA-binding capacity to SARS-CoV-2 N follows similar trends, although this determinant is not detected by the PL-NAb test. Therefore, we established using ELISA that the NIBSC standards show similar trends, in terms of binding to S (Table 2), to those that we observe with surrogate neutralization using PL-Nab (Figure 2).

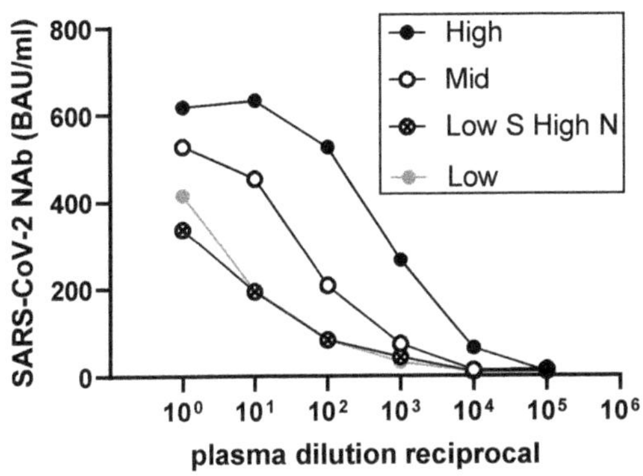

Figure 2. NIBSC standards display exceptional linearity when diluted and analysed via TRFIA. High, mid, low S high N, and low NIBSC standards were titrated in 10-fold steps and analysed for NAb using the PL-NAb test. Data are displayed as BAU/mL versus dilution and are represented by three independent determinations.

Table 2. ELISA titration endpoints from binding assay of NIBSC standard plasma (left column) to SARS-CoV-2 S and N proteins. Plasma was diluted until endpoint binding was achieved. Titres are shown as the first dilution of IgG and IgA to reach <0.2 OD450 nm.

	IgG Spike	IgA Spike	IgG Nucleocapsid	IgA Nucleocapsid
High (20/150)	1:25,600	1:800	>1:51,200	1:800
Mid (20/148)	1:6400	1:200	1:25,600	1:1600
Low S, High N (20/140)	1:1600	1:100	1:12,800	1:100
Low (20/140)	1:800	1:100	1:800	1:100
Negative (20/142)	1:100	1:100	1:200	1:100

Next, we correlated these NAb data obtained with the NIBSC standards to the reported titre of NAb (IU/mL) and anti-RBD (BAU/mL). Excellent correlations were observed with undiluted plasma (Figure 3A,B), plasma diluted 1:10 (Figure 3C,D), and plasma diluted 1:100 (Figure 3E,F). Correlations were consistently greater than R^2 of 0.90 when analysing the NAb data, except with undiluted plasma samples including the NIBSC negative when correlated with WHO values for NAb ($R^2 = 0.88$). Overall, the correlations observed were similar throughout, but tended to be stronger, such that greater R^2 values were seen when comparing PL-NAb data with anti-RBD antibody rather than NAb. These data suggest that the NAb test faithfully reproduces neutralisation data generated by PRNT with excellent sensitivity, and that anti-RBD antibodies in plasma are the likely neutralising component detected.

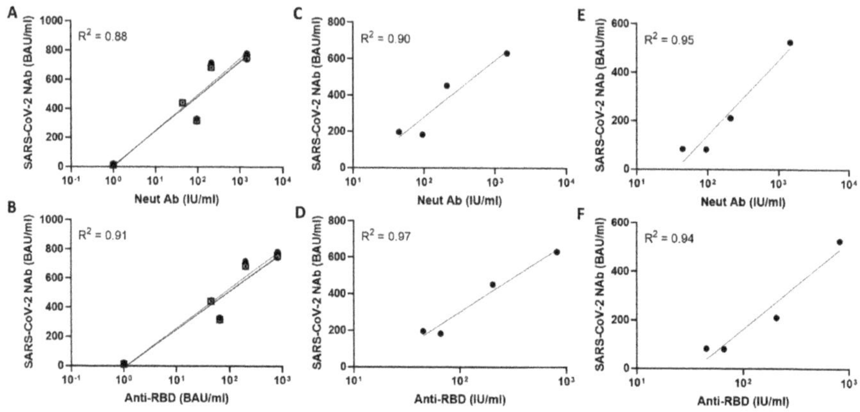

Figure 3. Correlations of levels of NIBSC standards NAb to SARS-CoV-2 by TRFIA with WHO data Neut Ab (**A**,**C**,**E**) and Anti-RBD (**B**,**D**,**F**). (**A**,**B**) represent undiluted NIBSC standards; (**C**,**D**) represent 1:10 dilution of NIBSC standards; and (**E**,**F**) represent 1:100 dilution of NIBSC standards. Correlations by linear regression are shown as R^2 values. Data are shown as the average of 3 independent analyses.

Lastly, we determined the PL-NAb assay specificity using 50 serum samples from pre-COVID-19 volunteers from a prior study [14]. Undiluted serum was analysed via a PL-NAb test and the LS-21100 analyser to produce scatter plots and determine the positive/negative cut-off value for BAU/mL that reflects functional NAb. Pre-COVID-19 serum samples produced mostly very low BAU/mL values at the LOD of 10 BAU/mL (Figure 4). Sample values ranged from the LOD to 68.99 BAU/mL, with a mean of 15.45 BAU/mL (standard deviation 16.07). Applying the mean + 2SD rule to these data provides a positive/negative cut-off value of 47.58 BAU/mL (dotted lines Figure 4). Out of the 50 samples, two were marginally above this cut-off, reflecting an assay specificity of 96%, which is consistent with prior observations that estimate the cut-off value to be 40 BAU/mL with 95.45% specificity (according to documents provided by the test manufacturers).

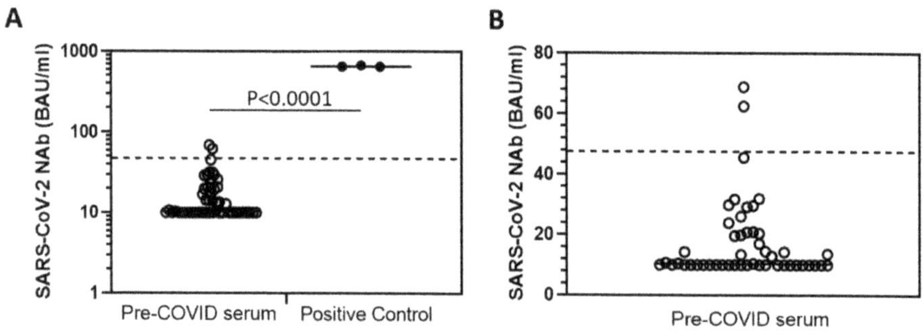

Figure 4. Scatterplots of NAb determination by TRFIA of Pre-COVID-19 serum samples (triplicate) to determine specificity and positive/negative cut-off value (dotted horizontal line = 47.58 BAU/mL). (**A**): displays scatter of Pre-COVID-19 serum data with respect to positive control (high NIBSC standard), significantly different $p < 0.0001$. (**B**): displays scatter range in values of Pre-COVID-19 serum data alone.

4. Discussion

Here, we have evaluated the performance of a new quantitative diagnostic test for SARS-CoV-2 NAb. The test is rapid; requires just 20 μL of plasma, serum, or whole blood; and provides a quantitative measure of the level of NAb in 15 min, with excellent similarity to traditional methods of detecting SARS-CoV-2 NAb. The test can therefore faithfully reflect levels of protective immunity to SARS-CoV-2 and could be a useful population screening tool to inform decisions regarding continued immunisation. Our analysis relied on the use of a set of standard immunoglobulin preparations promoted by the WHO as reagents that allow a standardised international reference panel to evaluate SARS-CoV-2 antibody diagnostics across regions and different tests. We also included a set of Pre-COVID-19 serum samples to estimate the diagnostic cut-off point at which a reliable measure of potential protective NAb could be expected.

The PL-NAb diagnostic test is a competitive immunoassay where the specific SARS-CoV-2 NAb inhibit the binding of the recombinant SARS-CoV-2 RBD, which has been fluorescently labelled, to recombinant human ACE-2. Detection is achieved by dry fluorescence in a time-resolved manner (TRFIA) following UV excitation of chelated lanthanide immunomicrospheres [16]. The assay allows for a stronger signal intensity to be detected at a slightly later timepoint compared with traditional immunofluorescence and improves the quality and accuracy of detection by minimising background emission interference. The data output determines NAb according to binding antibody units, a normalised measure that is based on the WHO standard, enabling comparisons to be made between different tests. The analytical measuring range is reported to be from 10 to 3000 BAU/mL, reflecting a significant breadth of detection spanning the likely sample variations. Linearity of the PL-NAb test was determined across the analytical measuring range using NIBSC standard immunoglobulins containing different levels of SARS-CoV-2 neutralising and binding antibodies. We did not observe BAU/mL >1000 with any of the standards, reflecting the levels of NAb in individuals that had recovered from a single SARS-CoV-2 infection early in the pandemic, rather than having received multiple immunisations with COVID-19 vaccines. Furthermore, at the upper detection range, saturation of detection was noted when assaying an undiluted plasma sample containing high levels of SARS-CoV-2 antibodies (NIBSC high). This phenomenon was not evident with immunoglobulin preparations containing lower levels of antibodies or with more dilute plasma. In fact, linearity of NAb detection was exceptional at these lower levels, reflecting good assay sensitivity and a low potential for false negatives. Importantly, we confirmed these data obtained with PL-Nab using alternative ELISA binding experiments.

Correlation of the PL-NAb test with NAb and RBD binding, according to the published WHO data for these preparations, was found to be very high. This reflects the fact that the majority of NAb to SARS-CoV-2 will interact with the RBD [17]. Of course, the assay will not determine other sites of NAb interaction with spike, such as those binding to the N-terminal domain (NTD), but these are likely to be of minimal contribution compared to RBD-binding NAb in the neutralising effect on SARS-CoV-2 [18]. For this reason, we did not correlate PL-NAb data with the other parameters for the standards supplied by NIBSC, such as anti-S1, anti-S and anti-N. However, inspecting the WHO data shown in Table 1 and based on our own analyses shown in Figure 3, it is likely that similar correlations to those we observed for NAb and RBD binding would be observed. Comparisons of antibody units, such as IU/mL for NAb and BAU/mL for binding, have been established, but are difficult to compare directly, particularly when different assays are used to define these parameters. However, it has been proposed that 1000 BAU/mL can be used to compare assays, particularly for the same class of immunoglobulin (e.g., IgG or IgA) and their specificity (e.g., S or N) [19]. Since the PL-NAb test determines NAb binding to RBD as a surrogate for neutralising, the use of BAU/mL values can be considered an appropriate measure to reflect the level of protective antibodies in a sample and therefore the level of protection afforded against SARS-CoV-2 in vivo. In fact, levels of anti-S IgG >2000 BAU/mL are routinely observed in samples measured using PL-NAb from fully vaccinated and COVID-19-experienced individuals. Approximately 90% of adults (age range 18–80) recorded levels above 1000 BAU/mL. Therefore, 1000 BAU/mL could be considered a benchmark for a decision regarding the timing of further immunisation against SARS-CoV-2 [20]. Despite a current lack of consensus on a robust immune correlate of protection for SARS-CoV-2, evidence does suggest that anti-S and NAb is a likely readout. However, translation of this to BAU/mL currently remains elusive, although values of 154 BAU/mL anti-S for ancestral SARS-CoV-2 and 171 BAU/mL anti-S for the alpha variant have been recorded as thresholds of protection [21], suggesting that 1000 BAU/mL with PL-Nab represents exceptional protection.

The lower BAU/mL values for NAb seen at assay lower LOD are more difficult to link to protective capabilities. Low positive or borderline samples have been classified as having SARS-CoV-2 S antibody levels <100 BAU/mL [22]. Our analysis of diluted NIBSC standards revealed that even the low standards produced BAU/mL values >100 when diluted 1:100, well below the dilution factor when analysing whole blood samples. The high NIBSC standard required dilution to 1:10,000 to reduce BAU/mL to <100, reflecting good assay sensitivity at the lower end of detection. Translation of this value to in vivo protection is not entirely possible, as it is just one component of protective immunity that is more complex, including a cell-mediated component [23,24]. Nevertheless, we could determine a lower-level cut-off of approximately 48 BAU/mL when analysing the range of values obtained using Pre-COVID-19 serum samples. Thus, one could envisage that the PL-NAb test would have the following conditions assigned based on currently available information.

PL-NAb (BAU/mL)	Outcome
0–50	Negative for NAb
51–100	Borderline or low positive for NAb (not protected)
101–1000	Moderately protective immunity
>1000	Fully protected (in terms of SARS-CoV-2 disease severity)

5. Conclusions

The PL-NAb test is a rapid and accurate alternative to PRNT for determining levels of NAb to SARS-CoV-2. It can assist with determinations of protective levels of immunity to SARS-CoV-2 at a population level and can assist with decisions regarding the further need for booster immunisations or not. The test does not provide information about the full spectrum of immunity and may require updating as new variants with altered RBD sequences emerge. However, it offers a snapshot of an essential protective immunity component that reflects immune status to this relatively new virus. We now know that having immunity to SARS-CoV-2 from prior infection or immunisation results in improved outcomes following infection, including reduced symptoms and disease severity and lower hospitalisation and death rates, even as new variants emerge [25]. At this late stage of the COVID-19 pandemic, when we are living with COVID-19, determinations of SARS-CoV-2 NAb levels at a population level could go some way towards allowing a return to pre-pandemic behaviour, speeding up the transition to a post-pandemic world based on immune protection status.

Author Contributions: Conceptualization, G.R.M. and A.M.; methodology, G.R.M., Y.Z., R.N. and J.W.; investigation, G.R.M., Y.Z., R.N. and J.W.; writing—original draft preparation, G.R.M.; writing—review and editing, G.R.M., Y.Z., R.N., A.M. and J.W. All authors have read and agreed to the published version of the manuscript.

Funding: This research received no external funding.

Institutional Review Board Statement: Ethical review and approval were waived for this study due to the standard immunoglobulin samples being obtained commercially and Pre-COVID-19 samples having undergone ethical approval according to the referenced prior study.

Informed Consent Statement: Patient consent was waived due to the standard immunoglobulin samples being obtained commercially and Pre-COVID-19 samples having undergone ethical approval according to the referenced prior study.

Data Availability Statement: Data are available upon request from the corresponding author.

Conflicts of Interest: G.R.M. and J.W. declare a potential conflict of interest as paid consultants for PremaLabs Diagnostics UK Ltd. Y.Z., R.N. and A.M. are employees of PremaLabs Diagnostics UK Ltd., which has a commercial interest in the outcomes of this study.

References

1. Zhou, P.; Yang, X.L.; Wang, X.G.; Hu, B.; Zhang, L.; Zhang, W.; Si, H.R.; Zhu, Y.; Li, B.; Huang, C.L.; et al. A pneumonia outbreak associated with a new coronavirus of probable bat origin. *Nature* **2020**, *579*, 270–273. [CrossRef] [PubMed]
2. Zhang, Y.; Huang, Z.; Zhu, J.; Li, C.; Fang, Z.; Chen, K.; Zhang, Y. An updated review of SARS-CoV-2 detection methods in the context of a novel coronavirus pandemic. *Bioeng. Transl. Med.* **2022**, e10356. [CrossRef] [PubMed]
3. Mallano, A.; Ascione, A.; Flego, M. Antibody Response against SARS-CoV-2 Infection: Implications for Diagnosis, Treatment and Vaccine Development. *Int. Rev. Immunol.* **2021**, *41*, 393–413. [CrossRef] [PubMed]
4. Devi, M.J.; Gaffar, S.; Hartati, Y.W. A review post-vaccination SARS-CoV-2 serological test: Method and antibody titer response. *Anal. Biochem.* **2022**, *658*, 114902. [CrossRef]
5. Ndzouboukou, J.-L.B.; Zhang, Y.-D.; Fan, X.-L. Recent Developments in SARS-CoV-2 Neutralizing Antibody Detection Methods. *Curr. Med. Sci.* **2021**, *41*, 1052–1064. [CrossRef]
6. Bonifacio, M.A.; Laterza, R.; Vinella, A.; Schirinzi, A.; Defilippis, M.; Di Serio, F.; Ostuni, A.; Fasanella, A.; Mariggiò, M.A. Correlation between In Vitro Neutralization Assay and Serological Tests for Protective Antibodies Detection. *Int. J. Mol. Sci.* **2022**, *23*, 9566. [CrossRef]
7. Von Rhein, C.; Scholz, T.; Henss, L.; Kronstein-Wiedemann, R.; Schwarz, T.; Rodionov, R.N.; Corman, V.M.; Tonn, T.; Schnierle, B.S. Comparison of potency assays to assess SARS-CoV-2 neutralizing antibody capacity in COVID-19 convalescent plasma. *J. Virol. Methods* **2021**, *288*, 114031. [CrossRef]
8. Wohlgemuth, N.; Whitt, K.; Cherry, S.; Roubidoux, E.K.; Lin, C.-Y.; Allison, K.J.; Gowen, A.; Freiden, P.; Allen, E.K.; Gaur, A.H.; et al. An Assessment of Serological Assays for SARS-CoV-2 as Surrogates for Authentic Virus Neutralization. *Microbiol. Spectr.* **2021**, *9*, e0105921. [CrossRef]
9. Pieri, M.; Infantino, M.; Manfredi, M.; Nuccetelli, M.; Grossi, V.; Lari, B.; Tomassetti, F.; Sarubbi, S.; Russo, E.; Amedei, A.; et al. Performance evaluation of four surrogate Virus Neutralization Tests (sVNTs) in comparison to the in vivo gold standard test. *Front. Biosci.* **2022**, *27*, 74. [CrossRef]
10. Jeremiah, S.S.; Miyakawa, K.; Ryo, A. Detecting SARS-CoV-2 neutralizing immunity: Highlighting the potential of split nanoluciferase technology. *J. Mol. Cell Biol.* **2022**, *14*, mjac023. [CrossRef]
11. Liu, J.; Mao, Q.; Wu, X.; He, Q.; Bian, L.; Bai, Y.; Wang, Z.; Wang, Q.; Zhang, J.; Liang, Z.; et al. Considerations for the Feasibility of Neutralizing Antibodies as a Surrogate Endpoint for COVID-19 Vaccines. *Front. Immunol.* **2022**, *13*, 814365. [CrossRef]
12. Chen, Z.; Zhang, Z.; Zhai, X.; Li, Y.; Lin, L.; Zhao, H.; Bian, L.; Li, P.; Yu, L.; Wu, Y.; et al. Rapid and Sensitive Detection of anti-SARS-CoV-2 IgG, Using Lanthanide-Doped Nanoparticles-Based Lateral Flow Immunoassay. *Anal. Chem.* **2020**, *92*, 7226–7231. [CrossRef]
13. Knezevic, I.; Mattiuzzo, G.; Page, M.; Minor, P.; Griffiths, E.; Nuebling, M.; Moorthy, V. WHO International Standard for evaluation of the antibody response to COVID-19 vaccines: Call for urgent action by the scientific community. *Lancet Microbe* **2022**, *3*, e235–e240. [CrossRef]
14. Mallia, P.; Footitt, J.; Sotero, R.; Jepson, A.; Contoli, M.; Trujillo-Torralbo, M.-B.; Kebadze, T.; Aniscenko, J.; Oleszkiewicz, G.; Gray, K.; et al. Rhinovirus Infection Induces Degradation of Antimicrobial Peptides and Secondary Bacterial Infection in Chronic Obstructive Pulmonary Disease. *Am. J. Respir. Crit. Care Med.* **2012**, *186*, 1117–1124. [CrossRef]
15. Siddiqui, S.; Hackl, S.; Ghoddusi, H.; McIntosh, M.R.; Gomes, A.C.; Ho, J.; Reeves, M.B.; McLean, G.R. IgA binds to the AD-2 epitope of glycoprotein B and neutralizes human cytomegalovirus. *Immunology* **2021**, *162*, 314–327. [CrossRef]
16. Dickson, E.F.G.; Pollak, A.; Diamandis, E.P. Ultrasensitive bioanalytical assays using time-resolved fluorescence detection. *Pharmacol. Ther.* **1995**, *66*, 207–235. [CrossRef]
17. Suthar, M.S.; Zimmerman, M.G.; Kauffman, R.C.; Mantus, G.; Linderman, S.L.; Hudson, W.H.; Vanderheiden, A.; Nyhoff, L.; Davis, C.W.; Adekunle, O.; et al. Rapid generation of neutralizing antibody responses in COVID-19 patients. *Cell Rep. Med.* **2020**, *1*, 100040. [CrossRef]
18. Chi, X.; Yan, R.; Zhang, J.; Zhang, G.; Zhang, Y.; Hao, M.; Zhang, Z.; Fan, P.; Dong, Y.; Yang, Y.; et al. A neutralizing human antibody binds to the N-terminal domain of the Spike protein of SARS-CoV-2. *Science* **2020**, *369*, 650–655. [CrossRef]
19. Kristiansen, P.A.; Page, M.; Bernasconi, V.; Mattiuzzo, G.; Dull, P.; Makar, K.; Plotkin, S.; Knezevic, I. WHO International Standard for anti-SARS-CoV-2 immunoglobulin. *Lancet* **2021**, *397*, 1347–1348. [CrossRef]
20. Borgonovo, F.; Stangalini, C.A.; Tinelli, C.; Mariani, C.; Mileto, D.; Cossu, M.V.; Abbati, L.; Bilardo, L.; Gagliardi, G.; Cutrera, M.; et al. Decay rate of antiS1/S2 IgG serum levels after 6 months of BNT162b2 vaccination in a cohort of COVID-19-naive and COVID-19-experienced subjects. *Hum. Vaccines Immunother.* **2022**, *18*, 2060018. [CrossRef]
21. Perry, J.; Osman, S.; Wright, J.; Richard-Greenblatt, M.; Buchan, S.A.; Sadarangani, M.; Bolotin, S. Does a humoral correlate of protection exist for SARS-CoV-2? A systematic review. *PLoS ONE* **2022**, *17*, e0266852. [CrossRef] [PubMed]
22. Egger, A.E.; Irsara, C.; Holzer, B.; Winkler, C.; Bellmann-Weiler, R.; Weiss, G.; Hartmann, B.; Prokop, W.; Hoermann, G.; Griesmacher, A.; et al. Borderline and weakly positive antibody levels against the S-protein of SARS-CoV-2 exhibit limited agreement with virus neutralization titres. *J. Clin. Virol. Plus* **2022**, *2*, 100058. [CrossRef] [PubMed]

23. Thevarajan, I.; Nguyen, T.H.O.; Koutsakos, M.; Druce, J.; Caly, L.; van de Sandt, C.E.; Jia, X.; Nicholson, S.; Catton, M.; Cowie, B.; et al. Breadth of concomitant immune responses prior to patient recovery: A case report of non-severe COVID-19. *Nat. Med.* **2020**, *26*, 453–455. [CrossRef] [PubMed]
24. Ni, L.; Ye, F.; Cheng, M.-L.; Feng, Y.; Deng, Y.-Q.; Zhao, H.; Wei, P.; Ge, J.; Gou, M.; Li, X.; et al. Detection of SARS-CoV-2-Specific Humoral and Cellular Immunity in COVID-19 Convalescent Individuals. *Immunity* **2020**, *52*, 971–977. [CrossRef]
25. McLean, G.; Kamil, J.; Lee, B.; Moore, P.; Schulz, T.F.; Muik, A.; Sahin, U.; Türeci, Ö.; Pather, S. The Impact of Evolving SARS-CoV-2 Mutations and Variants on COVID-19 Vaccines. *Mbio* **2022**, *13*, e0297921. [CrossRef]

vaccines

MDPI

Communication

Development of a Bioluminescent Imaging Mouse Model for SARS-CoV-2 Infection Based on a Pseudovirus System

Xi Wu [1], Nana Fang [2], Ziteng Liang [1,3], Jianhui Nie [1], Sen Lang [4], Changfa Fan [4], Chunnan Liang [4], Weijin Huang [1] and Youchun Wang [3,5,*]

1 Division of HIV/AIDS and Sex-Transmitted Virus Vaccines, Institute for Biological Product Control, National Institutes for Food and Drug Control (NIFDC), Beijing 102629, China; wuxi@nifdc.org.cn (X.W.); huangweijin@nifdc.org.cn (W.H.)
2 National Vaccine and Serum Institute, Beijing 101111, China; m18210828042@163.com
3 Chinese Academy of Medical Sciences & Peking Union Medical College, Beijing 100006, China
4 National Rodent Laboratory Animal Resources Center, Institute for Laboratory Animal Resources, National Institutes for Food and Drug Control (NIFDC), Beijing 102629, China
5 Institute of Medical Biology, Chinese Academy of Medical Sciences, Kunming 650118, China
* Correspondence: wangyc@imbcams.com.cn

Abstract: Coronavirus disease 2019 (COVID-19), caused by severe acute respiratory syndrome coronavirus 2 (SARS-CoV-2), remains widely pandemic around the world. Animal models that are sensitive to the virus are therefore urgently needed to evaluate potential vaccines and antiviral agents; however, SARS-CoV-2 requires biosafety level 3 containment. To overcome this, we developed an animal model using the intranasal administration of SARS-CoV-2 pseudovirus. As the pseudovirus contains the firefly luciferase reporter gene, infected tissues and the viral load could be monitored by in vivo bioluminescent imaging. We used the model to evaluate the protective efficacy of monoclonal antibodies and the tissue tropism of different variants. The model may also be a useful tool for the safe and convenient preliminary evaluation of the protective efficacy of vaccine candidates against SARS-CoV-2, as well as the treatment efficacy of anti-viral drugs.

Keywords: SARS-CoV-2 pseudovirus; mouse model; in vivo bioluminescent

1. Introduction

By March 2023, over 675 million confirmed cases of coronavirus disease 2019 (COVID-19), including over 6.87 million deaths, had been reported to the World Health Organization. The combination of high transmissibility and virulence of severe acute respiratory syndrome coronavirus 2 (SARS-CoV-2), the causative agent of COVID-19, underscores the urgent need to develop highly effective vaccines and antiviral agents, which can provide protection against rapidly evolving variants.

Both in vitro and in vivo methods have been developed to evaluate the efficacy of antiviral agents, among which mice are the most frequently used animal model. Mice offer the advantages of a low cost, high biological similarity to humans and a short lifecycle. However, mice cannot be infected by SARS-CoV-2 because of amino acid differences between mouse ACE2 (mACE2) and human ACE2 (hACE2) [1]. Multiple strategies have been applied to sensitize mice to SARS-CoV-2 infection, including expressing hACE2 via transgenic, knock-in (KI), and viral-vector transduction methods [2–5]. Other strategies include modifying the virus to adapt it to mice [5]. Among these models, a hACE2 KI mouse model was developed by replacing mACE2 with hACE2 cDNA. The hACE2 KI mice were infected intranasally and high viral loads were observed in the respiratory system, including the lungs and trachea [2]. However, research with SARS-CoV-2 must be performed in a biosafety level 3 (BSL-3) laboratory, which has hampered the development of antiviral therapeutic agents. Recently, pseudovirus-based neutralization assays have been established, which show a good correlation with human virus infection. Application of the pseudovirus allows for the evaluation of anti-SARS-CoV-2 agents in laboratories with a lower level of containment (BSL-2) [6,7].

Here, to facilitate the investigation of antiviral agents, we established a SARS-CoV-2 pseudovirus infection mouse model. Using pseudovirus harboring the luciferase gene, we optimized an in vivo bioluminescent model to mimic natural infection with SARS-CoV-2, which can be continuously monitored in live animals. The protection efficiencies of monoclonal neutralizing antibodies were evaluated and the infection efficiencies of different variants were detected using this model.

2. Materials and Methods

2.1. Cells

HEK293T (American Type Culture Collection (ATCC), CRL-3216) and Vero E6 (ATCC, CRL-1586) were cultured in high-glucose Dulbecco's modified Eagle's medium (DMEM; HyClone, Logan, UT, USA) supplemented with 10% fetal bovine serum (FBS; Gibco, Billings, MT, USA) and 1% penicillin–streptomycin (Gibco) in a humidified 5% CO_2 atmosphere at 37 °C.

2.2. Production of Single Cycle (sc) Vesicular Stomatitis Virus (VSV)-SARS-CoV-2

The Spike gene of SARS-CoV-2 was codon-optimized for human cells and expressed on plasmid pcDNA3.1. Plasmid expressing spike of the indicated variant was then transfected into 293T cells using Lipofectamine 3000 reagent (Invitrogen, L3000015, Carlsbad, CA, USA) following the manufacturer's instructions. At the same time, the transfected cells were infected with G gene-deleted (G*ΔG) VSV, which harbored a luciferase reporter gene, at a multiplicity of infection (MOI) of 4. Six hours after infection, the supernatants were removed. Cells were washed with 2% FBS/PBS three times, then fresh DMEM with 10% FBS was added. At 24 h post-infection, the supernatants containing scVSV-SARS-CoV2 were harvested, filtered through a 0.45 nm filter, and stored at −70 °C until use.

2.3. Generation of rVSV-SARS-CoV-2

rVSV-SARS-CoV-2 was rescued from cDNA clones. Briefly, a plasmid encoding the T7 promoter for the VSV antigenome modified by a replacement of the G gene with a 21-amino acid (aa)-deleted codon-optimized SARS-CoV-2 spike gene with the indicated mutations and a luciferase reporter gene was inserted into the genome as a separate transcriptional unit before the spike gene was constructed. Plasmid-based rescue of the replication-competent VSV-based pseudovirus was carried out as described previously [8]. Briefly, BHK21 cells were infected with vaccinia virus vTF7-3, which expresses T7 polymerase, for 2 h. Then, the supernatant was discarded and cells were transfected with plasmid encoding the VSV antigenome, along with helper plasmids encoding the T7 promoter for the VSV N, P, G, and L genes using Lipofectamine 3000 reagent. Cells were cultured for an additional 48 h and the supernatants were collected and filtered through a 0.22 nm filter and passaged in Vero E6 cells. The virus was further passaged and amplified every 2–3 days, three to four times. The viral RNA was then extracted and reverse transcription PCR was performed. The full-length spike gene was amplified and sequenced. The supernatant containing the virus was aliquoted and stored at −80 °C.

2.4. Animal Experiments

The study was approved by the Animal Care and Use Committee at the National Institute for Food and Drug Control (NIFDC). Three-week-old C57BL/6 and hACE2 KI mice were obtained from the Institute for Laboratory Animal Resources, NIFDC. The mice were infected with scVSV-SARS-CoV-2 or rVSV-SARS-CoV-2 harboring spike protein with the indicated mutation via intraperitoneal (IP), intravenous (IV), intragastric (IG), or intranasal (IN) injection. Two to four mice were used in each group. Three independent experiments were conducted and representative images are shown. The bioluminescent signals were monitored at the indicated timepoints post-infection.

2.5. Bioluminescent Imaging (BLI) Analysis

Mice were subjected to IP injection of the substrate, d-luciferin (50 mg/kg body weight), and were then anaesthetized with isoflurane at 10 min post-injection. In vivo bioluminescence was monitored using the IVIS Lumina Series III Imaging System. Images were acquired with an exposure time of 60 s. The images were displayed in color according to pseudotype, ranging from red to blue.

2.6. Viral Load Assay

Quantification of the virus was performed using real-time PCR. RNA was extracted from the lungs of control and monoclonal antibody (mAb)-treated mice. Reverse transcription was performed using the reverse transcription PCR kit (Thermo Fisher Scientific, Waltham, MA, USA). SYBR green quantitative real-time PCR was performed using a LightCycler 480 Real-Time PCR System (Roche). VSV-P mRNA was quantified. The sequences of primers used for this analysis were as follows: forward: 5′-GCAATTGACAGCTCTTCTGCTCA-3′, and reverse: 5′-GTCGTCAATCCTCCGGTACTATC ATCTG-3′.

2.7. Statistical Analysis

All data were analyzed and graphs were generated using GraphPad Prism 6.0 software (GraphPad, San Diego, CA, USA). p-values of <0.05 were considered statistically significant.

3. Results

3.1. Packing Optimization and Generation of Single Cycle and Replication-Competent VSV-SARS-CoV-2 Pseudovirus

To generate scVSV-SARS-CoV-2 pseudovirus, 293T cells were transfected with a plasmid expressing SARS-CoV-2 spike protein with the D614G mutation (S-D614G) and infected with G gene-deleted VSV harboring the firefly luciferase (Fluc) reporter, simultaneously (Figure 1A). To obtain a high titer of pseudotyped SARS-CoV-2 virus, the amount of spike expression vector, and the ratio of G*ΔG-VSV and transfection reagent were optimized. We observed that cells in the six-well plate transfected with 2 μg of spike-expressing vectors by Lipofectamine 3000 and infected with G*ΔG-VSV at a ratio of 1:1000, producing the highest titer of scVSV-SARS-CoV-2 (Figure 1B).

To generate a replication-competent VSV-SARS-CoV-2 (rVSV-SARS-CoV-2) pseudovirus, we synthetized a plasmid harboring the VSV antigenome with a luciferase reporter inserted between VSV-M and VSV-G. Then, the G gene was replaced by S-D614G with a 21-aa deletion at the C-terminus. This 21-aa sequence has been reported to be essential for the efficient replication of rVSV-SARS-CoV-2. Plasmid-based rescue of rVSV-SARS-CoV-2 was carried out as described previously [8] (Figure 1C) and generated a slowly replicating virus bearing the spike protein in the envelope. This virus was then amplified by passage in Vero E6 cells. Virus replication dynamics were characterized after low MOI (0.1) infection at 30 °C and 37 °C, respectively. We observed high viral titers 72–96 h following infection, and the viral titers were significantly higher at 30 °C (Figure 1D).

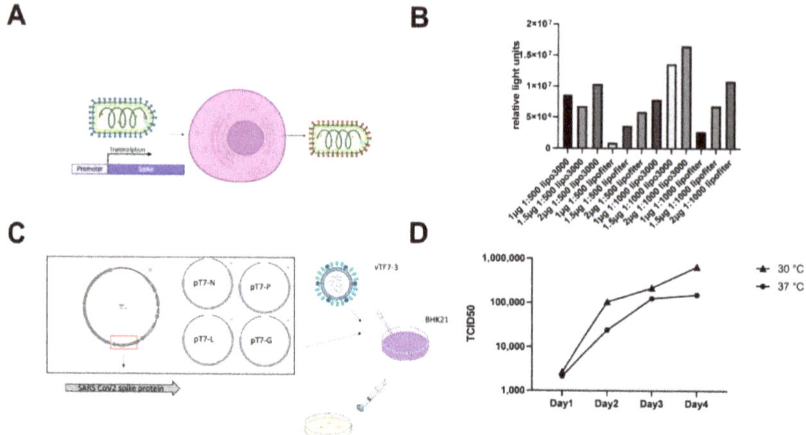

Figure 1. Packing optimization and rescue of scVSV-SARS-CoV-2 and rVSV-SARS-CoV-2. (**A**) A scheme for scVSV preparation. SARS-CoV-2 spike protein with a 21-amino acid deletion at the C-terminus was pseudotyped on the surface of VSV particles that carry the Fluc reporter gene. (**B**) Optimization of the ratio of spike expression vectors, G*ΔG-VSV, and transfection reagents. (**C**) A scheme for rVSV-SARS-CoV-2 rescue. BHK21 cells were infected with vaccinia virus vTF7-3, which expresses T7 polymerase for 2 h, then transfected with T7 promoter-driven VSV N, P, L, and G genes and the antigenome of VSV harboring a luciferase reporter. After 48 h of culture, the supernatants were collected and filtered through a 0.22 μm filter. Then, rVSV-SARS-CoV-2 was further amplified by passage in Vero cells. (**D**) The bioluminescence signals were monitored after infecting cells with rVSV-SARS-CoV-2 at the indicated timepoints and temperature.

3.2. Construction and Optimization of a Bioluminescent Imaging Mouse Model

To establish a robust, bioluminescent imaging SARS-CoV-2 mouse model, we first tested the infection dynamics of scVSV-SARS-CoV-2 and rVSV-SARS-CoV-2 in the mouse model. hACE2 KI mice were infected with single cycle or replication-competent VSV-SARS-CoV-2 pseudovirus, harboring S-D614G, intranasally at a dose of 2×10^5 $TCID_{50}$. Bioluminescence signals were monitored at 24, 48, and 72 h post-infection (Figure 2A). The signal intensity peaked at 24 h post-infection, then gradually decreased. Replication-competent pseudovirus-infected mice showed significantly higher signals than the single cycle virus, so replication-competent pseudovirus was selected for further infection studies and 24 h post-infection was selected as the optimal timepoint to carry out the assays. To further evaluate specificity and infection ability, different infection routes, including intraperitoneal (IP), intravenous (IV), intragastric (IG), and intranasal (IN) injection, were compared in hACE2 KI mice and C57BL/6 mice. To achieve an intensive infection process, mice were administrated with a maximum volume (1 mL through the IV, IP, and IG routes, 100 μL through the IN route) of virus at 6×10^6 $TCID_{50}$/mL concentration (Figure 2B). Bioluminescence signals were detected in the IV and IN groups, but not in the IP and IG groups. C57BL/6 mice also showed an enrichment of bioluminescence signals in the liver in the IV group, suggesting nonspecific infection through this route. By contrast, virus infected though the IN route was enriched in the nose and lungs in hACE2 KI mice, while signals were undetectable in C57BL/6 mice. These data showed that the dissemination of bioluminescence signals in the IN group were stable and more specific than in mice in the other groups.

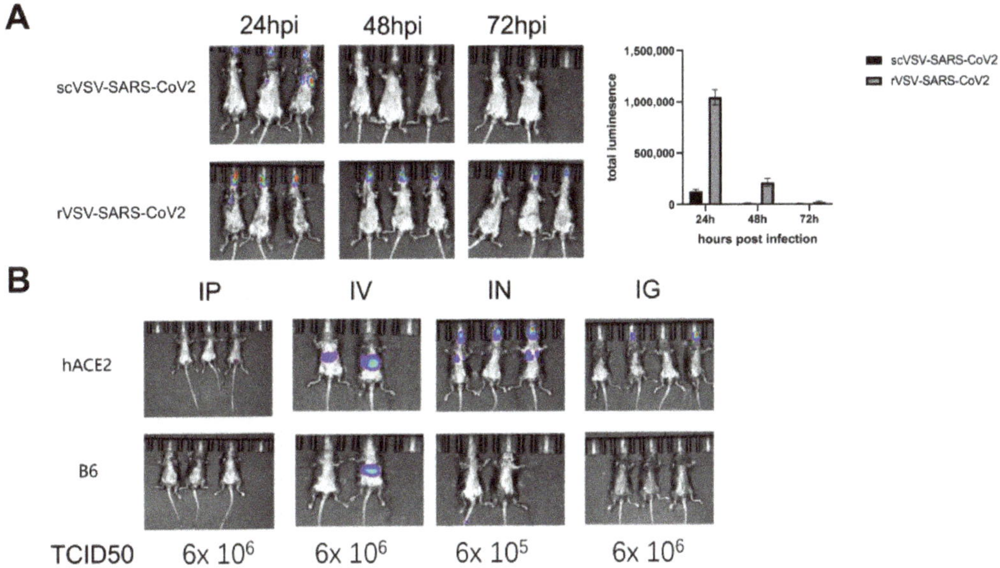

Figure 2. Construction and optimization of a bioluminescent imaging mouse model. (**A**) Comparison between scVSV-SARS-CoV-2 and rVSV-SARS-CoV-2 infected hACE2 KI mice. Mice were inoculated with the indicated pseudovirus harboring S-D614G via intranasal injection (2×10^5 $TCID_{50}$/mouse). The virus distribution and dissemination were monitored by bioluminescent imaging (BLI) on days 1, 2, and 3 post-infection (left panel) and quantified (right panel). (**B**) Comparison of different infection routes (IP, IV, IN, and IG) in hACE2 KI mice and C57BL/6 mice. Mice were inoculated with rVSV-SARS-CoV-2 harboring S-D614G at the indicated $TCID_{50}$. The virus distribution and infection intensity were monitored by the bioluminescent signal at 24 h post-infection.

3.3. *Evaluation of Monoclonal Neutralizing Antibodies Protection in the Mouse Model*

With the continuation of the COVID-19 pandemic around the world, mouse models to assess potential antiviral agents are urgently needed. BGB-DXP593 is a monoclonal neutralizing antibody identified from the B cells of convalescent COVID-19 patients [9] that can neutralize pseudovirus harboring S-D614G in vitro [10]. We used our infection model to assess the in vivo prevention efficacies of this antibody. Mice were administered two doses of the BGB-DXP593 mAb individually (20 mg/kg) via the intravenous (IV) route before pseudovirus exposure (days −2 and −1) and were then challenged with IN administration of rVSV-SARS-CoV-2 (day 0). Nearly complete protection was conferred by BGB-DXP593 mAb (Figure 3A,B). Quantitative PCR was performed using primers targeting the VSV-P gene, which resides in the backbone of the pseudovirus, and the data were normalized to that of the *Gapdh* gene (Figure 3C). The results confirmed that the viral load significantly decreased in the mAb-treated group. We further performed linear regression and found a good correlation between the relative VSV-P mRNA level and BLI (r^2 = 0.94), suggesting that bioluminescence intensity reflects the viral load in vivo (Figure 3D).

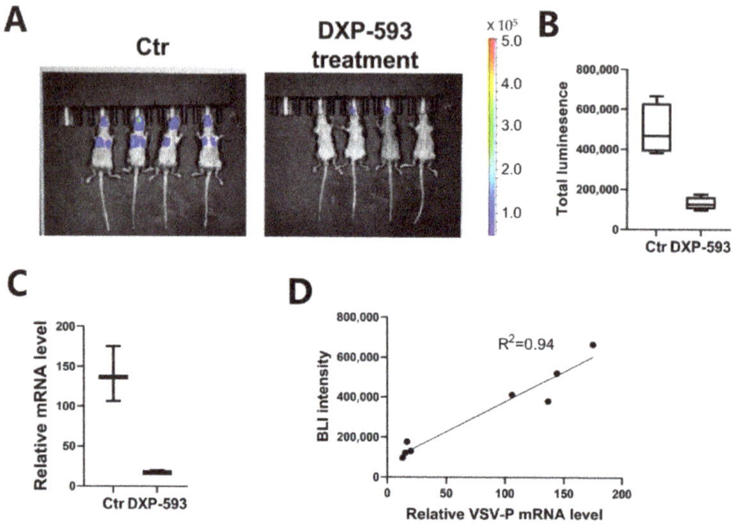

Figure 3. Evaluation of monoclonal neutralizing antibody protection efficiency in the mouse model. (**A**) Anti-SARS-CoV-2 mAbs were injected to evaluate their preventive efficacies. The mAbs were injected 2 days and 1 day before the inoculation of rVSV-SARS-CoV-2 to determine their preventive efficacies. Flux was measured at 1 day post-infection. (**B**) Quantification of the total flux for each mouse. (**C**) Quantification of VSV in the lungs of control and mAb-treated mice was performed by RT-PCR of VSV-P mRNA levels and were normalized to GAPDH levels. Values are shown as the means ± SEMs of three independent experiments. (**D**) Linear regression between the relative VSV-P mRNA levels in the lungs and the total bioluminescence of the lungs.

3.4. Evaluation of Tissue Tropism of Different Variants in the Mouse Model

Because of the highly variant nature of SARS-CoV-2, evaluation of the pathogenesis and virulence of different variants is important. Pseudoviruses enable different mutants to be easily constructed. We constructed rVSV-SARS-CoV-2 harboring the S gene from XBB strain. rVSV-SARS-CoV-2 (D614G) and rVSV-SARS-CoV-2 (XBB) reached similar titers on Vero cells and particle numbers were quantified by quantitative PCR. We then tested their infection ability in vivo. Mice infected with rVSV-SARS-CoV-2 (D614G) showed viral loads in the mouth and/or lungs (Figure 4A,B), while mice infected with rVSV-SARS-CoV-2 (XBB) showed significantly lower viral loads in the lungs (Figure 4C).

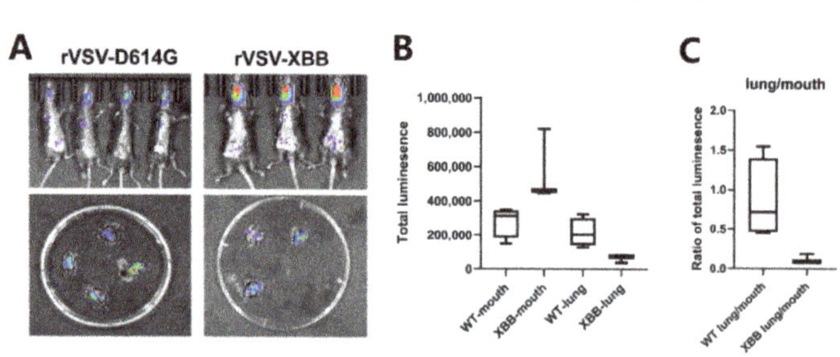

Figure 4. Evaluation of tissue tropism of different variants (**A**) Mice were inoculated with rVSV-SARS-CoV-2 harboring the indicated S variants by intranasal injection (2×10^5 $TCID_{50}$/mouse). Flux was measured at 1 day post-infection. Lungs were dissected for BLI. (**B**) Quantification of the total flux intensity in the mouths and lungs of each group. (**C**) The ratio of flux in the lung region to flux in the mouth region was determined for each mouse.

4. Discussion

The high transmissibility and virulence of SARS-CoV-2 dictate the need for BSL-3 containment, which presents a major obstacle to research. The objective of this study was to establish a safe and convenient model mimicking SARS-CoV-2 infection. A typical pseudovirus comprises the genomic backbone from one virus and the envelope protein from another virus. The genomic backbone is often modified and reporter genes may be added, which limits the replication and/or reduces the virulence of the virus. Pseudoviruses thereby allow researchers to safely work with a virus under BSL-2 containment. In this study, we used a VSV-based pseudovirus that could produce a high titer of SARS-CoV-2. Different spike variants of SARS-CoV-2 can be easily incorporated into the pseudovirus and studied without the need for live viruses.

In a previous study, lentivirus-based SARS-CoV-2 pseudoviruses were used to model SARS-CoV-2 infection following transduction with adenoviruses expressing hACE2 [11]. However, because of the replication-incompetent nature of the pseudovirus and inefficient transduction of hACE2 by adenovirus, the infection of SARS-CoV-2 pseudovirus was limited to the upper respiratory tract. We compared the infection efficiency of scVSV-SARS-CoV-2 and rVSV-SARS-CoV-2, and found the bioluminescence signal to be significantly higher in rVSV-SARS-CoV-2-infected mice than scVSV-SARS-CoV-2-infected mice, indicating that replication occurred in vivo. Furthermore, the rVSV-SARS-CoV-2 infection model exhibited lung infection, which appeared to mimic human SARS-CoV-2 infection. However, we noticed that the intensity of the bioluminescence signal was highest at 24 h post-infection, then gradually decreased, indicating that rVSV-SARS-CoV-2 did not continuously replicate in mice. The generation of hACE2 in an immunodeficient background may be useful to study the continuous infection process and the tissue distribution of rVSV-SARS-CoV-2.

It was previously reported that the SARS-CoV-2 variant Omicron BA.1 was associated with a lower risk of hospitalization and severe illness compared with the Delta variant [12]. Consistent with this observation, when hACE2 mice were infected with rVSV-SARS-CoV-2 harboring the spike protein of the XBB variant, the viral load in the lungs was much lower compared with those infected with wild-type rVSV-SARS-CoV-2. Some mutations have been suggested to influence cell surface entry pathways, which may affect tissue tropism. Further studies are needed to determine the crucial mutation site(s) responsible for the tissue tropism of different SARS-CoV-2 variants.

With the rapid evolution of SARS-CoV-2, the investigation and verification of new antibodies is urgently needed. We used a mouse model to evaluate the protective efficacy of mAbs administered via the IV route. Our model was effective in evaluating broad-acting neutralizing antibodies via the infection of pseudoviruses with different envelope proteins. Our model could be further expanded to evaluate the protective efficacy of vaccines [13,14] and to screen for anti-viral reagents targeting virus entry and fusion with the cell membrane.

In summary, we established an in vivo infection model using a replication-competent pseudovirus, which will facilitate the evaluation of vaccines, antibodies, and anti-viral reagents against SARS-CoV-2, and provide a safer and more flexible alternative to the use of live virus.

Author Contributions: Conceptualization, Y.W. and W.H.; methodology, J.N., C.F., C.L. and S.L.; validation, Y.W. and W.H.; data curation, X.W., N.F. and Z.L.; investigation, X.W., N.F. and Z.L.; writing—original draft preparation, X.W. and N.F.; writing—review and editing, Y.W. and W.H.; visualization, X.W. and Z.L.; supervision, Y.W. and W.H.; funding acquisition, Y.W. and W.H. All authors have read and agreed to the published version of the manuscript.

Funding: This work was funded by National Key Research and Development Program of China (2021YFC0863300) and a grant from National Key R&D Program of China (No. 2021YFC2301700).

Institutional Review Board Statement: The study was approved by the Ethics Committee of National Institute of Food and Drug Control.

Informed Consent Statement: Not applicable.

Data Availability Statement: The data presented in this study are available on request from the corresponding author. The data are not publicly available due to privacy.

Conflicts of Interest: The authors declare no conflict of interest.

References

1. Damas, J.; Hughes, G.M.; Keough, K.C.; Painter, C.A.; Persky, N.S.; Corbo, M.; Hiller, M.; Koepfli, K.P.; Pfenning, A.R.; Zhao, H.; et al. Broad host range of SARS-CoV-2 predicted by comparative and structural analysis of ACE2 in vertebrates. *Proc. Natl. Acad. Sci. USA* **2020**, *117*, 22311–22322. [CrossRef] [PubMed]
2. Sun, S.H.; Chen, Q.; Gu, H.J.; Yang, G.; Wang, Y.X.; Huang, X.Y.; Liu, S.S.; Zhang, N.N.; Li, X.F.; Xiong, R.; et al. A Mouse Model of SARS-CoV-2 Infection and Pathogenesis. *Cell Host Microbe* **2020**, *28*, 124–133 e124. [CrossRef] [PubMed]
3. Jiang, R.D.; Liu, M.Q.; Chen, Y.; Shan, C.; Zhou, Y.W.; Shen, X.R.; Li, Q.; Zhang, L.; Zhu, Y.; Si, H.R.; et al. Pathogenesis of SARS-CoV-2 in Transgenic Mice Expressing Human Angiotensin-Converting Enzyme 2. *Cell* **2020**, *182*, 50–58 e58. [CrossRef] [PubMed]
4. Hassan, A.O.; Case, J.B.; Winkler, E.S.; Thackray, L.B.; Kafai, N.M.; Bailey, A.L.; McCune, B.T.; Fox, J.M.; Chen, R.E.; Alsoussi, W.B.; et al. A SARS-CoV-2 Infection Model in Mice Demonstrates Protection by Neutralizing Antibodies. *Cell* **2020**, *182*, 744–753 e744. [CrossRef] [PubMed]
5. Gu, H.; Chen, Q.; Yang, G.; He, L.; Fan, H.; Deng, Y.Q.; Wang, Y.; Teng, Y.; Zhao, Z.; Cui, Y.; et al. Adaptation of SARS-CoV-2 in BALB/c mice for testing vaccine efficacy. *Science* **2020**, *369*, 1603–1607. [CrossRef] [PubMed]
6. Nie, J.; Li, Q.; Wu, J.; Zhao, C.; Hao, H.; Liu, H.; Zhang, L.; Nie, L.; Qin, H.; Wang, M.; et al. Establishment and validation of a pseudovirus neutralization assay for SARS-CoV-2. *Emerg. Microbes Infect.* **2020**, *9*, 680–686. [CrossRef] [PubMed]
7. Zettl, F.; Meister, T.L.; Vollmer, T.; Fischer, B.; Steinmann, J.; Krawczyk, A.; V'Kovski, P.; Todt, D.; Steinmann, E.; Pfaender, S.; et al. Rapid Quantification of SARS-CoV-2-Neutralizing Antibodies Using Propagation-Defective Vesicular Stomatitis Virus Pseudotypes. *Vaccines* **2020**, *8*, 386. [CrossRef] [PubMed]
8. Lawson, N.D.; Stillman, E.A.; Whitt, M.A.; Rose, J.K. Recombinant vesicular stomatitis viruses from DNA. *Proc. Natl. Acad. Sci. USA* **1995**, *92*, 4477–4481. [CrossRef] [PubMed]
9. Cao, Y.; Su, B.; Guo, X.; Sun, W.; Deng, Y.; Bao, L.; Zhu, Q.; Zhang, X.; Zheng, Y.; Geng, C.; et al. Potent Neutralizing Antibodies against SARS-CoV-2 Identified by High-Throughput Single-Cell Sequencing of Convalescent Patients' B Cells. *Cell* **2020**, *182*, 73–84 e16. [CrossRef] [PubMed]
10. Li, Q.; Wu, J.; Nie, J.; Zhang, L.; Hao, H.; Liu, S.; Zhao, C.; Zhang, Q.; Liu, H.; Nie, L.; et al. The Impact of Mutations in SARS-CoV-2 Spike on Viral Infectivity and Antigenicity. *Cell* **2020**, *182*, 1284–1294 e1289. [CrossRef] [PubMed]
11. Tseng, S.H.; Lam, B.; Kung, Y.J.; Lin, J.; Liu, L.; Tsai, Y.C.; Ferrall, L.; Roden, R.B.S.; Wu, T.C.; Hung, C.F. A novel pseudovirus-based mouse model of SARS-CoV-2 infection to test COVID-19 interventions. *J. Biomed. Sci.* **2021**, *28*, 34. [CrossRef] [PubMed]
12. Nyberg, T.; Ferguson, N.M.; Nash, S.G.; Webster, H.H.; Flaxman, S.; Andrews, N.; Hinsley, W.; Bernal, J.L.; Kall, M.; Bhatt, S.; et al. Comparative analysis of the risks of hospitalisation and death associated with SARS-CoV-2 omicron (B.1.1.529) and delta (B.1.617.2) variants in England: A cohort study. *Lancet* **2022**, *399*, 1303–1312. [CrossRef] [PubMed]
13. Nie, J.; Liu, L.; Wang, Q.; Chen, R.; Ning, T.; Liu, Q.; Huang, W.; Wang, Y. Nipah pseudovirus system enables evaluation of vaccines in vitro and in vivo using non-BSL-4 facilities. *Emerg. Microbes Infect.* **2019**, *8*, 272–281. [CrossRef] [PubMed]
14. Li, Q.; Liu, Q.; Huang, W.; Wu, J.; Nie, J.; Wang, M.; Zhao, C.; Zhang, L.; Wang, Y. An LASV GPC pseudotyped virus based reporter system enables evaluation of vaccines in mice under non-BSL-4 conditions. *Vaccine* **2017**, *35*, 5172–5178. [CrossRef] [PubMed]

vaccines

MDPI

Article

Antigen-Heterologous Vaccination Regimen Triggers Alternate Antibody Targeting in SARS-CoV-2-DNA-Vaccinated Mice

Anders Frische [1,2], Karen Angeliki Krogfelt [1,2,3], Anders Fomsgaard [1,4] and Ria Lassaunière [1,*]

1 Department of Virus & Microbiological Special Diagnostics, Statens Serum Institut, 2300 Copenhagen, Denmark; afi@ssi.dk (A.F.); karenak@ruc.dk (K.A.K.); afo@ssi.dk (A.F.)
2 Section of Molecular and Medicinal Biology, Department of Science and Environment, Roskilde University, 4000 Roskilde, Denmark
3 PandemiX, Center for Interdisciplinary Study of Pandemic Signatures, Institute for Science and Environment, Roskilde University, 4000 Roskilde, Denmark
4 Infectious Diseases Unit, Clinical Institute, University of Southern Denmark, 5230 Odense, Denmark
* Correspondence: mlas@ssi.dk

Abstract: An in-depth analysis of antibody epitopes following vaccination with different regimens provides important insight for developing future vaccine strategies. B-cell epitopes conserved across virus variants may be ideal targets for vaccine-induced antibodies and therapeutic drugs. However, challenges lie in identifying these key antigenic regions, and directing the immune system to target them. We previously evaluated the immunogenicity of two candidate DNA vaccines encoding the unmodified spike protein of either the SARS-CoV-2 Index strain or the Beta variant of concern (VOC). As a follow-on study, we characterized here the antibody binding profiles of three groups of mice immunized with either the DNA vaccine encoding the SARS-CoV-2 Index strain spike protein only, the Beta VOC spike protein only, or a combination of both as an antigen-heterologous prime-boost regimen. The latter induced an antibody response targeting overlapping regions that were observed for the individual vaccines but with additional high levels of antibody directed against epitopes in the SD2 region and the HR2 region. These heterologous-vaccinated animals displayed improved neutralization breadth. We believe that a broad-focused vaccine regimen increases neutralization breadth, and that the in-depth analysis of B-cell epitope targeting used in this study can be applied in future vaccine research.

Keywords: SARS-CoV-2; epitope mapping; microarray; neutralizing antibodies; DNA vaccine

1. Introduction

In December 2019, the SARS-CoV-2 Index strain emerged in Wuhan, China. Since then, the virus has evolved substantially, acquiring consecutively more amino acid changes in the major surface glycoprotein, the spike protein, which is the primary target for vaccine- and infection-induced neutralizing antibodies. Following the introduction of SARS-CoV-2 into Europe, the ancestral strain with the spike D614G substitution became dominant, followed by the emergence of a myriad of variants that include Alpha, Beta, Gamma, Delta, and several Omicron variants and recombinants. Immune pressure likely contributed to virus evolution that altered antigenicity, leading to escape of vaccine-induced immunity, natural immunity, and hybrid immunity. The pandemic has seen to an unprecedented and ongoing race to develop novel and updated vaccines that generate effective broadly neutralizing antibodies that are able to protect from severe disease against an everchanging virus.

SARS-CoV-2 vaccine development focused primarily on the spike protein due to its essential functions in virus entry, which makes it an ideal target for protective antibody responses. The spike protein is a trimer, and each monomer is built up by two non-covalently associated subunits S1 and S2. SARS-CoV-2 enters host cells by binding of the viral receptor-binding motif (RBM), located in the receptor-binding domain (RBD) of the spike protein, to the host cell receptor, angiotensin-converting enzyme 2 (ACE2). The RBD embedded in the S1 subunit shifts between a shielded 'down' conformation and an "open" receptor-binding conformation. Pre-cleavage at the S1/S2 cleavage site at the junction between S1 and S2 by furin or other cellular proteases promotes the RBD 'up' conformation, priming ACE2 binding [1–3]. In the 'up' position, the RBD can bind to the ACE2 receptor, leading to conformational changes in the spike and exposure of the S2′ cleavage site. To facilitate membrane fusion, S2′ is cleaved either by cell surface proteases such as TMPRSS2 or furin, or by cathepsin in the endosomal pathway, leading to a destabilization of the pre-fusion trimer and detachment of S1. Mediated by the S2 domain of the spike protein, the viral envelope fuses with the cell membrane through the insertion of the fusion peptide and the formation of a six-helix bundle by heptad repeat 1 and 2, pulling the two membranes together and resulting in fusion.

Rational vaccine design or monoclonal antibody therapy relies on knowledge about B-cell epitopes generating effective broadly neutralizing antibodies. As a host receptor engaging viral protein domain, the RBD is the primary target for neutralizing antibodies and a clear target for intervention strategies [4–6]. RBD-specific antibodies directed against ACE2 binding epitopes are classified as class 1 or 2 antibodies [7]. Class 1 antibodies bind only RBD 'up' epitopes and class 2 antibodies bind both RBD 'up' and 'down' epitopes. Conversely, RBD-specific antibodies of classes 3 and 4 are non-ACE2 blocking antibodies that bind 'up' and 'down' (class 3) or only 'up' (class 4). Multiple factors contribute to an enhanced frequency of mutations in the RBD, thereby influencing vaccine immunogenicity and monoclonal antibody recognition [8]. The N-terminal domain (NTD) upstream of the RBD in the S1 subunit, similarly known to elicit potent neutralizing antibodies [9–11], is also a region of highly mutated residues primarily centered around a supersite located in the N3 loop (residues 141–156) [9,11]. Deletions compromising neutralizing antibody efficiency are known to exclusively affect the S1 regions outside the RBD [12,13]. The key to effective broadly neutralizing antibodies is therefore to target epitopes that are conserved across variants. Protein function and structural constraints are believed to have a strong impact on the likeliness that a gene sequence will mutate [14]. Targeting conformationally "locked" regions could be crucial for effective neutralization.

Statens Serum Institut, Denmark, developed two candidate SARS-CoV-2 vaccines that encode either the unmodified spike protein of the Index strain or the spike protein of the Beta VOC (PANGO lineage B.1.351) [15,16]. Both vaccines are immunogenic in rabbits and mice, producing robust binding and neutralizing antibody responses. In studies conducted by Lassauniere et al. [15], where CB6F1 mice received a homologous regimen of three immunizations of either of these vaccines or a combination of both, the authors observed an increased breadth of neutralization responses when the vaccines were mixed. In this study, we aim to determine if this increased breadth is the result of different epitopes being targeted. To this end, we study serum from mice that received homologous or heterologous vaccine regimens. We use microarray technology with circular constrained peptides of the entire Index strain spike ectodomain to identify targeted B-cell epitopes.

2. Materials and Methods

2.1. Study Population

The DNA vaccines, vaccination strategies, and animal experiments are described in detail elsewhere [15,16]. In brief, eight-week-old female CB6F1 mice (Envigo, Horst, The Netherlands) were immunized with 50 μg of a DNA plasmid vector, encoding either the complete and unaltered spike protein of the SARS-CoV-2 Index strain (pNTC-Spike), derived from the Wuhan-Hu-1 strain (MN908947), or of the Beta (B.1.351) strain (pNTC-Spike.351). Nature Technology Corporation (Lincoln, NE, USA) subcloned human codon-optimized SARS-CoV-2 spike sequences synthesized by GeneArt (Thermo Fisher Scientific, Dreieich, Germany) into the NTC8685-eRNA41H vector backbone and produced the vaccines in NTC4862 E. coli cells (DH5α attλ::P5/6 6/6-RNA-IN-SacV, Cmr) using their antibiotic-free RNA-OUT selection procedure. The vaccine stocks were provided in a concentration of 10 mg/mL in phosphate-buffered saline (PBS) [15]. Immunizations were performed in weeks 0, 2, and 4. Groups of five mice were immunized intradermally with needle injection at the base of the tail according to three different vaccine regimens (Figure 1). The first group received only the pNTC-Spike vaccine at weeks 0, 2, and 4, while the second group received only the pNTC-Spike.351 vaccine at the same time points. These antigen-homologous groups are hereafter referred to as the Index and Beta groups, respectively. The third group received a combination of the two vaccines as the pNTC-Spike vaccine administered at weeks 0 and 2, and the pNTC-Spike.351 vaccine administered at week 4. The latter antigen-heterologous group is hereafter referred to as the mixed group. Blood samples were taken two weeks after the final vaccination. The animals were housed in facilities at Statens Serum Institut, Copenhagen, Denmark. All procedures were supervised by the laboratory animal veterinarians and complied with the Danish legislation, which is based on EU Directive 2010/63/EU on the protection of animals used for scientific purposes. The experiments received ethical approval from The Animal Experimentation Council, the National Competent Authority within this field (approval number 2017-15-0201-01322).

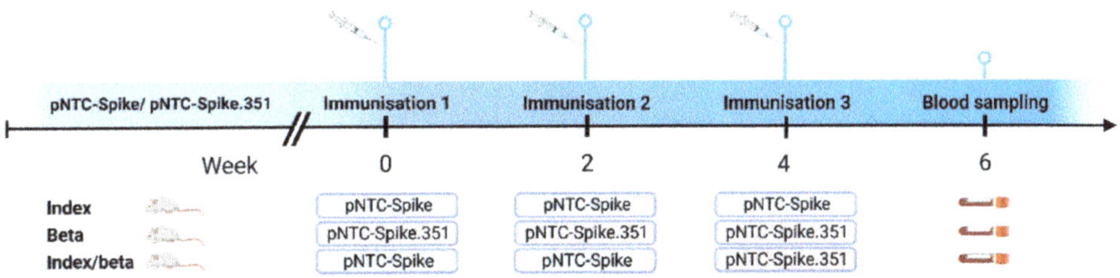

Figure 1. Vaccination regimens. Animals were immunized with either a pNTC DNA vector containing the full SARS-CoV-2 Index strain spike protein (pNTC-Spike) or the full SARS-CoV-2 Beta variant (pNTC-Spike.351) at two-week intervals. Immunizations were carried out according to the indicated 3 different regimens: Index, Beta, and mixed. Mice were immunized intradermally with a needle at the base of the tail, with a 50 µg dose. Blood samples were taken 2 weeks post final immunization. Vaccination regimens are further described in Lassauniere et al., 2021 [16].

2.2. Microarray Method

Epitope mapping was performed on pooled sera from each group of mice, according to vaccination regimen, using a custom-made microarray with 10′mer cyclic overlapping peptides spanning the full Index strain spike protein and 35 additional mutation sequences representing key mutations throughout the spike protein. To perform the epitope mapping, the peptide microarray obtained from PEPperPRINT (Heidelberg, Germany) was adjusted to room temperature and assembled in accordance with the manufacturer's instructions. Subsequently, a washing buffer comprising Dulbecco's phosphate-buffered saline (DPBS) with 0.05% Tween20, at a pH of 7.4, was added to each subarray. After incubation at room temperature for 15 min, the washing buffer was removed using a pipette and each subarray was blocked in 400 µL of Rockland Blocking Buffer (MB-070, Rockland Immunochemicals, Pottstown, PA, USA) at room temperature for 30 min. To evaluate non-specific reactions, we then pre-treated the peptide array with a secondary antibody/streptavidin solution diluted in staining buffer (DPBS with 0.005% Tween20 and 10% blocking buffer, pH 7.4) as described below and scanned. The microarray was then reinserted into the PEPperPRINT microarray cassette and incubated with staining buffer for 15 min to equilibrate. The staining buffer was then removed with a pipette. Diluted in staining buffer, 400 µL of each serum pool was pipetted into separate subarrays, and the microarray was left overnight at 2 to 8 °C on an orbital shaker at 140 rpm and in the dark. The following day, sample dilutions were removed by pipetting, followed by a two-times wash with 400 µL wash buffer for 10 s on an orbital shaker at 140 rpm, protected from light. Moreover, 400 µL of biotinylated goat anti-mouse IgG F(c) (cat. #31805, Invitrogen, Thermo Fisher Scientific, Waltham, MA, USA) was added to each subarray. This secondary antibody was diluted 1:500 in staining buffer and the microarray was subsequently incubated for 45 min at room temperature. The secondary antibody was removed by pipetting, followed by 2× wash with 400 μL wash buffer as previously described, and 400 μL streptavidin Alexa Fluor 647 conjugate (Thermo Fisher Scientific, Waltham, MA, USA) diluted 1:750 in staining buffer was added, and the peptide microarray was incubated at room temperature for 45 min. Each subarray was then washed two times with 400 μL wash buffer as previously described, and the entire array slide was submerged two times into dipping buffer (1 mM Tris buffer, pH 7.4), followed by a 1 min centrifugation at 1000 rpm to completely remove all dipping buffer. Microarray was scanned on a microarray scanner (SureScan from Agilent, Santa Clara, CA, USA) and analyzed using MAPIX Analyzer Software v. 9.1.0. The microarray was then re-inserted into the microarray cassette and equilibrated as previously described. After removal of the staining buffer using a pipette, the microarray was finally incubated with an anti-HA PEPperCHIP® control antibody from PEPperPRINT diluted 1:2000 in staining buffer and incubated at room temperature for 45 min. Subarrays were washed twice with 400 μL wash buffer, and, after disassembly, the entire array slide was dipped twice into dipping buffer, centrifuged, scanned, and analyzed as previously described. All RT incubations were carried out in the dark using an orbital shaker at 140 rpm.

2.3. Statistical Analysis and Calculations

Fluorescence (relative fluorescence units, RFU) was determined as fluorescence intensity at 635 nm with a subtraction of background determined at 635 nm from an area manually set within each subarray in MAPIX. Microarray fluorescence data were imported into Excel. Fluorescence was calculated as the mean of Δ-fluorescence from two identical peptides and normalized according to the mean of a positive HA control (n = 48) included on each subarray. Targeted epitopes were defined as the fluorescence intensities above the cut off. The cut off was defined as

$$\text{Cut off} = \overline{\text{X}} + \text{SD} \times \text{f}$$

where $\overline{\text{X}}$ is the mean of n negative control determinations (n = 96) included on each subarray, SD is the standard deviation, and f is the standard deviation multiplier with a confidence interval of 99.0% as provided by Frey et al. [17].

3. Results

3.1. Selection of Animals

To characterize the antibody binding profiles in DNA-vaccinated mice, we selected animals grouped according to the three different immunization regimens described in Section 2. Displaying quite distinct cross-neutralization profiles (Supplementary Figure S1) [15] and immunized with three different vaccine regimens, we hypothesized that these animals would target different epitopes on the spike protein. To study this and identify potentially broadly neutralizing epitopes, we assessed the overall antibody binding profiles from pools of serum from each group of animals having followed the same immunization regimen using microarray epitope mapping.

3.2. Antibody Binding Profiles

We designed a peptide microarray using cyclic 10′mer sequences, with eight overlapping amino acids and two amino acid shifts, spanning the full Index strain spike protein and 35 additional sequences of mutations representing important VOCs at the time of design. Because of a very limited amount of serum accessible for each animal, microarray analysis was carried out using the serum pools of the four to five individual mice included in each vaccination regimen group. The amount of individual mouse serum in each pool was normalized according to previously established levels of ELISA binding IgG titers (Supplementary Figure S1C), adding equal titers of spike-ectodomain-specific antibody to control for confounding variables, such as variable levels of overall spike IgG levels between animals. Each group of mice was analyzed in a 1:20 dilution and a 1:100 dilution, allowing assessment of low-affinity and high-affinity antibodies. Targeted (active) epitopes were defined as epitopes with relative fluorescence units (RFU) above a cut-off value calculated from a negative control set included on each subarray.

3.2.1. Targeting of RBD Epitopes

We observed moderate targeting and binding within the entire RBD, with maximum RFU values of 8087, 8529, and 7203 (diluted 1:20) in the Index-vaccinated group, the Beta-vaccinated group, and the mixed-vaccinated group, respectively (Figure 2). Maximum RBD targeting was found within the RBM in the Beta-vaccinated group, indicating an ACE2 blocking focus (antibody classes 1 and 2), and just outside the RBM in the remaining two groups. The Index-vaccinated group and the mixed-vaccinated group had the apex RBD peak at peptide 429, spanning residues 420–429. This region overlaps with two known monoclonal class 4 antibodies, CR3022 and COVA1-16 [18,19]. The apex RBD peak in the Beta-vaccinated group was found at peptide 473, spanning residues 464–473. This peptide, also targeted by the other two groups, overlaps with class 1 mAb GAR05 epitope [20,21]. An adjacent epitope, spanning residues 478–501, overlapping with another class 1 mAb, CB6 [19,22], similarly shows a generally higher level of antibody targeting in all three groups.

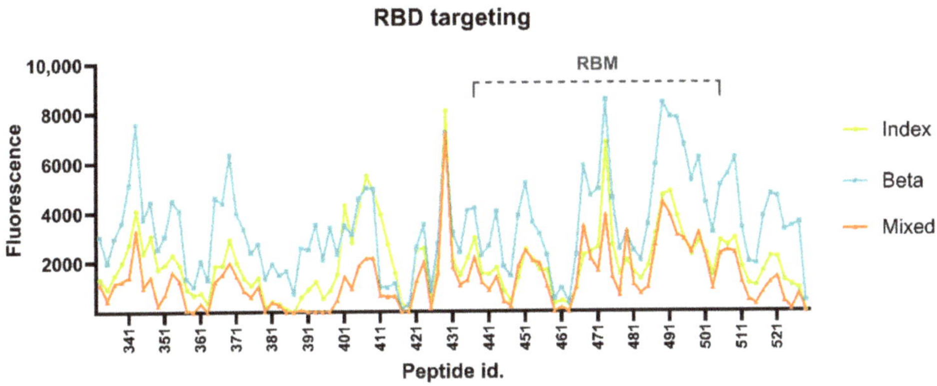

Figure 2. Antibody binding to the receptor-binding domain (RBD). Peptide id. on the x-axis refers to the position of the last amino acid of each 10′mer peptide on the microarray. The y-axis indicates the level of antibody binding to each peptide measured as the average of duplicate determinations of spot fluorescence minus background fluorescence normalized according to a positive control included on each subarray. RBM = receptor-binding motif.

3.2.2. Mutational Sensitivity

The ability to broadly neutralize different variants could be influenced by the capacity of antibodies to tolerate viral mutations. To test the animals in each vaccine regimen's resilience towards mutations, we included peptides of 35 of the most prominent mutations at the time of design of the microarray, including neutralization-resistant K417N and receptor-binding motif substitution E484K.

Substitutions C136F and D138Y and deletion DEL144 in the NTD supersite heavily affected binding in the Beta-vaccinated group and to a lesser extent the mixed-vaccinated group (Figure 3). In contrast, enhanced binding was observed in the Index-vaccinated group. These mutations resulted in a 58% reduction in total binding, from 188,843 RFU to 80,099 RFU, in the Beta-vaccinated group in serum diluted 1:20. In comparison, the total binding in the same region in the mixed-vaccinated group decreased 24% and increased 94% in the Index-vaccinated group. Although a moderate effect, E484K substitution leads to a general loss of antibody binding; in contrast, K417N results in a surprising increase in signal in all the groups. The N501Y substitution within the RBM, which is believed to enhance ACE2 binding affinity [23,24], does not seem to affect antibody binding in any of the groups. In contrast, Alpha variant substitution A570D in the SD1 region increased binding in the Index-vaccinated group and the Beta-vaccinated group, but it is nearly untargeted in the mixed-vaccinated group. The D614G substitution is generally low-targeted; however, the mixed-vaccinated group is the only group with enhanced targeting of the mutated form. Except for the D614G substitution, the groups vaccinated with the pNTC.Spike.351, in particular the Beta-vaccinated group, display enhanced antibody binding to all Beta mutations. The Index-vaccinated group, the group with the highest neutralizing antibody levels against the D614G variant and the Delta AY.4. variant, however, displays generally low binding to all the Beta mutation peptides and all the included Delta mutation peptides.

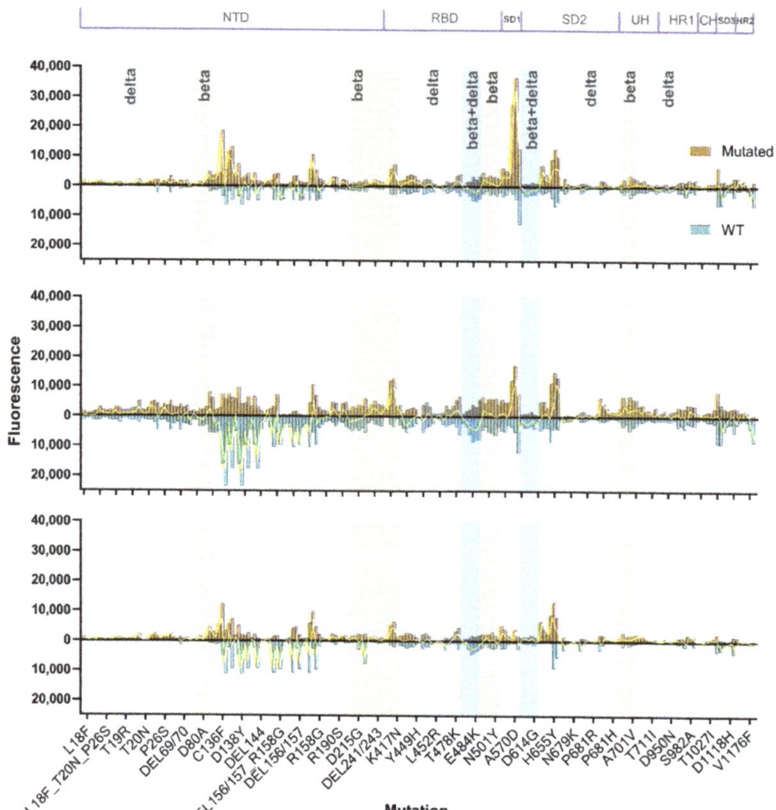

Figure 3. Comparative antibody binding to the non-mutated Index strain peptides mirrored to the mutated counterpart. The x-axis indicates the mutation; on the third peptide for each variant, the residue of interest is approximately in the center of the sequence. Substitutions make up 6 individual peptides, deletions make up 5 peptides, and deletions and substitutions make up 6 peptides. The y-axis indicates the level of antibody binding to each peptide measured as the average of duplicate determinations of spot fluorescence minus background fluorescence normalized according to a positive control included on each subarray. Values in gold indicate antibody binding to the mutated sequence and values in turquoise antibody binding to Index strain sequence. Yellow lines represent the antibody binding to the mutated peptide minus antibody binding to the corresponding Index strain peptide. Beta and Delta mutations are marked in colored bars.

3.2.3. Consistently Targeted Regions

To identify highly immunogenic regions on the spike protein targeted by DNA-vaccine-induced antibodies in mice, we evaluated the epitopes consistently targeted between the three groups. At a serum dilution of 1:20, all three groups consistently targeted multiple epitopes within the region spanning residues 130–179 around the N-terminal domain supersite, with maximum intensity at peptide 141 (EFQFCNDPFL) consisting of amino acids 132–141 (Figure 4). This site, along with the SD3 site spanning residues 1144–1153, displays moderate to very high targeting, with a minimum fluorescence exceeding 6000 RFU. The neighboring epitopes adjacent to these hotspots, spanning residues 148–163, 1138–1151, and 1199–1217, were similarly consistently targeted with low to high expression (low 1835 RFU; high 9117 RFU). The RBD site at peptide 429 (DYNYKLPDDF) spanning residues 420–429, and peptides 1263 and 1265 spanning the area of residues 1254–1265 in the connector domain in the periphery of the protruding spike, equally showed moderate binding levels (>6000 RFU) in all three groups. At a higher serum dilution of 1:100, peptides 141, 1263, and 1265 also displayed notable antibody binding, indicating high-affinity/high-abundance antibodies (Supplementary Figure S2).

3.2.4. Non-Consistently Targeted Regions

In addition to the consistently targeted regions, the Index-vaccinated group and the Beta-vaccinated group both had high binding to peptide 579 (ADTTDAVRDP), spanning residues 570–579 in the SD1 region, with RFUs exceeding 12,000 (12,733, 12,081), and peptide 1187 (NIQKEIDRLN; RFU 6285, 8163) in the HR2 region in a 1:20 serum dilution. Diluted 1:100, peptide 579 is the apex peak in the Index-vaccinated group, with a 2-fold (1.9) higher antibody binding signal than peptide 1113 (Supplementary Figure S2). These two peptides are targeted with equal fluorescence at a serum dilution of 1:20. This SD1 epitope is significantly less targeted in the mixed-vaccinated group (RFU = 2743). The Beta group displayed a markedly higher (3.6-fold) level of antibodies targeting NTD peptide 141 compared to the lowest level observed in the Index-vaccinated group, and a 4.2-fold higher level of the 1265 peptide (Figure 4B,C). Apart from the consistently targeted regions shared between all the groups, the mixed-vaccinated group showed unique high levels of antibody binding to peptides 625 and 627 spanning residues 616–627 very close to the 630-loop in the SD2 region, with a maximum RFU of 21,676 compared to the Index-vaccinated group and the Beta-vaccinated group, with low intensities of 1688 and 3404, respectively. Peptide 1169 (TSPDVDLGDI) in the HR2 region similarly showed very high antibody binding in the mixed-vaccinated group (RFU = 29,320) compared to 4283 and 6584 RFU as observed in the Index and in the Beta-vaccinated groups, respectively. The Beta-vaccinated group showed high general antibody levels covering most parts of the spike protein and focusing less on individual epitopes, in contrast to the more focused response observed for the mixed-vaccinated group (Figure 4C). The serum antibodies of the mixed-vaccinated group do, however, seem to target regions in the NTD, RBD, CR, and CH also observed for the Beta-vaccinated group, which were absent to low in the Index-vaccinated group (Figure 4C). Of note, none of the groups displayed noticeably high levels of binding to the S1/S2 cleavage site, the S2′ cleavage site, or the fusion peptide.

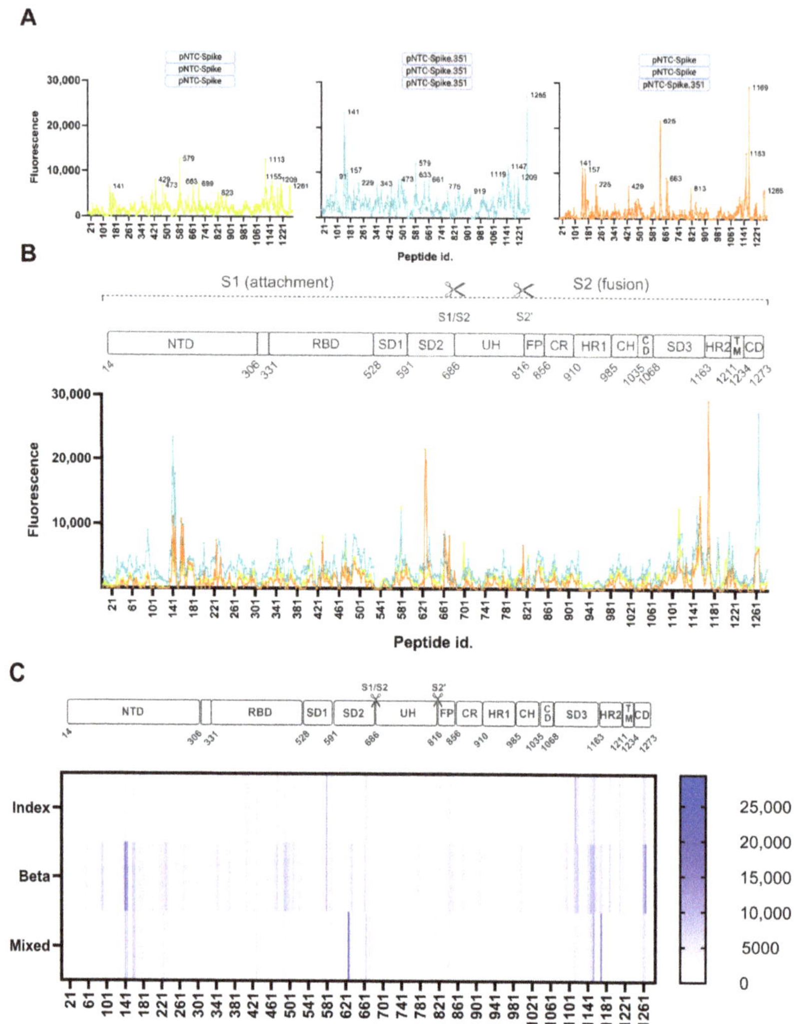

Figure 4. Serum-binding antibody profiles. Antibody binding to 10′mer overlapping peptides covering the entire SARS-CoV-2 Index strain spike ectodomain. The *x*-axis refers to the position of the last amino acid of each 10′mer peptide on the microarray. The *y*-axis (**A**,**B**) indicates the level of antibody binding to each peptide measured as the average of duplicate determinations of spot fluorescence minus background fluorescence normalized according to a positive control included on each subarray. (**A**) Individual binding profiles from antibody binding in pools of serum samples stratified according to vaccination regimen. Major peaks are labeled. (**B**) Full overview of antibody binding in all three groups relative to the different SARS-CoV-2 spike protein domains. S1 subunit (residues 14–685); S2 subunit (residues 686–1273); NTD = N-terminal domain; RBD = receptor-binding domain; SD1, SD2, SD3 = sub-domain 1, 2, 3; S1/S2 = furin cleavage site; UH = upstream helix; S2′ cleavage site; FP = fusion peptide; CR = connecting region; HR1 and HR2 = heptad repeat sequence 1 and 2; CH = central helix; CD = connector domain; TM = transmembrane helix; CD = connector domain. (**C**) Heatmap showing serum antibody binding relative to the different SARS-CoV-2 spike protein domains in the Index-strain-vaccinated group, the Beta-VOC-vaccinated group, and the mixed-vaccinated group.

4. Discussion

This study applied peptide microarray technology to map B-cell epitopes targeted by SARS-CoV-2 spike-specific antibodies induced by three different regimens of DNA vaccines to aid rational vaccine development in the future. It is a follow-up study of the development and pre-clinical immunogenicity studies of the SARS-CoV-2 DNA vaccines described in Lassauniere et al. [15,16]. These studies found an enhanced breadth in heterologous-vaccinated mice that received a combination of a DNA plasmid encoding the full Index strain spike protein sequence and an identical plasmid vector encoding the full Beta VOC spike protein. Being one of the first variants bearing immune escape capabilities such as the E484K and K417N/T substitutions, the Beta VOC represented a highly antibody-resistant variant compared to the ancestral Index strain. To determine if the observed improved ability to cross-neutralize different VOCs in the heterologous-vaccinated mice is the result of different antibody targeting, we established B-cell epitope profiles from polyclonal serum from groups of mice representing each vaccination regimen. The animals following the antigen-heterologous vaccination regimen, combining both of the two used DNA vaccine constructs leading to a slightly improved cross-neutralizing capacity, showed the most distinct binding profile.

The ideal vaccine should induce a broad neutralizing antibody response, conferring protection against present and emerging virus strains. Our results indicate that using an antigen-homologous vaccine strategy directs the immune system into a focused strain-specific response. In contrast, applying an antigen-heterologous vaccination regime resulted in broader cross-neutralization [15]. We observed an antigen-heterologous vaccine regimen antibody profile targeting common regions in the NTD, RBD, and SD3 equally targeted by the antigen homologous regimens, with the addition, however, of a high level of antibody binding to an SD2 region spanning residues 616–627 and an HR2 region spanning residues 1160–1169. This group displayed a more focused antibody response with markedly lower omnipresent binding, and lower binding to mutational spots, although with a potent antibody response at the targeted regions. This diverging antibody binding profile was associated with a slightly improved ability to cross-neutralize variants (Supplementary Figure S1).

A SARS-CoV-2 vaccine design using a native spike protein, as used for this study cohort, may induce a different RBD-binding antibody repertoire compared to that induced by a pre-fusion-stabilized spike protein. The pre-fusion-stabilized S-2P spike used in the mRNA-1273 vaccine [25] is believed to adopt a more open conformation with one or more monomers in the 'up' state in comparison with the native spike protein [3,26,27]. With a maximum RFU of 8529 (Beta-vaccinated group), none of the vaccinated groups tend to focus their antibody repertoire against the RBD. The antibodies in all three groups were mainly targeting class 4 and class 1 epitopes, both of which target RBD 'up' conformations. These epitopes are not accessible in the closed formation. Surprisingly, all three vaccinated groups targeted an epitope containing the glycan-shielded N343 with considerable intensity (Figure 2). The N343 RBD glycan is, besides shielding, suspected to interact with the ACE2 backbone and stabilize ACE2-RBD binding [4,28]. The mAb S309 (classified as a class 3 antibody) does not bind this exact epitope; however, N343 is sandwiched between the mAb's binding sites, and the binding of this antibody possibly compromises ACE2-RBD binding stability, thereby inhibiting infectivity [28]. The observed antibody binding to the peptide containing N343 in all the groups of vaccinated mice may behave similarly. Despite the usage of a native spike, supposedly adopting a more 'closed' spike conformation compared to the pre-fusion-stabilized spike, the main antibody targeting is directed against epitopes only accessible in the 'open' conformation. Antibodies directed against the down state are regarded as being more potent, and, because of a more conserved area of the exposed surface of the closed conformation, they are also believed to have a higher level of cross-neutralization [10,27]. The observed weak antibody binding to closed conformation epitopes, combined with overall moderate RBD targeting, point towards effective antibody repertoires targeting regions outside the receptor-binding domain.

As opposed to moderate RBD antibody targeting, we observed high antibody binding just outside the RBD, in the SD1 region, most noticeably observed in the Index-vaccinated group. The non-mutated sequence of peptide 579 displayed high antibody binding even at a 1:100 dilution compared to the remaining targeted peptides in this group, indicating a higher proportion of high-affinity antibodies targeting this peptide (Supplementary Figure S2). For serum diluted 1:20, this peptide had an RFU of 12,733. However, a single substitution, A570D, significantly increased the antibody binding to the neighboring mutated peptide spanning residues 569–578 (IDDTTDAVRD) to an RFU of 36,706 (3-fold). The A570D substitution has been reported to increase viral entry into target cells [29]. The targeted peptide is overlapping with a known broadly neutralizing epitope in the SD1 [8,30–33] that presumably functions as a hinge of the open/closed state of RBD. Vaccination with the native spike, used for this study cohort [16], is known to elicit antibodies targeting this region [34]. In contrast, this SD1 region is occluded when using a vaccine design with a pre-fusion-stabilized spike protein. Being immunized with the Index strain without the A570D substitution, it is surprising that the elicited antibodies bind the peptide with the substitution with 3-fold greater intensity. However, despite high-resolution amino acid coverage on the microarray, an amino acid shift of two amino acids could have some limitations. Thus, an exact comparison of an identical peptide frame differentiating only in the substitution was not possible. The enhanced antibody fit could therefore partly be explained by the slight frame shift varying two terminal amino acids. It is unlikely, however, to be the sole reason for the increased signal as the terminal amino acids are positioned close to the linker attaching the 10′mer circular peptide to the glass plate; hence, the minimal epitope is likely unaffected by these amino acid changes.

The S1 subdomain, and in particular the RBD, has been the primary target for research within neutralizing antibodies. However, the more conserved regions of S1 and S2 are speculated to have a higher contribution to cross-reactivity [35,36]. Residues 616–627 of the S2 region are parts of the 630 loop stretching from residue 617 to residue 644 [37]. The 630 loop is, along with the fusion peptide, believed to play a critical role in stabilizing the spike protein and serves a function in retaining the RBD in the 'down' mode [37–39]. Targeting this region with antibodies could possibly disturb the structural rearrangements necessary for opening up the RBD, leaving it in a locked 'down' position. The stability of the 630 loop is affected by the neighboring H655Y substitution; however, all three vaccinated groups share enhanced binding to the mutated form (Figure 3) [37]. HR2, which interacts with HR1 in the formation of the 6-helix bundle during membrane fusion [39,40], similarly reveals a broadly neutralizing epitope that becomes accessible after ACE2 binding [40–42]. Heavily targeting of this epitope, as observed in the mixed-vaccinated group, is likely to interfere with membrane fusion. Binding to HR2 has been reported to correlate with disease severity of SARS-CoV-2 [43].

5. Conclusions

Our results indicate that a heterologous vaccine regimen does not simply elicit an antibody response targeting the sum of antibodies induced by each individual vaccine. In contrast, a broader focused regimen induces an immune response tweaked to fit regions that could serve as common denominators, increasing neutralization breadth. Vaccinating with different variants might direct the immune system into a conserved region-directed response, targeting areas serving critical functions that are less likely to mutate and are evolutionarily "fixed". Including a third vaccine, based on an additional variant such as the Omicron variant, or reversing the order of vaccination priming with the Beta vaccine and boosting with the Index vaccine, could broaden our understanding of this phenomenon. Furthermore, combining a heterologous vaccination regimen with a stepwise functional antibody characterization could provide interesting insight into the transition towards a conserved region-directed immune response.

Supplementary Materials: The following supporting information can be downloaded at: https://www.mdpi.com/article/10.3390/vaccines12030218/s1, Figure S1: Levels of cross-neutralization; Figure S2: Serum-binding low-affinity antibody profiles.

Author Contributions: Conceptualization, R.L., A.F. (Anders Fomsgaard), K.A.K. and A.F. (Anders Frische); methodology, R.L., A.F. (Anders Fomsgaard), K.A.K. and A.F. (Anders Frische); formal analysis, R.L. and A.F. (Anders Frische); investigation, R.L., K.A.K. and A.F. (Anders Frische); resources, R.L. and A.F. (Anders Frische); data curation, R.L. and A.F. (Anders Frische); writing—original draft preparation, A.F. (Anders Frische); writing—review and editing, A.F. (Anders Frische), R.L., K.A.K. and A.F. (Anders Fomsgaard); visualization, A.F. (Anders Frische) and R.L.; supervision, R.L., K.A.K. and A.F. (Anders Fomsgaard); project administration, R.L. and A.F. (Anders Fomsgaard); funding acquisition, K.A.K., A.F. (Anders Fomsgaard) and R.L. All authors have read and agreed to the published version of the manuscript.

Funding: The work was partly supported by funding from the Danish Ministry of Education and Research, grant number 0237-00006B.

Institutional Review Board Statement: The experiment received approval by The Animal Experimentation Council, the National Competent Authority within this field (approval number 2017-15-0201-01322; approval date 29 September 2017) and the Internal Animal Welfare committee at the animal facility at Statens Serum Institute (approval date 10 December 2021).

Informed Consent Statement: Not applicable.

Data Availability Statement: The raw data supporting the conclusions of this article will be made available by the authors on request.

Acknowledgments: We would like to acknowledge the skilled technical assistance of Birgit Knudsen.

Conflicts of Interest: Anders Fomsgaard is a co-inventor on a patent application covering a SARS-CoV-2 DNA vaccine; all rights to the vaccine have been assigned to Statens Serum Institut (SSI), a Danish national not-for-profit governmental public health institute. Other authors declare no competing interests. The funders had no role in the design of the study; in the collection, analyses, or interpretation of data; in the writing of the manuscript; or in the decision to publish the results.

References

1. Takeda, M. Proteolytic activation of SARS-CoV-2 spike protein. *Microbiol. Immunol.* **2022**, *66*, 15–23. [CrossRef]
2. Tang, T.; Jaimes, J.A.; Bidon, M.K.; Straus, M.R.; Daniel, S.; Whittaker, G.R. Proteolytic Activation of SARS-CoV-2 Spike at the S1/S2 Boundary: Potential Role of Proteases beyond Furin. *ACS Infect. Dis.* **2021**, *7*, 264–272. [CrossRef] [PubMed]
3. Walls, A.C.; Park, Y.J.; Tortorici, M.A.; Wall, A.; McGuire, A.T.; Veesler, D. Structure, Function, and Antigenicity of the SARS-CoV-2 Spike Glycoprotein. *Cell* **2020**, *181*, 281–292.e6. [CrossRef]
4. Barnes, C.O.; Jette, C.A.; Abernathy, M.E.; Dam, K.-M.A.; Esswein, S.R.; Gristick, H.B.; Malyutin, A.G.; Sharaf, N.G.; Huey-Tubman, K.E.; Lee, Y.E.; et al. SARS-CoV-2 neutralizing antibody structures inform therapeutic strategies. *Nature* **2020**, *588*, 682–687. [CrossRef] [PubMed]
5. Yin, W.; Xu, Y.; Xu, P.; Cao, X.; Wu, C.; Gu, C.; He, X.; Wang, X.; Huang, S.; Yuan, Q.; et al. Structures of the Omicron spike trimer with ACE2 and an anti-Omicron antibody. *Science* **2022**, *375*, 1048–1053. [CrossRef] [PubMed]
6. Wang, Z.; Muecksch, F.; Cho, A.; Gaebler, C.; Hoffmann, H.H.; Ramos, V.; Zong, S.; Cipolla, M.; Johnson, B.; Schmidt, F.; et al. Analysis of memory B cells identifies conserved neutralizing epitopes on the N-terminal domain of variant SARS-Cov-2 spike proteins. *Immunity* **2022**, *55*, 998–1012.e8. [CrossRef] [PubMed]
7. Barnes, C.O.; Jette, C.A.; Abernathy, M.E.; Dam, K.-M.A.; Esswein, S.R.; Gristick, H.B.; Malyutin, A.G.; Sharaf, N.G.; Huey-Tubman, K.E.; Lee, Y.E.; et al. Structural classification of neutralizing antibodies against the SARS-CoV-2 spike receptor-binding domain suggests vaccine and therapeutic strategies. *bioRxiv* **2020**. [CrossRef]
8. Liu, L.; Casner, R.G.; Guo, Y.; Wang, Q.; Iketani, S.; Chan, J.F.-W.; Yu, J.; Dadonaite, B.; Nair, M.S.; Mohri, H.; et al. Antibodies that neutralize all current SARS-CoV-2 variants of concern by conformational locking. *bioRxiv* **2023**. [CrossRef]
9. Cerutti, G.; Guo, Y.; Zhou, T.; Gorman, J.; Lee, M.; Rapp, M.; Reddem, E.R.; Yu, J.; Bahna, F.; Bimela, J.; et al. Potent SARS-CoV-2 neutralizing antibodies directed against spike N-terminal domain target a single supersite. *Cell Host Microbe* **2021**, *29*, 819–833.e7. [CrossRef]
10. Liu, L.; Wang, P.; Nair, M.S.; Yu, J.; Rapp, M.; Wang, Q.; Luo, Y.; Chan, J.F.W.; Sahi, V.; Figueroa, A.; et al. Potent neutralizing antibodies against multiple epitopes on SARS-CoV-2 spike. *Nature* **2020**, *584*, 450–456. [CrossRef]
11. Klinakis, A.; Cournia, Z.; Rampias, T. N-terminal domain mutations of the spike protein are structurally implicated in epitope recognition in emerging SARS-CoV-2 strains. *Comput. Struct. Biotechnol. J.* **2021**, *19*, 5556–5567. [CrossRef]

12. Papanikolaou, V.; Chrysovergis, A.; Ragos, V.; Tsiambas, E.; Katsinis, S.; Manoli, A.; Papouliakos, S.; Roukas, D.; Mastronikolis, S.; Peschos, D.; et al. From delta to Omicron: S1-RBD/S2 mutation/deletion equilibrium in SARS-CoV-2 defined variants. *Gene* **2022**, *814*, 146134. [CrossRef] [PubMed]
13. Bal, A.; Destras, G.; Gaymard, A.; Stefic, K.; Marlet, J.; Eymieux, S.; Regue, H.; Semanas, Q.; d'Aubarede, C.; Billaud, G.; et al. Two-step strategy for the identification of SARS-CoV-2 variant of concern 202012/01 and other variants with spike deletion H69–V70, France, August to December 2020. *Eurosurveillance* **2021**, *26*, 2100008. [CrossRef] [PubMed]
14. Forni, D.; Cagliani, R.; Pontremoli, C.; Mozzi, A.; Pozzoli, U.; Clerici, M.; Sironi, M. Antigenic variation of SARS-CoV-2 in response to immune pressure. *Mol. Ecol.* **2021**, *30*, 3548–3559. [CrossRef]
15. Lassaunière, R.; Polacek, C.; Linnea Tingstedt, J.; Fomsgaard, A. Preclinical evaluation of a SARS-CoV-2 variant B.1.351-based candidate DNA vaccine. *Vaccine* **2023**, *41*, 6505–6513. [CrossRef] [PubMed]
16. Lassaunière, R.; Polacek, C.; Gram, G.J.; Frische, A.; Tingstedt, J.L.; Krüger, M.; Dorner, B.G.; Cook, A.; Brown, R.; Orekov, T.; et al. Preclinical evaluation of a candidate naked plasmid DNA vaccine against SARS-CoV-2. *Npj Vaccines* **2021**, *6*, 156. [CrossRef] [PubMed]
17. Frey, A.; Di Canzio, J.; Zurakowski, D. A statistically defined endpoint titer determination method for immunoassays. *J. Immunol. Methods* **1998**, *221*, 35–41. [CrossRef] [PubMed]
18. Jette, C.A.; Cohen, A.A.; Gnanapragasam, P.N.P.; Muecksch, F.; Lee, Y.E.; Huey-Tubman, K.E.; Schmidt, F.; Hatziioannou, T.; Bieniasz, P.D.; Nussenzweig, M.C.; et al. Broad cross-reactivity across sarbecoviruses exhibited by a subset of COVID-19 donor-derived neutralizing antibodies. *Cell Rep.* **2021**, *36*, 109760. [CrossRef]
19. Chen, Y.; Zhao, X.; Zhou, H.; Zhu, H.; Jiang, S.; Wang, P. Broadly neutralizing antibodies to SARS-CoV-2 and other human coronaviruses. *Nat. Rev. Immunol.* **2023**, *23*, 189–199. [CrossRef]
20. Rouet, R.; Henry, J.Y.; Johansen, M.D.; Sobti, M.; Balachandran, H.; Langley, D.B.; Walker, G.J.; Lenthall, H.; Jackson, J.; Ubiparipovic, S.; et al. Broadly neutralizing SARS-CoV-2 antibodies through epitope-based selection from convalescent patients. *Nat. Commun.* **2023**, *14*, 3–6. [CrossRef]
21. Chu, X.; Ding, X.; Yang, Y.; Lu, Y.; Li, T.; Gao, Y.; Zheng, L.; Xiao, H.; Yang, T.; Cheng, H.; et al. Mechanism of an RBM-targeted rabbit monoclonal antibody 9H1 neutralizing SARS-CoV-2. *Biochem. Biophys. Res. Commun.* **2023**, *660*, 43–49. [CrossRef] [PubMed]
22. Shi, R.; Shan, C.; Duan, X.; Chen, Z.; Liu, P.; Song, J.; Song, T.; Bi, X.; Han, C.; Wu, L.; et al. A human neutralizing antibody targets the receptor-binding site of SARS-CoV-2. *Nature* **2020**, *584*, 120–124. [CrossRef] [PubMed]
23. Grabowski, F.; Preibisch, G.; Giziński, S.; Kochańczyk, M.; Lipniacki, T. SARS-CoV-2 Variant of Concern 202012/01 Has about Twofold Replicative Advantage and Acquires Concerning Mutations. *Viruses* **2021**, *13*, 392. [CrossRef] [PubMed]
24. Liu, Y.; Liu, J.; Plante, K.S.; Plante, J.A.; Xie, X.; Zhang, X.; Ku, Z.; An, Z.; Scharton, D.; Schindewolf, C.; et al. The N501Y spike substitution enhances SARS-CoV-2 transmission. *bioRxiv* **2021**. [CrossRef]
25. Anderson, E.J.; Rouphael, N.G.; Widge, A.T.; Jackson, L.A.; Roberts, P.C.; Makhene, M.; Chappell, J.D.; Denison, M.R.; Stevens, L.J.; Pruijssers, A.J.; et al. Safety and Immunogenicity of SARS-CoV-2 mRNA-1273 Vaccine in Older Adults. *N. Engl. J. Med.* **2020**, *383*, 2427–2438. [CrossRef]
26. Wrapp, D.; Wang, N.; Corbett, K.S.; Goldsmith, J.A.; Hsieh, C.L.; Abiona, O.; Graham, B.S.; McLellan, J.S. Cryo-EM structure of the 2019-nCoV spike in the prefusion conformation. *Science* **2020**, *367*, 1260–1263. [CrossRef]
27. Juraszek, J.; Rutten, L.; Blokland, S.; Bouchier, P.; Voorzaat, R.; Ritschel, T.; Bakkers, M.J.G.; Renault, L.L.R.; Langedijk, J.P.M. Stabilizing the closed SARS-CoV-2 spike trimer. *Nat. Commun.* **2021**, *12*, 244. [CrossRef]
28. Hsu, Y.P.; Frank, M.; Mukherjee, D.; Shchurik, V.; Makarov, A.; Mann, B.F. Structural remodeling of SARS-CoV-2 spike protein glycans reveals the regulatory roles in receptor-binding affinity. *Glycobiology* **2023**, *33*, 126–137. [CrossRef]
29. Peng, Q.; Zhou, R.; Liu, N.; Wang, H.; Xu, H.; Zhao, M.; Yang, D.; Au, K.K.; Huang, H.; Liu, L.; et al. Naturally occurring spike mutations influence the infectivity and immunogenicity of SARS-CoV-2. *Cell. Mol. Immunol.* **2022**, *19*, 1302–1310. [CrossRef]
30. Seow, J.; Khan, H.; Rosa, A.; Calvaresi, V.; Graham, C.; Pickering, S.; Pye, V.E.; Cronin, N.B.; Huettner, I.; Malim, M.H.; et al. A neutralizing epitope on the SD1 domain of SARS-CoV-2 spike targeted following infection and vaccination. *Cell Rep.* **2022**, *40*, 111276. [CrossRef]
31. Poh, C.M.; Carissimo, G.; Wang, B.; Amrun, S.N.; Lee, C.Y.P.; Chee, R.S.L.; Fong, S.W.; Yeo, N.K.W.; Lee, W.H.; Torres-Ruesta, A.; et al. Two linear epitopes on the SARS-CoV-2 spike protein that elicit neutralising antibodies in COVID-19 patients. *Nat. Commun.* **2020**, *11*, 2806. [CrossRef] [PubMed]
32. Xu, S.; Wang, Y.; Wang, Y.; Zhang, C.; Hong, Q.; Gu, C.; Xu, R.; Wang, T.; Yang, Y.; Zang, J.; et al. Mapping cross-variant neutralizing sites on the SARS-CoV-2 spike protein. *Emerg. Microbes Infect.* **2022**, *11*, 351–367. [CrossRef] [PubMed]
33. Guenthoer, J.; Lilly, M.; Starr, T.N.; Dadonaite, B.; Lovendahl, K.N.; Croft, J.T.; Stoddard, C.I.; Chohan, V.; Ding, S.; Ruiz, F.; et al. Identification of broad, potent antibodies to functionally constrained regions of SARS-CoV-2 spike following a breakthrough infection. *bioRxiv* **2023**. [CrossRef] [PubMed]
34. Frische, A.; Gunalan, V.; Krogfelt, K.A.; Fomsgaard, A.; Lassaunière, R. A Candidate DNA Vaccine Encoding the Native SARS-CoV-2 Spike Protein Induces Anti-Subdomain 1 Antibodies. *Vaccines* **2023**, *11*, 1451. [CrossRef] [PubMed]
35. Grobben, M.; van der Straten, K.; Brouwer, P.J.; Brinkkemper, M.; Maisonnasse, P.; Dereuddre-Bosquet, N.; Appelman, B.; Lavell, A.A.; van Vught, L.A.; Burger, J.A.; et al. Cross-reactive antibodies after SARS-CoV-2 infection and vaccination. *Elife* **2021**, *10*, e70330. [CrossRef] [PubMed]

36. Huang, Y.; Nguyen, A.W.; Hsieh, C.-L.; Silva, R.; Olaluwoye, O.S.; Wilen, R.E.; Kaoud, T.S.; Azouz, L.R.; Qerqez, A.N.; Le, K.C.; et al. Identification of a conserved neutralizing epitope present on spike proteins from all highly pathogenic coronaviruses. *bioRxiv* **2021**. Available online: https://www.biorxiv.org/content/10.1101/2021.01.31.428824v2.abstract (accessed on 28 April 2023).
37. Wang, X.; Chen, X.; Tan, J.; Yue, S.; Zhou, R.; Xu, Y.; Lin, Y.; Yang, Y.; Zhou, Y.; Deng, K.; et al. 35B5 antibody potently neutralizes SARS-CoV-2 Omicron by disrupting the N-glycan switch via a conserved spike epitope. *Cell Host Microbe* **2022**, *30*, 887–895.e4. [CrossRef]
38. Wang, Y.; Liu, C.; Zhang, C.; Wang, Y.; Hong, Q.; Xu, S.; Li, Z.; Yang, Y.; Huang, Z.; Cong, Y. Structural basis for SARS-CoV-2 Delta variant recognition of ACE2 receptor and broadly neutralizing antibodies. *Nat. Commun.* **2022**, *13*, 871. [CrossRef]
39. Zhang, J.; Xiao, T.; Cai, Y.; Chen, B. Structure of SARS-CoV-2 spike protein. *Curr. Opin. Virol.* **2021**, *50*, 173–182. [CrossRef]
40. Jiang, S.; Zhang, X.; Du, L. Therapeutic antibodies and fusion inhibitors targeting the spike protein of SARS-CoV-2. *Expert Opin. Ther. Targets* **2021**, *25*, 415–421. [CrossRef]
41. Bianchini, F.; Crivelli, V.; Abernathy, M.E.; Guerra, C.; Palus, M.; Muri, J.; Marcotte, H.; Piralla, A.; Pedotti, M.; De Gasparo, R.; et al. Human neutralizing antibodies to cold linear epitopes and subdomain 1 of the SARS-CoV-2 spike glycoprotein. *Sci. Immunol.* **2023**, *8*, eade0958. [CrossRef]
42. Elshabrawy, H.A.; Coughlin, M.M.; Baker, S.C.; Prabhakar, B.S. Human Monoclonal Antibodies against Highly Conserved HR1 and HR2 Domains of the SARS-CoV Spike Protein Are More Broadly Neutralizing. *PLoS ONE* **2012**, *7*, e50366. [CrossRef]
43. Garrido, J.L.; Medina, M.A.; Bravo, F.; McGee, S.; Fuentes-Villalobos, F.; Calvo, M.; Pinos, Y.; Bowman, J.W.; Bahl, C.D.; Barria, M.I.; et al. IgG targeting distinct seasonal coronavirus- conserved SARS-CoV-2 spike subdomains correlates with differential COVID-19 disease outcomes. *Cell Rep.* **2022**, *39*, 110904. [CrossRef] [PubMed]

vaccines

MDPI

Article

A Candidate DNA Vaccine Encoding the Native SARS-CoV-2 Spike Protein Induces Anti-Subdomain 1 Antibodies

Anders Frische [1,2], Vithiagaran Gunalan [1], Karen Angeliki Krogfelt [1,2], Anders Fomsgaard [1,3] and Ria Lassaunière [1,*]

1 Department of Virus & Microbiological Special Diagnostics, Statens Serum Institut, 2300 Copenhagen, Denmark; afi@ssi.dk (A.F.); vigu@ssi.dk (V.G.); kak@ssi.dk (K.A.K.); afo@ssi.dk (A.F.)
2 Section of Molecular and Medicinal Biology, Department of Science and Environment, Roskilde University, 4000 Roskilde, Denmark
3 Infectious Diseases Unit, Clinical Institute, University of Southern Denmark, 5230 Odense, Denmark
* Correspondence: mlas@ssi.dk

Abstract: The ideal vaccine against viral infections should elicit antibody responses that protect against divergent strains. Designing broadly protective vaccines against SARS-CoV-2 and other divergent viruses requires insight into the specific targets of cross-protective antibodies on the viral surface protein(s). However, unlike therapeutic monoclonal antibodies, the B-cell epitopes of vaccine-induced polyclonal antibody responses remain poorly defined. Here we show that, through the combination of neutralizing antibody functional responses with B-cell epitope mapping, it is possible to identify unique antibody targets associated with neutralization breadth. The polyclonal antibody profiles of SARS-CoV-2 index-strain-vaccinated rabbits that demonstrated a low, intermediate, or high neutralization efficiency of different SARS-CoV-2 variants of concern (VOCs) were distinctly different. Animals with an intermediate and high cross-neutralization of VOCs targeted fewer antigenic sites on the spike protein and targeted one particular epitope, subdomain 1 (SD1), situated outside the receptor binding domain (RBD). Our results indicate that a targeted functional antibody response and an additional focus on non-RBD epitopes could be effective for broad protection against different SARS-CoV-2 variants. We anticipate that the approach taken in this study can be applied to other viral vaccines for identifying future epitopes that confer cross-neutralizing antibody responses, and that our findings will inform a rational vaccine design for SARS-CoV-2.

Keywords: SARS-CoV-2; DNA vaccine; epitope mapping; microarray; neutralizing antibodies

1. Introduction

On 12 December 2019, severe acute respiratory syndrome virus 2 (SARS-CoV-2) emerged, presumably through a zoonotic spillover from an animal in Wuhan, China [1]. The early release of the viral whole genome sequence and the characterization of the viral entry mediated by the spike protein facilitated critical research within SARS-CoV-2 spike/ACE2 interactions, pathogenesis and possible intervention strategies [2]. The spike protein occurs on the viral surface as homotrimers comprising the non-covalently associated subunits S1 and S2. The receptor binding domain (RBD), positioned at the S1 subunit, can, through a hinge-like structure, attend either a 'down', conformational masked mode for immune evasion, or an 'up' receptor-accessible conformation [3–7]. To enter a cell, the spike protein must bind to the host receptor ACE2 through the RBD, followed by cleavage at the S2′ cleavage site by cell surface proteases TMPRSS2, furin, or cathepsin (endosomal pathway) to release the fusion peptide [3–5,7–12]. The prefusion trimer is then destabilized and the S1 subunit is disconnected. This positions the fusion peptide on the S2 subunit to be inserted as a wedge into the cell membrane, followed by the formation of a six-helix bundle through heptad repeats 1 and 2, which brings the viral envelope and the host cell membrane into proximity, resulting in membrane fusion [3,5,13]. Considering the functional significance of the RBD, the S2′ cleavage site, and fusion elements, these spike protein domains are important targets for vaccine-induced antibodies to block/neutralize virus entry into the host cell.

Neutralizing antibodies may block viral entry into the host cell through a direct binding to the receptor binding motif, the steric hindrance of receptor binding, the locking of the RBD in a 'down' position, a spike-conformational change disturbance, or membrane fusion interference [6,14,15]. The RBD is a known target for infection-induced and vaccine-induced neutralizing antibodies. However, the RBD of SARS-CoV-2 maintains a 'down' mode at a higher frequency compared to SARS-CoV [3], leaving it less accessible for neutralizing antibodies. In addition, S1, and in particular the RBD, are the primary regions for mutations in the spike, leaving effective RBD neutralizing antibodies as being more strain-specific. For these reasons, the more preserved regions of the S1 and the highly conserved S2 domain (91% similarity to SARS-CoV [13]) could potentially be more prone to broadly neutralizing antibodies [13]. Studies have reported broadly neutralizing B-cell epitopes within the conformational active regions of the S1, such as the SD1 and the N-terminal domain [16–23]. In the S2 subunit, essential fusion elements such as the fusion peptide and the heptad repeats are also well-known targets for neutralizing antibodies and therapeutic drugs [5,13,24]. Defining the exact binding epitope of strong neutralizing antibodies is a key to rational vaccine design and antibody drug development.

Epitope mapping is a powerful approach that provides valuable insights into antibody binding patterns. Peptide microarrays can identify antigenic sites and immunogenic hotspots on proteins such as the SARS-CoV-2 spike protein. B-cells or secreted antibodies recognize these antigenic determinants, hereafter referred to as B-cell epitopes. Characterizing B-cell epitopes of monoclonal antibodies, and polyclonal responses in animal models and clinical cohorts, can be valuable knowledge for the rational vaccine design to obtain vaccines that induce broadly neutralizing antibodies [25–27], define the binding of therapeutic monoclonal antibodies to viral variants [28], and identify antibodies generating undesired clinical outcomes such as antibody-dependent enhancement [29,30]. It can also improve specificity in serological assays [27,29,31,32]. Immunization with peptides, binding strongly neutralizing antibodies, is believed to elicit equally strongly neutralizing antibodies [30]. Defining the B-cell epitopes of broadly protective anti-SARS-CoV-2 antibodies is essential for the next generation of intervention strategies.

During the SARS-CoV-2 pandemic, we developed a DNA vaccine that encodes the native spike protein of the SARS-CoV-2 index strain [33]. This plasmid DNA vaccine induced anti-spike binding IgG and neutralizing antibodies in mice, rabbits, and rhesus macaques, as well as inducing robust Th1-dominant cellular responses in small animals [33]. In the present study, we aim to elucidate the underlying antibody targets associated with broad cross-neutralization observed in a proportion of intramuscular, needle-free-vaccinated rabbits during the preclinical evaluation of this vaccine. Here, we used peptide microarray technology, with circular constrained peptides, to define immunogenic hotspots and identify B-cell epitopes associated with the ability to cross-neutralize different variants, often referred to as neutralization breadth.

2. Materials and Methods

2.1. DNA Vaccination and Study Population

We developed and performed a preclinical evaluation of a candidate DNA vaccine in different animal models, including rabbits [33]. Five ten-week-old New Zealand white rabbits each received 125 µg of a nanoplasmid vector three times at two-week intervals. This DNA vaccine encodes the full, unmodified SARS-CoV-2 index strain spike protein derived from the Wuhan-Hu-1 strain (MN908947). GeneArt (Thermo Fisher Scientific, Dreieich, Germany) synthesized the human codon optimized SARS-CoV-2 spike sequence with EcoRI and XhoI restriction enzyme sites for cloning into the NTC8685-eRNA41H vector backbone (Nature Technology Corporation, Lincoln, NE, USA). The Nature Technology Corporation produced the pNTC-Spike plasmid using their antibiotic-free RNA-OUT selection procedure in NTC4862 *E. coli* cells (DH5α attλ::P5/6 6/6-RNA-IN-SacV, Cmr). Vaccine stocks were supplied at 10 mg/mL in phosphate buffered saline (PBS).

The rabbits were immunized intramuscularly using a PharmaJet® Stratis IM device. Serum samples taken 2 weeks after the last vaccination were included for analysis in the present study. No control animals were used in this study. The rabbits were kept in a temperature-regulated, pathogen-free environment at the Statens Serum Institut in Copenhagen, Denmark. The animals had continuous access to water and regular pellet food. Their cages included bedding and elements for environmental stimulation. All care and processes involving the animals adhered to Danish laws, which are aligned with the EU Directive 2010/63/EU 107 concerning the safeguarding of animals utilized in scientific research. The experiments were supervised by the laboratory animal veterinarians at Statens Serum Institut. The experiment received approval by The Animal Experimentation Council, the National Competent Authority within this field (approval number 2017-15-0201-01322; approval date 29 September 2017) and the Internal Animal Welfare committee at the animal facility at Statens Serum Institute (approval date 10 December 2021).

2.2. Microneutralization

The SARS-CoV-2 live virus microneutralization assay was performed as described using different SAS-CoV-2 variants [34,35]. Heat-inactivated serum samples were 2-fold diluted and incubated with 300 × $TCID_{50}$ for 1 h at 37 °C, 5% CO_2. Tissue culture plates that were seeded with a Vero E6 cell monolayer (10,000 cells/well) the preceding day were washed twice with 100 μL of Dulbecco's phosphate buffered saline (DPBS) and overlaid with 100 μL of the serum–virus mixture. The inoculated cells were incubated at 37 °C, 5% CO_2, for 24 to 32 h, depending on the SARS-CoV-2 variant. Quadruplicate wells with cells containing either 100 μL of 300 × $TCID_{50}$ SARS-CoV-2 virus (no serum) or 100 μL of the virus diluent only were included on each microneutralization plate as the virus control and cell control, respectively. Microneutralization plates were washed twice with 100 μL of DPBS, and the SARS-CoV-2-infected monolayers were subsequently fixed with cold 80% (v/v) acetone in DPBS for 10 min. The replicating virus was quantified in an ELISA, targeting the nucleocapsid protein of SARS-CoV-2 using primary mouse anti-SARS-CoV-2 nucleocapsid protein mAb clone 7E1B, diluted at 1:4000 (cat. #BSM-41414M, Bioss Inc, Woburn, MA, USA) and a secondary antibody, a goat anti-mouse IgG peroxidase-conjugated polyclonal antibody, diluted at 1:10,000 (cat. #A16078, Invitrogen, Thermo Fisher, Waltham, MA, USA). The incubations with the primary and secondary antibodies were carried out for five minutes on an orbital shaker (300 rpm) at room temperature, followed by a one-hour incubation at 37 °C. The plates were washed three times with wash buffer (1% (v/v) Triton-X100, 0.2% Tween20 in DPBS) for 30 s before the primary and secondary antibody incubations. After the secondary antibody incubation, the plates were washed 5 times with the wash buffer and 3 times with deionized water (no soak) to remove all washing buffer debris. Then, 100 μL of TMB ONE substrate (Kem-en-tec, cat. #4380) was added to each well and the plates were incubated in the dark for 10 min at room temperature. The reaction was stopped with 100 μL of 0.2 M H_2SO_4 and the absorbance was read at 450 nm, with 620 nm as a reference, on a FluoStar Omega plate reader (BMG Labtech, Offenburg, Germany).

2.3. Microarray

The custom cyclic constrained peptide microarray, spanning the full-length spike protein of the SARS-CoV-2 index strain, was designed and manufactured by PEPperPRINT (Heidelberg, Germany). The serum antibody binding to the peptides on the microarray was performed according to the manufacturer's instructions. In brief, the microarray was equilibrated with a washing buffer (DPBS with 0.05% Tween20, pH 7.4) for 15 min. The subarrays were subsequently blocked with a blocking buffer (Rockland Blocking Buffer MB-070, Rockland Immunochemicals, Pottstown, PA, USA) for 30 min. To assess unspecific reactions, the peptide array was pre-stained with a biotinylated secondary antibody and fluorophore-conjugated streptavidin, as described below, and scanned. The microarray was equilibrated for 15 min in a staining buffer (DPBS with 0.005% Tween20 and 10% blocking buffer, pH 7.4). The serum, diluted in the staining buffer, was added into the individual subarrays and incubated overnight at 4 °C. The following day, each subarray was washed twice with the wash buffer for 10 s, and a biotin-conjugated goat anti-rabbit IgG F(c) secondary antibody (cat. #A16122, Invitrogen, Thermo Fisher), diluted at 1:500 in the staining buffer, was added to each subarray and incubated for 45 min. The subarrays were washed twice as before and incubated for 45 min with streptavidin Alexa Fluor 647 conjugate (Thermo Fisher Scientific, USA), diluted at 1:750. The subarrays were washed as before, with the microarray slide dipped twice in a dipping buffer (1 mM Tris buffer, pH 7.4), dried using a centrifuge at 1000 rpm for 1 min, and scanned on an Agilent SureScan microarray scanner. The subarrays were equilibrated with the staining buffer for 15 min and subsequently stained with a PEPperCHIP® anti-HA control antibody, diluted at 1:2000 in the staining buffer, and incubated for 45 min. The subarrays were washed, dipped twice in the dipping buffer, dried via centrifugation and scanned as described above. All volumes added to the subarrays were 400 µL, as appropriate for a microarray glass slide with four subarrays. All incubation steps were carried out in the provided cassette from PEPperPRINT, ensuring dark conditions, on an orbital shaker at 140 rpm. With the exception of the overnight incubation of the serum at 4 °C, all other incubations were at room temperature. The microarrays were analyzed using the MAPIX Analyzer Software v. 9.1.0.

2.4. Spike and RBD Indirect ELISA

White Nunc™ MaxiSorp™ plates were coated with 100 µL of 1 µg/mL recombinant SARS-CoV-2 spike ectodomain or RBD proteins, and incubated overnight at 4 °C. The recombinant proteins included SARS-CoV-2 D614G Spike ectodomain (Cat. #10587-CV), SARS-CoV-2 B.1.351 Spike ectodomain (Cat. #10777-CV), SARS-CoV-2 Spike RBD (Cat. #10500-CV), or SARS-CoV-2 B.1.351 Spike RBD (Cat. #10735-CV). All proteins were sourced from R&D Systems. The following day, the coated plates were emptied and, without washing, 150 µL of blocking buffer (Dilution Buffer pH 7.2, Cat #1322, SSI Diagnostica (Hillerød, Denmark) + 2% bovine serum albumin, and 0.1% Tween20) was added. The plates were subsequently incubated for 5 min on an orbital shaker (300 rpm) at room temperature, and then for 1 h at 37 °C. The plates were washed three times in a washing buffer (PBS with 0.05% Tween20), allowing for a 30-s soak each time. Then, 100 µL of serially diluted rabbit serum, diluted from 1:20 to 1:81,920 in the dilution buffer, were added, and the plates were incubated as described above. The plates were washed three times in the washing buffer with a 30 s soak, and 100 µL of a 1:2000 dilution of horse radish peroxidase-conjugated mouse-anti-rabbit IgG antibody (Cat. #A1949 Sigma-Aldrich, Darmstadt, Germany) was added. The plates were incubated as described above, followed by being washed five times with the wash buffer and three times with deionized water, with no soak. Lastly, the plates were treated with 100 µL of BM Chemiluminescent Substrate (Cat. # 11582950001, Sigma-Aldrich, Germany) and incubated for 10 min in a dark environment at room temperature. The resulting luminescence was recorded using a FLUOstar Microplate Reader (BMG LABTECH, Germany).

2.5. Total Antibody RBD ELISA

The Wantai SARS-CoV-2 Ab ELISA was performed according to the manufacturers' instructions. The Wantai ELISA measures the total antibodies of all classes in a double antigen sandwich format. Firstly, 100 µL of the serially diluted rabbit sera (1:20 to 1:81,920) were added to wells that were pre-coated with the recombinant SARS-CoV-2 RBD antigen, and incubated for 30 min at 37 °C. The wells were washed five times and the horse radish peroxidase-conjugated SARS-CoV-2 antigen was added and incubated for 30 min at 37 °C. The wells were washed five times, followed by the addition of a chromogen solution, and incubated for 15 min at 37 °C. The reaction was stopped and the OD values were read on an ELISA reader.

2.6. Statistics and Calculations

2.6.1. Microneutralization

Neutralization titers were calculated as the interception of a 4-parameter logistic curve fit of the OD values obtained from the 2-fold titration of the samples, starting from a 1:10 dilution to 1:1280, with a cut off. The 50% virus infection cut off was calculated from the virus control wells and the cell control wells as follows:

$$50\%\ virus\ infection\ cutoff = \frac{(average\ OD\ of\ virus\ control\ wells) + (average\ OD\ of\ cell\ control\ wells)}{2}$$

Curve fitting was carried out using GraphPad Prism and the cut off calculations and titer calculations were carried out using Excel. The final titers are based on triplicate determinations.

2.6.2. Microarray

The individual determinations were the fluorescence intensity at 635 nm minus the background at 635 nm. The area for background determination was manually set in Mapix when placing the grid. The fluorescence calculations were calculated as the mean of Δ-fluorescence from two identical peptides. The calculations were normalized according to the mean of a positive HA control ($n = 48$) included on each subarray. Active epitopes were defined as the fluorescence intensities above the cut off. The cut off was defined as

$$Cutoff = \overline{X} + SD \times f$$

where $\overline{X}$ is the mean of n negative control determinations ($n = 96$) on each subarray, SD is the standard deviation, and f is the standard deviation multiplier provided by Frey et al. [36]. The confidence interval was set to 99.0%. The negative control was 96 negative spots included on each subarray.

2.6.3. Spike ELISA and RBD ELISA

The endpoint titers were calculated as the interception of a 4-parameter logistic curve fit of OD values obtained from the 4-fold titration of the samples, starting from a 1:20 dilution to 1:81,920, with a cut off. The cut off was defined as

$$Cutoff = \overline{X} + SD \times f$$

where $\overline{X}$ is the mean of n blank determinations ($n = 6$), SD is the standard deviation, and f is the standard deviation multiplier provided by Frey et al. [36]. The confidence interval was set to 99.0%. The final titers were based on triplicate determinations. The curve fitting was carried out using GraphPad Prism and the cut off calculations and titer calculations were carried out using Excel.

3. Results

3.1. Interindividual Differences in Cross-Neutralization of SARS-CoV-2 Variants of Concern (VOCs)

To define the ability of individual DNA-vaccinated rabbit sera to cross-neutralize diverse SARS-CoV-2 variants, we first determined the neutralization titer for each of the five rabbits against the ancestral strain D614G as well as eight World Health Organization-defined SARS-CoV-2 VOCs (Figure 1) [33]. Based on the median neutralization titers calculated for all VOCs, the five rabbits could be categorized broadly into three groups: low (J044890: median VOC titer = 15), intermediate (J044904: median VOC titer = 33; J044906: median VOC titer = 51), and high (J044907: median VOC titer = 143; J044908: median VOC titer = 174) cross-neutralization. Despite having comparable anti-spike IgG levels (Supplementary Figure S1), the five animals had distinct cross-neutralization profiles that would suggest a role for inter-individual-specific and/or dominant epitopes targeted within the polyclonal serum. To address this, a representative animal from each category was selected for an in-depth analysis of the B-cell epitopes targeted in the polyclonal sera, i.e., J044890, J044906, and J044908 (hereafter referred to as Rabbits 1, 2, and 3).

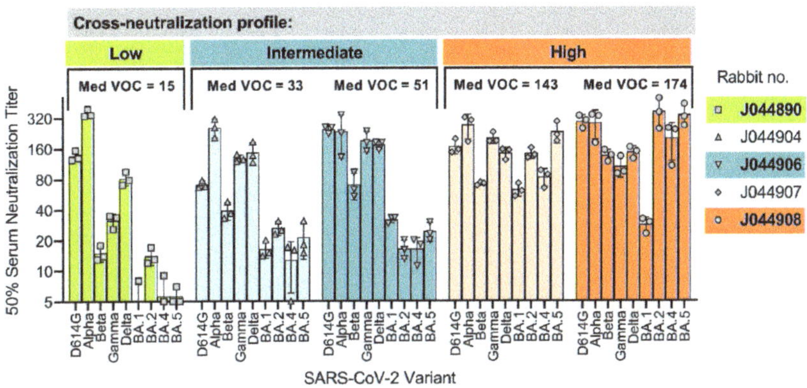

Figure 1. Cross-neutralization of serum antibodies from DNA-vaccinated rabbits against SARS-CoV-2 variants of concern. Rabbits received a DNA vaccine encoding the SARS-CoV-2 index strain spike protein three times with 2-week intervals [33]. Two weeks after final vaccination, 50% virus neutralization titers were determined for SARS-CoV-2 VOCs. Bars represent the mean virus neutralization titer for triplicate measurements against each variant and lines indicate standard deviation. Animals were categorized into low, intermediate, or high levels of cross-neutralization depending on their median neutralization titers against all VOCs. Med VOC was calculated as the median of all VOC neutralization titers for each rabbit. Animals selected for further analysis were J044890, J044906, and J044908 (hereafter referred to as Rabbits 1, 2, and 3).

3.2. Polyclonal Antibody Binding Profiles of Serum with Variable Cross-Neutralization of the SARS-CoV-2 VOCs

We next characterized the antibody binding profiles of individual rabbit sera using a custom-made peptide microarray, presenting the cyclic constrained peptides spanning the complete index strain spike protein. The cyclic peptides may, to some extent, represent conformational epitopes better than linear peptides and were therefore selected for the analysis. To adjust for a potential confounding effect of inter-individual anti-spike IgG levels, an equivalent amount of total anti-spike IgG was tested for each rabbit, as determined from the ELISA binding IgG titers. We evaluated two different serum dilutions on the peptide microarray, at a 1:20 dilution to represent the peptide binding of low-affinity and/or low-abundance antibodies, and a 1:100 dilution to represent high-affinity and/or high-abundance antibodies.

As expected for the inter-individual polyclonal antibody responses, the animals had different serum antibody binding profiles (Figure 2A). At a serum dilution of 1:20, Rabbits 1, 2, and 3 targeted 210, 102, and 62 individual peptides, respectively. At a 5-fold higher serum dilution of 1:100, the three rabbits had fewer positive peptide detections at 77, 11, and 8, respectively. For both serum dilutions, Rabbit 2 and Rabbit 3, which had an intermediate and high cross-neutralization against the VOCs, targeted significantly fewer peptides and had distinct peptides in common (Figure 2B).

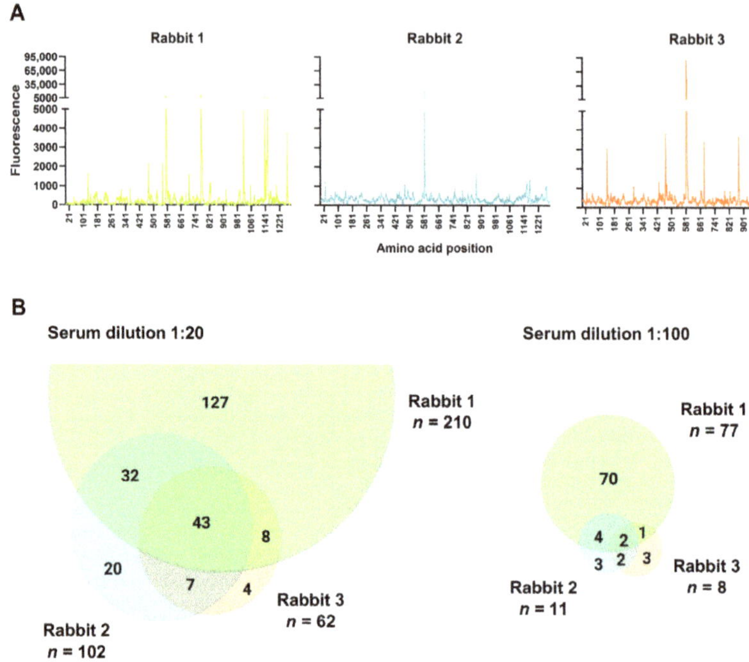

Figure 2. Serum antibody binding profiles targeting the SARS-CoV-2 spike protein. (**A**) Antibody binding of 1:20 diluted serum samples to 10-mer overlapping peptides spanning the ectodomain of the SARS-CoV-2 index strain spike protein. The *x*-axis represents the spike protein amino acid, as in the last position of the individual peptides. Fluorescence levels on the *y*-axis indicate the average level of antibody binding to each peptide measured in duplicate and are normalized according to a positive control included on each subarray. (**B**) The number of targeted peptides with fluorescent signals above a 99.9% confidence level cut-off value calculated from a negative control on each subarray. The Venn diagrams show the number of targeted peptides shared between the different animals and number specific to each animal.

3.3. Dominant Targeting of an SD1 Epitope in Animals with Intermediate and High Cross-Neutralization of VOCs

The higher serum dilution of 1:100, which likely represents a detection of more abundant and/or higher affinity antibodies, identified two regions common between the two rabbits with VOC cross-neutralization, but not Rabbit 1, which had a low-to-undetectable neutralization against neutralization-resistant VOCs. These include a peptide in the SD1 within the S1 subunit, spanning amino acids 572–581 (TTDAVRDPQT), and a peptide in the connecting region of the S2 subunit, spanning amino acids 862–871 (PPLLTDEMIA) (Figure 3A). For Rabbit 3, the antibody binding to the peptide spanning residues 572–581 (TTDAVRDPQT) peaked above the assay upper limit of quantification at the 1:20 serum dilution (Figure 3B). At a 5-fold lower serum input on the microarray (1:100), the fluorescence intensity was near the upper limit of quantification (97.7% saturated), thus enabling a quantitative intra- and inter-individual comparison. Rabbit 2 similarly targeted the 572–581 peptide, but with a markedly lower intensity (27-fold; relative fluorescence units (RFUs) 3286 vs. 88,324), while Rabbit 1 did not target peptide 572–581 (RFU = 14), but rather an adjacent sequence spanning three peptides and residues 563–577 (QFGRDIADTTDAVR; RFUs = 922–2146). The level of antibody binding to peptide 572–581 correlated with the level of cross-neutralization of the VOCs (Figure 3C).

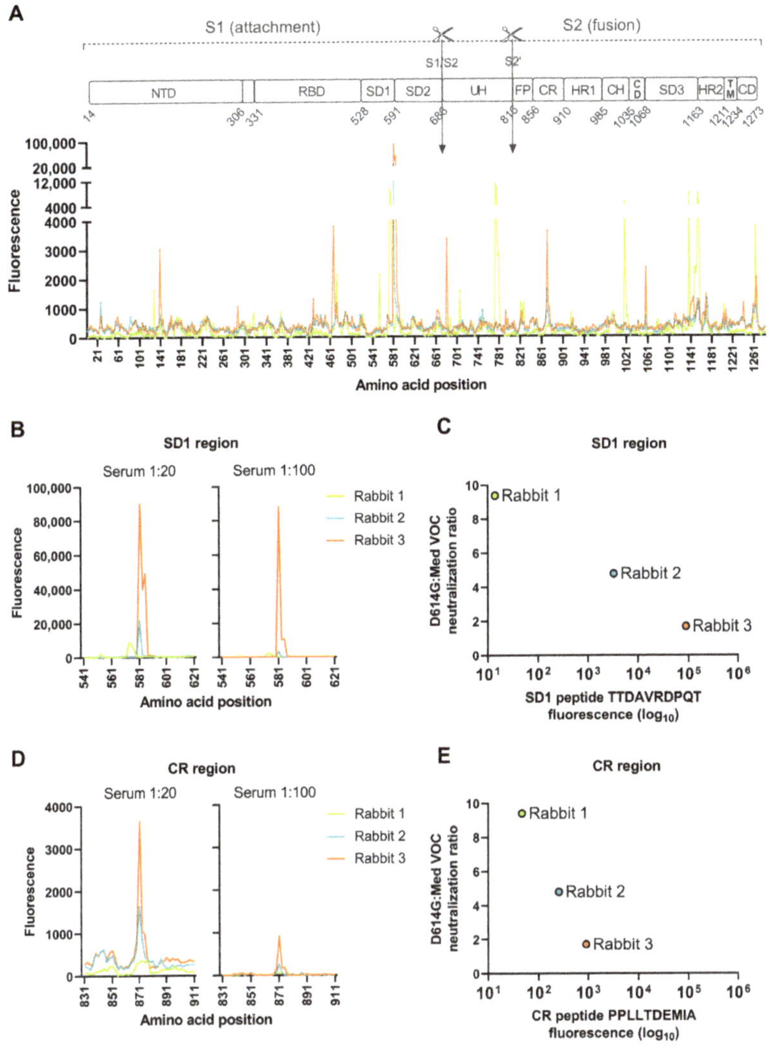

Figure 3. Dominant antibody binding to an SD1 epitope in animals with cross-neutralization of the VOCs. (**A**) Comparative serum antibody binding for the three animals relative to the different SARS-CoV-2 spike protein domains. S1 subunit (residues 14–685); S2 subunit (residues 686–1273); N-terminal domain (NTD); receptor binding domain (RBD); sub-domain 1, 2, 3 (SD1, SD2, SD3), S1/S2 furin cleavage site; upstream helix (UH); S2′ cleavage site; fusion peptide (FP); connecting region (CR); heptad repeat sequence 1 and 2 (HR1, HR2); central helix (CH); connector domain (CD); transmembrane helix (TM); and connector domain (CD). Antibody binding with a sample dilution of 1:20 and 1:100 to the SD1 region (**B**) and correlation analysis of antibody binding to the peptide TTDAVRDPQT (amino acids 572–581 on the index strain) to cross-neutralization capacity of the VOCs (**C**). Antibody binding with a sample dilution of 1:20 and 1:100 to the CR region (**D**) and correlation analysis of antibody binding to the peptide PPLLTDEMIA (amino acids 862–871 on the index strain) to cross-neutralization capacity of the VOCs (**E**). Antibody binding is expressed as fluorescence intensity measured for the duplicate spots of the peptide. Cross-neutralization capacity is calculated as the ratio of neutralization titers determined for the D614G strain to the median of neutralization titers of the VOCs (Med VOC).

The other spike ectodomain peptide common between Rabbit 3 and Rabbit 2 includes the peptide spanning residues 862–871 (PPLLTDEMIA; RFUs at serum 1:100 = 911 and 255, respectively) (Figure 3D). The binding to this peptide was 97-fold (Rabbit 3) and 13-fold (Rabbit 2) lower compared to the binding measured for the aforementioned 572–581 peptide. Rabbit 1 had a low-level binding to peptide 862–871 (RFUs = 42). As observed for peptide 572–581, the level of antibody binding correlated with the level of the cross-neutralization of the VOCs (Figure 3E).

In summary, Rabbit 3, with the highest cross-neutralization of VOCs, shares common epitopes with Rabbit 2 that has a detectable, yet lower, cross-neutralization of neutralization-resistant VOCs. The level of antibody binding to both peptides (TTDAVRDPQT and PPLLTDEMIA) correlate with the level of cross-neutralization. However, proportionally, the magnitude of antibody binding to the peptide spanning residues 572–581 (TTDAVRDPQT) dominated the responses for both Rabbit 3 and Rabbit 2 (Figure 4). This SD1 epitope is highly conserved between all SARS-CoV-2 variants (Figure 5). In contrast to this targeted response, Rabbit 1 demonstrated a more diffuse antibody binding profile across the entire spike ectodomain. Beyond the shared epitopes between Rabbit 3 and Rabbit 2, Rabbit 3 furthermore had binding to a peptide in the RBD (residues 458–467; KSNLKPFERD; RFUs at serum 1:100 = 491) and immediately upstream of the S1/S2 furin cleavage site (residues 672–681; ASYQTQTNSP; RFUs = 841).

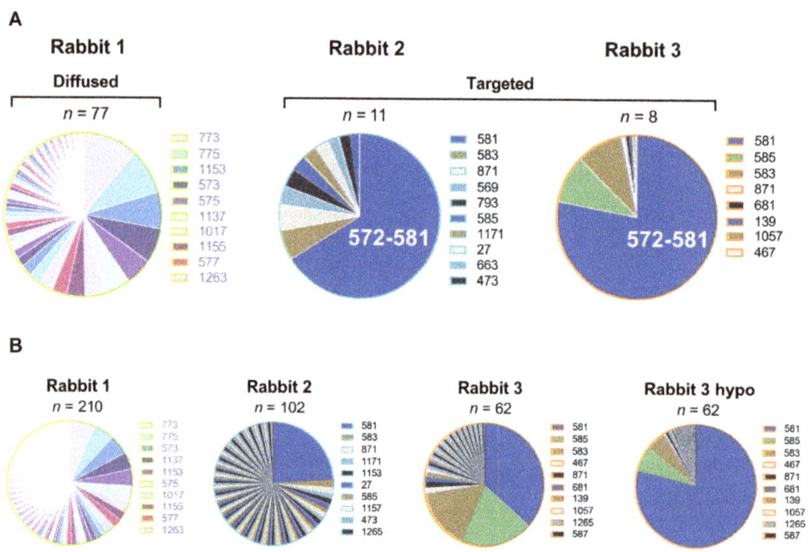

Figure 4. Proportionate binding of serum IgG to overlapping peptides spanning the SARS-CoV-2 index strain spike ectodomain for each rabbit. The pie charts consider the level of antibody binding, expressed as relative fluorescent units, to each peptide as a proportion of the sum of relative fluorescent units measured for binding to all overlapping peptides. (**A**) At a serum dilution of 1:100, Rabbit 2 and Rabbit 3 show dominant, targeted binding to the peptide spanning residues 572–581 (TTDAVRDPQT), while Rabbit 1 lacks binding to this peptide and show a more diffuse antibody binding profile. (**B**) At a serum dilution of 1:20, the dominance of peptide 572–581 remains for Rabbit 2 and Rabbit 3. Rabbit 3 hypo is the hypothetical proportion with no limiting spot saturation.

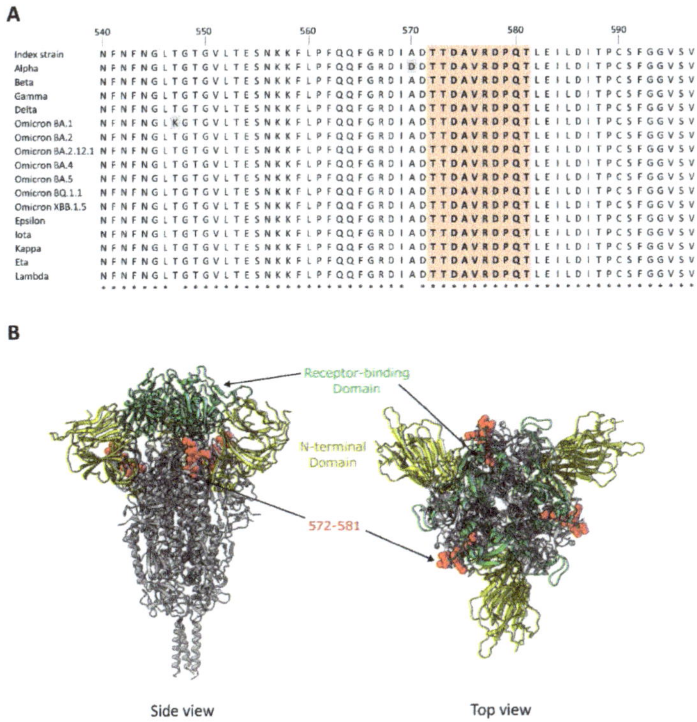

Figure 5. Sequence conservation of peptide 572–581 (TDAVRDPQT) and the position of the epitope in the SD1 regions of the full index strain spike protein trimer. (**A**) An amino acid sequence alignment of the partial SARS-CoV-2 spike protein encompassing the SD1 and peptide (highlighted in peach) targeted in animals with cross-neutralization capacity. (**B**) The position of the SD1 peptide on the crystal structure of the spike protein indicated in red spheres. The receptor binding domain is in green and the N-terminal domain is in beige. Asterisk (*) refers to identical amino acids in all sequences. Substitutions are indicated in grey.

3.4. Epitopes Targeted within the Spike RBD

The RBD is a frequent target of neutralizing antibodies, including broadly cross-neutralizing antibodies such as S309 and CR3022. Anti-RBD antibodies are grouped into different classes (class 1–4) based on their blocking of the direct binding of ACE2 (class 1 and class 2), non-ACE2 blocking (class 3 and class 4), and binding to the specific conformation of RBD, i.e., up (class 1–4) or down (class 2 and 3) [37]. At a serum dilution of 1:20, we observed 32, 20, and 14 positive peptides within the RBD for Rabbits 1, 2 and 3, respectively. For the targets within the receptor binding motif (class 1 and class 2 antibodies), there were 15, 12, and 13 positive peptides with maximum RFUs of 2134, 1078 and 3809 for Rabbits 1, 2 and 3, respectively. For the targets of the class 3 and 4 antibodies, at a serum dilution of 1:20, there were 17, 8, and 1 positive peptides that included a maximum RFU of 854 observed in Rabbit 1 spanning amino acids 358–367 (ISNCVADYSV), 744 observed in Rabbit 2 spanning amino acids 508–517 (YRVVVLSFEL), and 747 observed in Rabbit 3 spanning amino acids 334–343 (NLCPFGEVFN). At a serum dilution of 1:100, we found 9, 1, and 1 positive peptides within the RBD for Rabbits 1, 2 and 3, respectively. For the targets within the receptor binding motif (class 1 and class 2 antibodies), there were 6, 1, and 1 positive peptides with maximum RFUs of 219, 111, and 491 for Rabbits 1, 2 and 3, respectively. For the targets of the class 3 and 4 antibodies, at a serum dilution of 1:100, there were three positive peptides in Rabbit 1 that included a maximum RFU of 257 spanning amino acids 358–367 (ISNCVADYSV).

3.5. The Effect of Beta VOC-Specific Spike Mutations on Peptide Binding

Each subarray included peptides with specific amino acid changes within the Beta spike protein. These include amino acid substitutions D80A, D215G, K417N, E484K, N501Y, D614G, and A701V, and deletion LAL241–243del. Of these, mutations K417N and E484K are known to induce neutralization resistance [38–42]. In comparison to the dominant antibody binding sites in the spike ectodomain, the antibody binding level to peptides spanning the aforementioned residues ranged from undetectable to low for both the index strain and Beta VOC (Figure 6). In all three animals, there was a low-level antibody binding to the index strain peptides spanning amino acid 484, which was reduced notably by the mutation in the Beta VOC. A very low binding to the K417N mutation was detected, with a modestly increased binding to the mutated peptides in all three animals. The antibody binding to D80A, LAL241–243del, N501Y, and A701V remained mostly unaffected by the amino acid changes. The analysis of Omicron-specific amino acid changes is precluded, since the microarray was designed and produced prior to the emergence of Omicron.

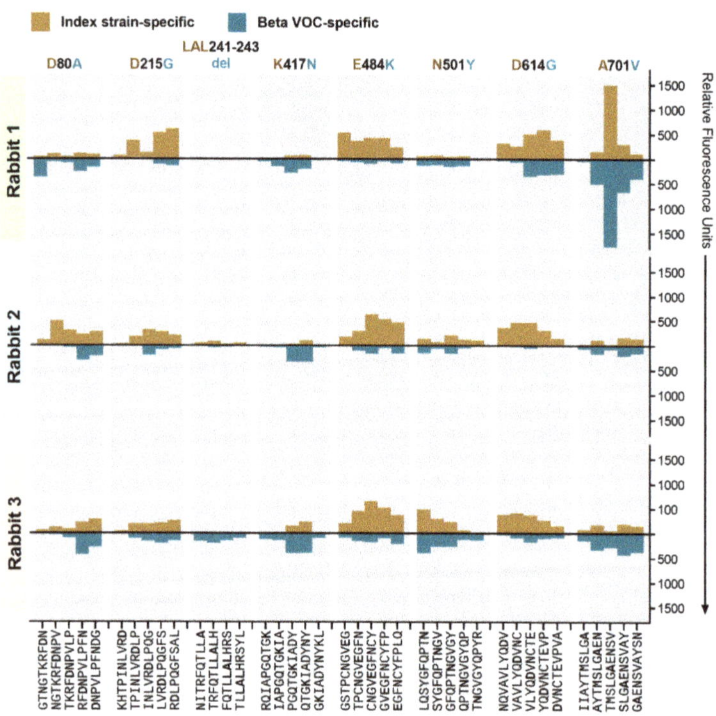

Figure 6. Antibody binding to overlapping peptides bearing SARS-CoV-2 Beta VOC spike protein variations relative to the index strain. Mirrored serum antibody binding to index strain peptide and their counterpart containing Beta-specific amino acid substitutions D80A, D215G, K417N, E484K, N501Y, D614G, and A701V, and deletion LAL241–243del. The x-axis indicates the 10-mer peptide sequence with the indicated amino acid variation at different positions; the third peptide for each variant bears the residue of interest approximately in the center of the sequence. The y-axis presents the mean fluorescence intensity minus background for duplicate peptide spots on the microarray. Values in gold indicate antibody binding to the index strain sequence and values in turquoise indicate antibody binding to the Beta VOC sequence.

3.6. Binding Antibody Levels against the Spike Ectodomain and RBD

Since higher levels of antibody titers can lead to a cross-neutralization of resistant SARS-CoV-2 variants, we determined the levels of the anti-spike ectodomain IgG and the anti-RBD IgG relative to the levels of neutralizing antibodies against the D614G strain and the Beta VOC (Figure 7). Intriguingly, Rabbit 3, with the highest cross-neutralization, had the lowest spike ectodomain-binding IgG titers against the D614G spike (endpoint titer = 4584 vs. 7776 and 6294), and an even lower level of binding IgG titers against the Beta VOC spike ectodomain, although comparable to the other rabbits (endpoint titer = 2535 vs. 2747 and 4322) (Figure 7B). Anti-RBD IgG titers followed a similar pattern with endpoint titers against the D614G strain of 5087, 4377, and 3442 for Rabbits 1, 2, and 3, respectively (Figure 7C). Notably, Rabbit 3 together with Rabbit 1 had the lowest anti-RBD IgG binding titers for the Beta VOC (endpoint titers = 774 and 786, respectively, vs. Rabbit 2 = 2779) (Figure 7C).

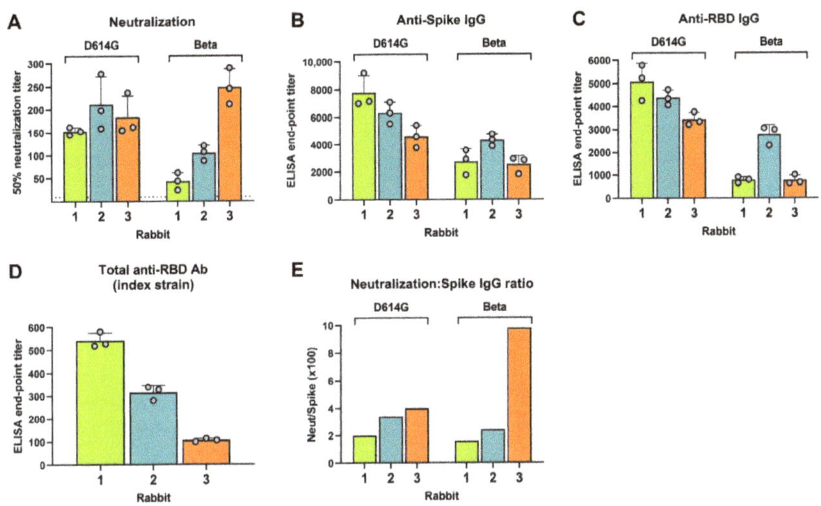

Figure 7. Virus neutralization and antibody binding of the SARS-CoV-2 D614G strain and Beta VOC. (**A**) The 50% live virus neutralization titers against the D614G strain and Beta VOC. Dotted line represents the lowest serum dilution tested. (**B**) Endpoint titers of binding IgG specific for the spike ectodomain of the D614G strain and Beta VOC, as measured in an indirect ELISA. (**C**) Endpoint titers of binding IgG specific for the spike protein receptor binding domain (RBD) of the D614G strain and the Beta VOC, as measured in an indirect ELISA. (**D**) Endpoint titers of total antibodies of all isotypes binding to the RBD of the index strain, as measured in a double antigen sandwich ELISA. (**E**) Virus neutralization normalized for IgG-binding antibody levels calculated as the neutralization titers divided by the anti-spike IgG titers. All error bars are standard deviation error bars.

A different ELISA format—the antigen–antibody–antigen sandwich ELISA—that detects the total antibodies of all isotypes against the index strain RBD, similarly showed that Rabbit 3 had the lowest RBD-specific total antibody levels (Figure 7D). To remove the possible confounding effect of high antibody titers on cross-neutralization, we normalized the virus neutralization titers using the binding IgG titers, which revealed that Rabbit 3 had the highest cross-neutralization capacity per unit of IgG compared to the other two rabbits (Figure 7E). In particular, compared to Rabbit 1, Rabbit 3 showed a 2.0-fold higher normalized neutralization capacity for the D614G strain and a 6.1-fold higher ratio for the Beta VOC.

4. Discussion

This study defines the serum antibody profiles of SARS-CoV-2 DNA-vaccinated rabbits with divergent cross-neutralization of SARS-CoV-2 VOCs using a peptide microarray displaying cyclic constrained peptides spanning the SARS-CoV-2 spike protein. The findings demonstrate that the cross-neutralization of variants is not necessarily the function of higher total-antibody responses, since the rabbit with the greatest cross-neutralization had the lowest anti-spike IgG binding antibody levels against the spike ectodomain and RBD. Furthermore, cross-neutralization did not appear to associate with an increased tolerance of amino acid changes known to induce neutralization resistance. Rather, we observed an IgG antibody response dominated by targeting an SD1 epitope in both animals with a moderate-to-high neutralization of VOCs, but not in an animal that lacked this functional antibody property.

On the spike S1 subunit, SD1 forms part of the hinge region and facilitates the RBD upward rotation, a conformational change required for the binding of the RBD to the ACE2 host receptor [43]. Despite the location outside the RBD, SD1 is a known target of neutralizing SARS-CoV-2 antibodies induced by infection and vaccination [16–18,22,23]. Evidence suggests that the mechanism of anti-SD1 antibody neutralization likely occurs through the conformational locking of the RBD in the 'down' conformation, and destabilizing spike trimers, rather than inducing S1 shedding or the stearic hindrance of ACE2 engagement [16,23]. In the present study, rabbit sera with a cross-neutralization of neutralization-resistant VOCs predominantly targeted an SD1 epitope (TTDAVRDPQT), which overlaps with that of two broadly neutralizing monoclonal antibodies [23]. The breadth of neutralization associated with anti-SD1 antibodies is attributable to this region of the spike protein being highly conserved between SARS-CoV-2 variants (Supplementary Figure S2). Residues within this epitope (572–581) comprise one of the most conserved regions within the S1 subunit. The very low conservation scores indicate a highly conserved area among SARS and MERS-related viruses containing multiple residues that are believed to be well-suited for drug intervention [44,45]. It is therefore plausible that the prominent targeting of an SD1 epitope by the serum antibodies may have contributed to the observed cross-neutralization in Rabbit 3 and Rabbit 2. To validate the role of SD1 peptide 572–581 as a broadly neutralizing epitope, an appropriate approach would be to remove the antibodies targeting this site. In theory, this is achievable through antibody competition or antibody depletion. Antibody competition with a peptide, in this case a 10-mer peptide, does, however, pose a challenge, as the binding affinity is generally too low to compete with the binding to the full immunogen. Antibody depletion, using the circular 10-mer peptide to mimic the binding conditions present on the microarray, similarly pose challenges that could distort the interpretation. Slight changes in the epitope presentation, binding accessibility, linker chemistry, etc., could cause a drift in the targeted antibodies, resulting in an uncontrolled depletion and an incorrect evaluation of the effect on the neutralization capacity. In summary, our data show a correlation between SD1 antibody targeting and neutralization breadth. However, a validation of the contribution of these antibodies to the cross-neutralization capacity, supporting the findings in this study, is warranted.

Neutralizing anti-SD1 antibody epitopes may be structurally occluded in the prefusion state of the spike protein [16]. This brings into question the efficiency by which SARS-CoV-2 vaccines using the prefusion-stabilized spike protein, e.g., mRNA vaccines, induce neutralizing antibodies targeting this region. Of note, the majority of anti-SD1 neutralizing monoclonal antibodies were isolated from convalescent donors or those who experienced a breakthrough infection after mRNA vaccination, as well as a donor vaccinated with the native spike-encoding AZD1222 vaccine [16,18,22,23]. It is therefore of consequence that the SARS-CoV-2 pNTC-Spike DNA vaccine construct that induced the anti-SD1 antibodies in the rabbits also encodes a native spike protein, rather than a prefusion-stabilized protein [33]. It could be argued that expressing the native spike in vivo using a DNA vaccine is more representative of a natural virus infection compared to an mRNA vaccine using a prefusion-stabilized spike protein. It should be noted, though, that pre-fusion stabilization of the glycoprotein is regarded as being immunogenicity-enhancing, presumably because the stabilized glycoprotein cannot convert to the post-fusion structure or misfold, thereby preserving the immunogenic conformation [46–48].

We cannot exclude a possible contribution of antibodies targeting other spike regions sensitive to neutralization, such as the RBD, N-terminal domain, and fusion peptide. On the peptide microarray, the binding to peptides spanning the RBD was generally low, with the exception of residues 458–467 (KSNLKPFERD). The monoclonal antibody C98C7 that binds these residues has previously been reported to broadly neutralize VOCs including Omicron BA.1 [49]. RBD class 3 and 4 monoclonal antibodies C309 and CR3022 that bind epitopes immediately upstream of the receptor binding motif have similarly been shown to broadly neutralize VOCs including Omicron BA.1, BA.1.1, and BA.2 [50–53]. We found only a weak binding to this region, with a maximum RFU of 854 in all three rabbits. The general low-level binding to the entire RBD detected in this study may be attributable to the conformational structure of RBD that may be poorly represented on the peptide microarray [3], or to the effective glycan shielding limiting the antibody targeting of sites on the RBD [54]. Nonetheless, the low level of the anti-RBD IgG specific for the Beta VOC, and the total antibody levels to the D614G strain observed in Rabbit 3—the animal with the greatest level of cross-neutralization—would suggest that the antibody target that is responsible for the broad neutralization resides outside the RBD. The serum antibodies of the rabbit lacking cross-neutralization (Rabbit 1) targeted regions in the upstream helix and sub-domain 3, close to the heptad repeat 2. These two regions have previously been associated with neutralization breadth [51], although in Rabbit 1, specific peptides targeted in this region did not associate with the cross-neutralization of VOCs.

Taken together, this study demonstrates the combination of functional antibody data with epitope mapping using peptide microarray technology as a powerful approach to unveiling key antigenic sites with and without functional significance. A cyclic constrained peptide array, instead of a linear peptide array, may better mimic native protein folding, improving the comparison of array data with ELISA and live-virus neutralization assay data [55]. Discontinuous epitopes brought together from distant amino acid positions are, however, difficult to recapitulate with individual peptides, introducing a limitation of this approach. Nonetheless, with a small but well-selected group of animal subjects, it was possible in the polyclonal serum responses to identify a known, broadly neutralizing antibody target validated by studies of monoclonal antibodies targeting the same site. Those studies showed that this epitope reported to bind broadly cross-reactive monoclonal antibody sd1.040 [56,57], and 12–16 and 12–19 [23] are in fact discontinuous, giving indications that high-resolution cyclic peptide microarrays can identify antibodies targeting discontinuous conformational epitopes.

In our initial selection of the study population, we saw a relatively clear distinction in the cross-neutralization capacities and median of the VOC neutralization titers for each rabbit. Among the five rabbits tested, the two animals that were excluded from the epitope mapping, one with intermediate cross-neutralization and one with high cross-neutralization, had a high resemblance to the two animals included in each segment. We do, however, anticipate that there would be heterogeneity in the profiles of the antibody responses elicited from the remaining two rabbits, as we can see within the group of three rabbits. Including additional animals would highly likely show variations in antigenic targeting as well. Though we performed antibody profiling of only three animals, with each animal representing different categories of cross-neutralization efficiency, this study population displayed highly diverse antibody repertoires, unveiling key features that might be valuable to support future vaccine strategies.

This study contributes to the growing interest within functional and applied antibody research that supports findings on recently discovered SARS-CoV-2 broadly neutralizing epitopes. Combining the application of a DNA vaccine, encoding the full native spike protein with a high-resolution conformational peptide microarray spanning the full index strain spike protein adds a valuable novel perspective to the field, and provides insight into natural infection immunology. The identification of potential immunogenic epitopes that associate with broadly neutralizing antibodies may aid rational vaccine development, such as with the design of effective peptide-based vaccines. Vaccines containing multiple peptides triggering both a B-cell and a T-cell response, or DNA/RNA vectors containing synthetic genes encoding immunogenic epitopes, have promising perspectives and can be produced at a relatively low cost [58]. Peptide-based vaccines are, however, known to be less immunogenic, and the inclusion of an adjuvant to stimulate the immune response is therefore essential [58,59]. Peptide-based SARS-CoV-2 vaccines using the adjuvant lipopeptide synthetic TLR1/2 ligand XS15 [60] or aluminum phosphate (Adju-Phos®) [61] have been reported to induce potent cellular responses in neutralizing antibodies. Future analysis using this approach, including additional animals and other validation methods, e.g., vaccination challenge studies with the SD1 epitope, will further improve our understanding of these initial findings.

Supplementary Materials: The following supporting information can be downloaded at https://www.mdpi.com/article/10.3390/vaccines11091451/s1. Figure S1: ELISA IgG titers against SARS-CoV-2 D614G spike ectodomain for the complete group of rabbits immunized intramuscularly with a candidate DNA vaccine encoding the index strain spike protein of SARS-CoV-2; and Figure S2: Structural Conservation of the SD1 Epitope in BA.1 and Alpha lineages.

Author Contributions: Conceptualization, R.L. and A.F. (Anders Frische); methodology, R.L. and A.F. (Anders Frische); formal analysis, V.G., R.L. and A.F. (Anders Frische); investigation, R.L. and A.F. (Anders Frische); resources, V.G., R.L. and A.F. (Anders Frische); data curation, R.L. and A.F. (Anders Frische); writing—original draft preparation, A.F. (Anders Frische); writing—review and editing, R.L.; visualization, A.F. (Anders Frische), V.G., and R.L.; supervision, R.L., K.A.K., and A.F. (Anders Fomsgaard); project administration, R.L. and A.F. (Anders Fomsgaard); funding acquisition, A.F. (Anders Fomsgaard) and R.L. All authors have read and agreed to the published version of the manuscript.

Funding: This work was partly supported by funding from the Danish Ministry of Education and Research, grant number 0237-00006B.

Institutional Review Board Statement: This experiment received approval by The Animal Experimentation Council, the National Competent Authority within this field (approval number 2017-15-0201-01322; approval date 29 September 2017) and the Internal Animal Welfare committee at the animal facility at Statens Serum Institute (approval date 10 December 2021).

Informed Consent Statement: Not applicable.

Data Availability Statement: The datasets presented in this study can be found in online repositories. The name of the repository and accession number can be found at ArrayExpress; E-MTAB-13040.

Acknowledgments: We would like to thank the skilled technical assistance of Birgit Knudsen.

Conflicts of Interest: Anders Fomsgaard is a co-inventor on a patent application covering a SARS-CoV-2 DNA vaccine; all rights to the vaccine have been assigned to Statens Serum Institut (SSI), a Danish national not-for-profit governmental public health institute. Other authors declare no competing interests. The funders had no role in the design of the study; in the collection, analyses, or interpretation of data; in the writing of the manuscript; or in the decision to publish the results.

References

1. Holmes, E.C.; Goldstein, S.A.; Rasmussen, A.L.; Robertson, D.L.; Crits-Christoph, A.; Wertheim, J.O.; Anthony, S.J.; Barclay, W.S.; Boni, M.F.; Doherty, P.C.; et al. The origins of SARS-CoV-2: A critical review. *Cell* **2021**, *184*, 4848–4856. [CrossRef] [PubMed]
2. Zhou, P.; Yang, X.L.; Wang, X.G.; Hu, B.; Zhang, L.; Zhang, W.; Si, H.R.; Zhu, Y.; Li, B.; Huang, C.L.; et al. A pneumonia outbreak associated with a new coronavirus of probable bat origin. *Nature* **2020**, *579*, 270–273. [CrossRef] [PubMed]
3. Shang, J.; Wan, Y.; Luo, C.; Ye, G.; Geng, Q.; Auerbach, A.; Li, F. Cell entry mechanisms of SARS-CoV-2. *Proc. Natl. Acad. Sci. USA* **2020**, *117*, 11727–11734. [CrossRef] [PubMed]
4. Wrapp, D.; Wang, N.; Corbett, K.S.; Goldsmith, J.A.; Hsieh, C.L.; Abiona, O.; Graham, B.S.; McLellan, J.S. Cryo-EM structure of the 2019-nCoV spike in the prefusion conformation. *Science* **2020**, *367*, 1260–1263. [CrossRef]

5. Huang, Y.; Yang, C.; Xu, X.; Xu, W.; Liu, S. wen Structural and functional properties of SARS-CoV-2 spike protein: Potential antivirus drug development for COVID-19. *Acta Pharmacol. Sin.* **2020**, *41*, 1141–1149. [CrossRef]
6. Barnes, C.O.; Jette, C.A.; Abernathy, M.E.; Dam, K.M.A.; Esswein, S.R.; Gristick, H.B.; Malyutin, A.G.; Sharaf, N.G.; Huey-Tubman, K.E.; Lee, Y.E.; et al. SARS-CoV-2 neutralizing antibody structures inform therapeutic strategies. *Nature* **2020**, *588*, 682–687. [CrossRef]
7. Cai, Y.; Zhang, J.; Xiao, T.; Peng, H.; Sterling, S.M.; Walsh, R.M.; Rawson, S.; Rits-Volloch, S.; Chen, B. Distinct conformational states of SARS-CoV-2 spike protein. *Science* **2020**, *369*, 1586–1592. [CrossRef]
8. Hoffmann, M.; Kleine-Weber, H.; Pöhlmann, S. A Multibasic Cleavage Site in the Spike Protein of SARS-CoV-2 Is Essential for Infection of Human Lung Cells. *Mol. Cell* **2020**, *78*, 779–784.e5. [CrossRef]
9. Hoffmann, M.; Kleine-Weber, H.; Schroeder, S.; Krüger, N.; Herrler, T.; Erichsen, S.; Schiergens, T.S.; Herrler, G.; Wu, N.-H.; Nitsche, A.; et al. SARS-CoV-2 Cell Entry Depends on ACE2 and TMPRSS2 and Is Blocked by a Clinically Proven Protease Inhibitor. *Cell* **2020**, *181*, 271–280. [CrossRef]
10. Peacock, T.P.; Goldhill, D.H.; Zhou, J.; Baillon, L.; Frise, R.; Swann, O.C.; Kugathasan, R.; Penn, R.; Brown, J.C.; Sanchez-David, R.Y.; et al. The furin cleavage site in the SARS-CoV-2 spike protein is required for transmission in ferrets. *Nat. Microbiol.* **2021**, *6*, 899–909. [CrossRef]
11. Johnson, B.A.; Xie, X.; Bailey, A.L.; Kalveram, B.; Lokugamage, K.G.; Muruato, A.; Zou, J.; Zhang, X.; Juelich, T.; Smith, J.K.; et al. Loss of furin cleavage site attenuates SARS-CoV-2 pathogenesis. *Nature* **2021**, *591*, 293–299. [CrossRef] [PubMed]
12. Essalmani, R.; Jain, J.; Susan-Resiga, D.; Andréo, U.; Evagelidis, A.; Derbali, R.M.; Huynh, D.N.; Dallaire, F.; Laporte, M.; Delpal, A.; et al. Distinctive Roles of Furin and TMPRSS2 in SARS-CoV-2 Infectivity. *J. Virol.* **2022**, *96*, e00128-22. [CrossRef] [PubMed]
13. Koppisetti, R.K.; Fulcher, Y.G.; Van Doren, S.R. Fusion Peptide of SARS-CoV-2 Spike Rearranges into a Wedge Inserted in Bilayered Micelles. *J. Am. Chem. Soc.* **2021**, *143*, 13205–13211. [CrossRef] [PubMed]
14. Gruell, H.; Vanshylla, K.; Weber, T.; Barnes, C.O.; Kreer, C.; Klein, F. Antibody-mediated neutralization of SARS-CoV-2. *Immunity* **2022**, *55*, 925–944. [CrossRef] [PubMed]
15. Barnes, C.O.; West, A.P.; Huey-Tubman, K.E.; Hoffmann, M.A.G.; Sharaf, N.G.; Hoffman, P.R.; Koranda, N.; Gristick, H.B.; Gaebler, C.; Muecksch, F.; et al. Structures of Human Antibodies Bound to SARS-CoV-2 Spike Reveal Common Epitopes and Recurrent Features of Antibodies. *Cell* **2020**, *182*, 828–842.e16. [CrossRef]
16. Seow, J.; Khan, H.; Rosa, A.; Calvaresi, V.; Graham, C.; Pickering, S.; Pye, V.E.; Cronin, N.B.; Huettner, I.; Malim, M.H.; et al. A neutralizing epitope on the SD1 domain of SARS-CoV-2 spike targeted following infection and vaccination. *Cell Rep.* **2022**, *40*, 111276. [CrossRef]
17. Xu, S.; Wang, Y.; Wang, Y.; Zhang, C.; Hong, Q.; Gu, C.; Xu, R.; Wang, T.; Yang, Y.; Zang, J.; et al. Mapping cross-variant neutralizing sites on the SARS-CoV-2 spike protein. *Emerg. Microbes Infect.* **2022**, *11*, 351–367. [CrossRef]
18. Guenthoer, J.; Lilly, M.; Starr, T.N.; Dadonaite, B.; Lovendahl, K.N.; Croft, J.T.; Stoddard, C.I.; Chohan, V.; Ding, S.; Ruiz, F.; et al. Identification of broad, potent antibodies to functionally constrained regions of SARS-CoV-2 spike following a breakthrough infection. *bioRxiv* **2023**. [CrossRef]
19. Cerutti, G.; Guo, Y.; Zhou, T.; Gorman, J.; Lee, M.; Rapp, M.; Reddem, E.R.; Yu, J.; Bahna, F.; Bimela, J.; et al. Potent SARS-CoV-2 neutralizing antibodies directed against spike N-terminal domain target a single supersite. *Cell Host Microbe* **2021**, *29*, 819–833.e7. [CrossRef]
20. Liu, L.; Wang, P.; Nair, M.S.; Yu, J.; Rapp, M.; Wang, Q.; Luo, Y.; Chan, J.F.W.; Sahi, V.; Figueroa, A.; et al. Potent neutralizing antibodies against multiple epitopes on SARS-CoV-2 spike. *Nature* **2020**, *584*, 450–456. [CrossRef]
21. Klinakis, A.; Cournia, Z.; Rampias, T. N-terminal domain mutations of the spike protein are structurally implicated in epitope recognition in emerging SARS-CoV-2 strains. *Comput. Struct. Biotechnol. J.* **2021**, *19*, 5556–5567. [CrossRef] [PubMed]
22. Meng, P.C.; Carissimo, G.; Wang, B.; Naqiah, A.S.; Lee, C.; Chee, R.; Fong, S.W.; Kim-Wah, Y.N.; Lee, W.H.; Torres-Ruesta, A.; et al. Two linear epitopes on the SARS-CoV-2 spike protein that elicit neutralising antibodies in COVID-19 patients. *Nat. Commun.* **2020**, *11*, 2806. [CrossRef]
23. Liu, L.; Casner, R.G.; Guo, Y.; Wang, Q.; Iketani, S.; Chan, J.F.-W.; Yu, J.; Dadonaite, B.; Nair, M.S.; Mohri, H.; et al. Antibodies that neutralize all current SARS-CoV-2 variants of concern by conformational locking. *bioRxiv* **2023**. Available online: https://www.biorxiv.org/content/10.1101/2023.04.08.536123v1 (accessed on 30 August 2023).
24. Fan, X.; Cao, D.; Kong, L.; Zhang, X. Cryo-EM analysis of the post-fusion structure of the SARS-CoV spike glycoprotein. *Nat. Commun.* **2020**, *11*, 3618. [CrossRef] [PubMed]
25. Zhou, P.; Yuan, M.; Song, G.; Beutler, N.; Shaabani, N.; Huang, D.; He, W.; Zhu, X.; Callaghan, S.; Yong, P.; et al. A human antibody reveals a conserved site on beta-coronavirus spike proteins and confers protection against SARS-CoV-2 infection. *Sci. Transl. Med.* **2022**, *14*, eabi9215. [CrossRef]
26. Haynes, W.A.; Kamath, K.; Bozekowski, J.; Baum-Jones, E.; Campbell, M.; Casanovas-Massana, A.; Daugherty, P.S.; Dela Cruz, C.S.; Dhal, A.; Farhadian, S.F.; et al. High-resolution epitope mapping and characterization of SARS-CoV-2 antibodies in large cohorts of subjects with COVID-19. *Commun. Biol.* **2021**, *4*, 1317. [CrossRef]
27. Vengesai, A.; Kasambala, M.; Mutandadzi, H.; Mduluza-Jokonya, T.L.; Mduluza, T.; Naicker, T. Scoping review of the applications of peptide microarrays on the fight against human infections. *PLoS ONE* **2022**, *17*, e0248666. [CrossRef]
28. Miller, N.L.; Raman, R.; Clark, T.; Sasisekharan, R. Complexity of Viral Epitope Surfaces as Evasive Targets for Vaccines and Therapeutic Antibodies. *Front. Immunol.* **2022**, *13*, 904609. [CrossRef]

29. Shrock, E.; Fujimura, E.; Kula, T.; Timms, R.T.; Lee, I.H.; Leng, Y.; Robinson, M.L.; Sie, B.M.; Li, M.Z.; Chen, Y.; et al. Viral epitope profiling of COVID-19 patients reveals cross-reactivity and correlates of severity. *Science* **2020**, *370*, eabd4250. [CrossRef]
30. Gershoni, J.M.; Roitburd-Berman, A.; Siman-Tov, D.D.; Freund, N.T.; Weiss, Y. Epitope mapping: The first step in developing epitope-based vaccines. *BioDrugs* **2007**, *21*, 145–156. [CrossRef]
31. Musicò, A.; Frigerio, R.; Mussida, A.; Barzon, L.; Sinigaglia, A.; Riccetti, S.; Gobbi, F.; Piubelli, C.; Bergamaschi, G.; Chiari, M.; et al. SARS-CoV-2 epitope mapping on microarrays highlights strong immune-response to n protein region. *Vaccines* **2021**, *9*, 35. [CrossRef] [PubMed]
32. Wang, H.; Wu, X.; Zhang, X.; Hou, X.; Liang, T.; Wang, D.; Teng, F.; Dai, J.; Duan, H.; Guo, S.; et al. SARS-CoV-2 Proteome Microarray for Mapping COVID-19 Antibody.pdf. *ACS Cent. Sci.* **2020**, *6*, 2238–2249. [CrossRef]
33. Lassaunière, R.; Polacek, C.; Gram, G.J.; Frische, A.; Tingstedt, J.L.; Krüger, M.; Dorner, B.G.; Cook, A.; Brown, R.; Orekov, T.; et al. Preclinical evaluation of a candidate naked plasmid DNA vaccine against SARS-CoV-2. *NPJ Vaccines* **2021**, *6*, 156. [CrossRef] [PubMed]
34. Frische, A.; Brooks, P.T.; Gybel-Brask, M.; Sækmose, S.G.; Jensen, B.A.; Mikkelsen, S.; Bruun, M.T.; Boding, L.; Strandh, C.P.; Jørgensen, C.S.; et al. Optimization and evaluation of a live virus SARS-CoV-2 neutralization assay. *PLoS ONE* **2022**, *17*, e0272298. [CrossRef] [PubMed]
35. Khan, K.; Karim, F.; Ganga, Y.; Bernstein, M.; Jule, Z.; Reedoy, K.; Cele, S.; Lustig, G.; Amoako, D.; Wolter, N.; et al. Omicron BA.4/BA.5 escape neutralizing immunity elicited by BA.1 infection. *Nat. Commun.* **2020**, *13*, 4686. [CrossRef] [PubMed]
36. Frey, A.; Di Canzio, J.; Zurakowski, D. A statistically defined endpoint titer determination method for immunoassays. *J. Immunol. Methods* **1998**, *221*, 35–41. [CrossRef]
37. Barnes, C.O.; Jette, C.A.; Abernathy, M.E.; Dam, K.-M.A.; Esswein, S.R.; Gristick, H.B.; Malyutin, A.G.; Sharaf, N.G.; Huey-Tubman, K.E.; Lee, Y.E.; et al. Structural classification of neutralizing antibodies against the SARS-CoV-2 spike receptor-binding domain suggests vaccine and therapeutic strategies. *bioRxiv* **2020**. [CrossRef]
38. Lassaunière, R.; Polacek, C.; Fonager, J.; Bennedbæk, M.; Boding, L.; Rasmussen, M.; Fomsgaard, A. Neutralisation of the SARS-CoV-2 Delta sub-lineage AY.4.2 and B.1.617.2 + E484K by BNT162b2 mRNA vaccine-elicited sera. *medRxiv* **2021**. [CrossRef]
39. Kuzmina, A.; Khalaila, Y.; Voloshin, O.; Keren-Naus, A.; Boehm-Cohen, L.; Raviv, Y.; Shemer-Avni, Y.; Rosenberg, E.; Taube, R. SARS-CoV-2 spike variants exhibit differential infectivity and neutralization resistance to convalescent or post-vaccination sera. *Cell Host Microbe* **2021**, *29*, 522–528.e2. [CrossRef]
40. Wu, L.; Peng, C.; Yang, Y.; Shi, Y.; Zhou, L.; Xu, Z.; Zhu, W. Exploring the immune evasion of SARS-CoV-2 variant harboring E484K by molecular dynamics simulations. *Brief. Bioinform.* **2022**, *23*, bbab383. [CrossRef]
41. Muruato, A.E.; Fontes-Garfias, C.R.; Ren, P.; Garcia-Blanco, M.A.; Menachery, V.D.; Xie, X.; Shi, P.Y. A high-throughput neutralizing antibody assay for COVID-19 diagnosis and vaccine evaluation. *Nat. Commun.* **2020**, *11*, 4059. [CrossRef] [PubMed]
42. Garcia-beltran, W.F.; Lam, E.C.; Denis, K.S.; Iafrate, A.J.; Naranbhai, V.; Balazs, A.B.; Garcia-beltran, W.F.; Lam, E.C.; Denis, K.S.; Nitido, A.D.; et al. Multiple SARS-CoV-2 variants escape neutralization by vaccine-induced humoral immunity ll Article Multiple SARS-CoV-2 variants escape neutralization by vaccine-induced humoral immunity. *Cell* **2021**, *184*, 2372–2383.e9. [CrossRef] [PubMed]
43. Henderson, R.; Edwards, R.J.; Mansouri, K.; Janowska, K.; Stalls, V.; Gobeil, S.M.C.; Kopp, M.; Li, D.; Parks, R.; Hsu, A.L.; et al. Controlling the SARS-CoV-2 spike glycoprotein conformation. *Nat. Struct. Mol. Biol.* **2020**, *27*, 925–933. [CrossRef] [PubMed]
44. Verkhivker, G.M. Molecular Simulations and Network Modeling Reveal an Allosteric Signaling in the SARS-CoV-2 Spike Proteins. *J. Proteome Res.* **2020**, *19*, 4587–4608. [CrossRef] [PubMed]
45. Trigueiro-Louro, J.; Correia, V.; Figueiredo-Nunes, I.; Gíria, M.; Rebelo-de-Andrade, H. Unlocking COVID therapeutic targets: A structure-based rationale against SARS-CoV-2, SARS-CoV and MERS-CoV Spike. *Comput. Struct. Biotechnol. J.* **2020**, *18*, 2117–2131. [CrossRef]
46. Hsieh, C.-L.; Goldsmith, J.A.; Schaub, J.M.; DiVenere, A.M.; Kuo, H.-C.; Javanmardi, K.; Le, K.C.; Wrapp, D.; Lee, A.G.; Liu, Y.; et al. Structure-based design of prefusion-stabilized SARS-CoV-2 spikes. *Science* **2020**, *369*, 1501–1505. [CrossRef]
47. Pallesen, J.; Wang, N.; Corbett, K.S.; Wrapp, D.; Kirchdoerfer, R.N.; Turner, H.L.; Cottrell, C.A.; Becker, M.M.; Wang, L.; Shi, W.; et al. Immunogenicity and structures of a rationally designed prefusion MERS-CoV spike antigen. *Proc. Natl. Acad. Sci. USA* **2017**, *114*, E7348–E7357. [CrossRef]
48. Crank, M.C.; Ruckwardt, T.J.; Chen, M.; Morabito, K.M.; Phung, E.; Costner, P.J.; Holman, L.S.A.; Hickman, S.P.; Berkowitz, N.M.; Gordon, I.J.; et al. A proof of concept for structure-based vaccine design targeting RSV in humans. *Science* **2019**, *365*, 505–509. [CrossRef]
49. Windsor, I.W.; Tong, P.; Lavidor, O.; Moghaddam, A.S.; McKay, L.G.A.; Gautam, A.; Chen, Y.; MacDonald, E.A.; Yoo, D.K.; Griffths, A.; et al. Antibodies induced by an ancestral SARS-CoV-2 strain that cross-neutralize variants from Alpha to Omicron BA.1. *Sci. Immunol.* **2022**, *7*, eabo3425. [CrossRef]
50. Case, J.B.; Mackin, S.; Errico, J.M.; Chong, Z.; Madden, E.A.; Whitener, B.; Guarino, B.; Schmid, M.A.; Rosenthal, K.; Ren, K.; et al. Resilience of S309 and AZD7442 monoclonal antibody treatments against infection by SARS-CoV-2 Omicron lineage strains. *Nat. Commun.* **2022**, *13*, 3824. [CrossRef]
51. Chen, Y.; Zhao, X.; Zhou, H.; Zhu, H.; Jiang, S.; Wang, P. Broadly neutralizing antibodies to SARS-CoV-2 and other human coronaviruses. *Nat. Rev. Immunol.* **2023**, *23*, 189–199. [CrossRef] [PubMed]

52. Magnus, C.L.; Hiergeist, A.; Schuster, P.; Rohrhofer, A.; Medenbach, J.; Gessner, A.; Peterhoff, D.; Schmidt, B. Targeted escape of SARS-CoV-2 in vitro from monoclonal antibody S309, the precursor of sotrovimab. *Front. Immunol.* **2022**, *13*, 966236. [CrossRef] [PubMed]
53. Yuan, M.; Wu, N.C.; Zhu, X.; Lee, C.C.D.; So, R.T.Y.; Lv, H.; Mok, C.K.P.; Wilson, I.A. A highly conserved cryptic epitope in the receptor binding domains of SARS-CoV-2 and SARS-CoV. *Science* **2020**, *368*, 630–633. [CrossRef] [PubMed]
54. Casalino, L.; Gaieb, Z.; Goldsmith, J.A.; Hjorth, C.K.; Dommer, A.C.; Harbison, A.M.; Fogarty, C.A.; Barros, E.P.; Taylor, B.C.; Mclellan, J.S.; et al. Beyond shielding: The roles of glycans in the SARS-CoV-2 spike protein. *ACS Cent. Sci.* **2020**, *6*, 1722–1734. [CrossRef] [PubMed]
55. Schwarz, T.; Heiss, K.; Mahendran, Y.; Casilag, F.; Kurth, F.; Sander, L.E.; Wendtner, C.M.; Hoechstetter, M.A.; Müller, M.A.; Sekul, R.; et al. SARS-CoV-2 Proteome-Wide Analysis Revealed Significant Epitope Signatures in COVID-19 Patients. *Front. Immunol.* **2021**, *12*, 629185. [CrossRef]
56. Bianchini, F.; Crivelli, V.; Abernathy, M.E.; Guerra, C.; Palus, M.; Muri, J.; Marcotte, H.; Piralla, A.; Pedotti, M.; De Gasparo, R.; et al. Human neutralizing antibodies to cold linear epitopes and subdomain 1 of the SARS-CoV-2 spike glycoprotein. *Sci. Immunol.* **2023**, *8*, eade0958. [CrossRef]
57. Zhou, D.; Ren, J.; Fry, E.E.; Stuart, D.I. Broadly neutralizing antibodies against COVID-19. *Curr. Opin. Virol.* **2023**, *61*, 101332. [CrossRef]
58. Lim, H.X.; Lim, J.; Jazayeri, S.D.; Poppema, S.; Poh, C.L. Development of multi-epitope peptide-based vaccines against SARS-CoV-2. *Biomed. J.* **2021**, *44*, 18–30. [CrossRef]
59. Azmi, F.; Ahmad Fuaad, A.A.H.; Skwarczynski, M.; Toth, I. Recent progress in adjuvant discovery for peptide-based subunit vaccines. *Hum. Vaccin. Immunother.* **2014**, *10*, 778–796. [CrossRef]
60. Heitmann, J.S.; Bilich, T.; Tandler, C.; Nelde, A.; Maringer, Y.; Marconato, M.; Reusch, J.; Jäger, S.; Denk, M.; Richter, M.; et al. A COVID-19 peptide vaccine for the induction of SARS-CoV-2 T cell immunity. *Nature* **2022**, *601*, 617–622. [CrossRef]
61. Guirakhoo, F.; Kuo, L.; Peng, J.; Huang, J.-H.; Kuo, B.-S.; Lin, F.; Liu, Y.-J.; Liu, Z.; Wu, G.; Ding, S.; et al. A Novel SARS-CoV-2 Multitope Protein/Peptide Vaccine Candidate is Highly Immunogenic and Prevents Lung Infection in an AAV hACE2 Mouse Model and non-human primates. *bioRxiv* **2021**. [CrossRef]

vaccines

MDPI

Article

The Immunogenicity of CpG, MF59-like, and Alum Adjuvant Delta Strain Inactivated SARS-CoV-2 Vaccines in Mice

Kangwei Xu [1], Jing Li [2], Xu Lu [1], Xiaoqin Ge [2], Kaiqin Wang [1], Jiahao Wang [2], Zhizhong Qiao [1], Yaru Quan [1] and Changgui Li [1,*]

1 National Institutes for Food and Drug Control, NHC Key Laboratory of Research on Quality and Standardization of Biotech Products, NMPA Key Laboratory for Quality Research and Evaluation of Biological Products, No. 2, Tiantan Xili, Dongcheng District, Beijing 100050, China; xukw@nifdc.org.cn (K.X.)

2 Sinovac Life Sciences Co., Ltd., No. 21, Tianfu St., Daxing Biomedicine Industrial Base of Zhongguancun Science Park, Daxing District, Beijing 100050, China

* Correspondence: changguili@aliyun.com

Abstract: The continuous evolution and mutation of SARS-CoV-2 have highlighted the need for more effective vaccines. In this study, CpG, MF59-like, and Alum adjuvant Delta strain inactivated SARS-CoV-2 vaccines were prepared, and the immunogenicity of these vaccines in mice was evaluated. The Delta + MF59-like vaccine group produced the highest levels of S- and RBD-binding antibodies and live Delta virus neutralization levels after one shot of immunization, while mice in the Delta + Alum vaccine group had the highest levels of these antibodies after two doses, and the Delta + MF59-like and Delta + Alum vaccine groups produced high levels of cross-neutralization antibodies against prototype, Beta, and Gamma strain SARS-CoV-2 viruses. There was no significant decrease in neutralizing antibody levels in any vaccine group during the observation period. CpG, MF59-like, and Alum adjuvant Delta strain inactivated SARS-CoV-2 vaccines excited different antibody subtypes compared with unadjuvanted vaccines; the Delta + CpG vaccine group had a higher proportion of IgG2b antibodies, indicating bias towards Th1 immunity. The proportions of IgG1 and IgG2b in the Delta + MF59-like vaccine group were similar to those of the unadjuvanted vaccine. However, the Delta + Alum vaccine group had a higher proportion of IgG1 antibodies, indicating bias towards Th2 immunity. Antigen-specific cytokine secretion CD4/8^{+} T cells were analyzed. In conclusion, the results of this study show differences in the immune efficacy of CpG, MF59-like, and Alum adjuvant Delta strain inactivated SARS-CoV-2 vaccines in mice, which have significant implications for the selection strategy for vaccine adjuvants.

Keywords: SARS-CoV-2; inactivated vaccines; Delta; immunogenicity; adjuvants

1. Introduction

Severe Acute Respiratory Syndrome Coronavirus 2 (SARS-CoV-2), the causative agent of the disease COVID-19, was first identified in human populations in 2019 and rapidly escalated into a global pandemic. The disease manifests in a spectrum of severity in patients, with most individuals experiencing mild symptoms and recovering within a few weeks. However, some develop serious complications such as pneumonia or acute respiratory distress syndrome which, unfortunately, have resulted in fatalities in some instances [1–3]. SARS-CoV-2 is categorized as an enveloped single-stranded RNA virus, falling under the genus Beta coronavirus within the Coronavirus family. It produces four structural proteins: nucleocapsid (N), glycoprotein (S), membrane (M), and envelope (E) proteins [4,5]. The S protein is composed of two subunits, S1 and S2. The receptor-binding domain (RBD) situated on the S1 subunits attaches to the angiotensin-converting enzyme 2 (ACE2) receptor in the host, while the S2 subunits facilitate the virus penetration into the host cells. Therefore, despite the N protein being the most abundant, the S protein is the primary target for the action of neutralizing antibodies [6–9].

In response to the urgent public health crisis presented by the COVID-19 pandemic, there has been a concerted global effort to expedite research and development processes for a vaccine. This international cooperation has yielded several types of vaccines that have undergone rigorous clinical trials and have subsequently been approved for use. The World Health Organization has sanctioned the emergency use of four primary types of vaccines: inactivated virus, RNA, non-replicating viral vector, and protein subunit vaccines [10]. Among these, inactivated vaccines, which contain a virus that has been rendered non-pathogenic, have a long history of use and are considered among the safest and most commonly utilized types of vaccines. These vaccines, distributed in billions of doses, have demonstrated their effectiveness in safeguarding against a wide variety of viral and bacterial diseases.

COVID-19 inactivated vaccines such as CoronaVac, BBIBP-CorV, and Covaxin have been widely used. These vaccines are prepared by inactivating cell culture viruses using chemical reagents such as β-propiolactone [11]. The vaccine maintains the integrity of the virus particle test and uses the whole virus as an immunogen. Compared with subunit vaccines of S protein or RBS fragments, a broader immune response can be induced [12,13]. Clinical trial results have shown that the overall adverse effects of inactivated vaccines are low, indicating a good safety profile [14–17]. They also have good safety among different populations, such as children and adolescents, the elderly, cancer patients, pregnant women, immunocompromised people, those with comorbidities, and HIV-infected people [18].

Adjuvants are often added to vaccines to improve the efficiency of the immune response, increase the duration of protection, reduce the amount of antigen used, and provide cross-protection by enhancing the breadth of the immune response to multiple antigens. At present, more than 10 adjuvants have been used in human vaccines, including aluminum salt, MF59, CpG, AS01, AS03, AS04, liposomes, etc. Aluminum salts are the most widely used adjuvant, used in billions of vaccine doses [19,20]. The inactivated vaccines CoronaVac and BBIBP-CorV both use aluminum hydroxide as adjuvants, while Covaxin uses aluminum hydroxide with toll-like receptor 7/8 agonist molecule (IMDG) as adjuvants [11,21]. In this study, the Delta strain virus was purified and inactivated after in vitro culture, and combined with an aluminum adjuvant, a MF59-like adjuvant, or a CpG adjuvant, and the immune effects of the different adjuvant inactivated vaccines were evaluated in mice.

2. Materials and Methods

2.1. Preparation of Inactivated Delta Vaccine with Different Adjuvants

The Delta (B.1.617.2) strain virus of SARS-CoV-2 was isolated from a COVID-19 patient and an adaptive subculture on Vero cells was cultured. Preparation of the inactivated Delta vaccine bulk was consistent with previous reports [9,11]. Briefly, the Delta strain virus was propagated in Vero cells (Obtained from WHO) for large-scale culture, and β propiolactone was added to the culture product at a ratio of 1:4000 for virus inactivation. After purification via ultrafiltration, ion exchange, molecular sieving, and other purification steps, the inactivated Delta vaccine bulk was obtained. Delta vaccines were prepared by combining the inactivated Delta vaccine bulk with different adjuvants. A total of five groups of vaccines and a control were prepared in this study (Table 1). The choice of an inactivated Delta virus and aluminum adjuvant immunization dose was based on our previous study [11], and the CpG and MF59-like adjuvants were used at the manufacturer's recommended dose. The Delta vaccine was prepared by diluting inactivated Delta vaccine bulk to 6 μg/mL of protein using PBS. The Delta + CpG vaccine was prepared by diluting inactivated Delta vaccine bulk to 12 μg/mL of protein using PBS and mixed with CpG adjuvants (Synbio Tech, Chengdu, China) with an equal volume concentration of 40 μg/mL. The Delta + MF59-like vaccine was prepared by diluting inactivated Delta vaccine bulk to 12 μg/mL of protein using PBS and then mixed with an equal volume of MF59-like adjuvant (InvivoGen, Toulouse, French) before use. The Delta + Alum vaccine was prepared by diluting inactivated Delta vaccine bulk to 12 μg/mL protein content using PBS and mixed with aluminum hydroxide adjuvant (SINOVAC, Beijing, China) with an equal volume concentration of 1 mg/mL, letting it stand for 24 h. PBS was used as a negative control. Each dose of the vaccine preparation was 0.5 mL.

Table 1. Vaccine formulations and study groups.

Vaccine	Antigen Content	Adjuvant
PBS	w/o	w/o
Delta	3 μg/dose	w/o
Delta + CpG	3 μg/dose	CpG 10 μg/dose
Delta + MF59-like	3 μg/dose	MF59-like 0.25 mL/dose
Delta + Alum	3 μg/dose	Aluminum hydroxide 250 μg/dose

2.2. *Mouse Immunization*

This study was approved by the Institutional Animal Care and Use Committee at the National Institutes for Food and Drug Control, China. Female SPF-level Balb/c mice, 6 weeks old, were selected as experimental animals, and 10 mice were immunized in each group. Immunization was performed twice via intraperitoneal injection of 0.5 mL of vaccine on days 0 and 14. Blood was collected and serum was isolated 7, 14, 21, 28, 35, and 42 days after the first immunization; ELISA and neutralizing antibody tests were performed. The mouse spleen was removed at 42 days for flow cytometry assay (Figure 1).

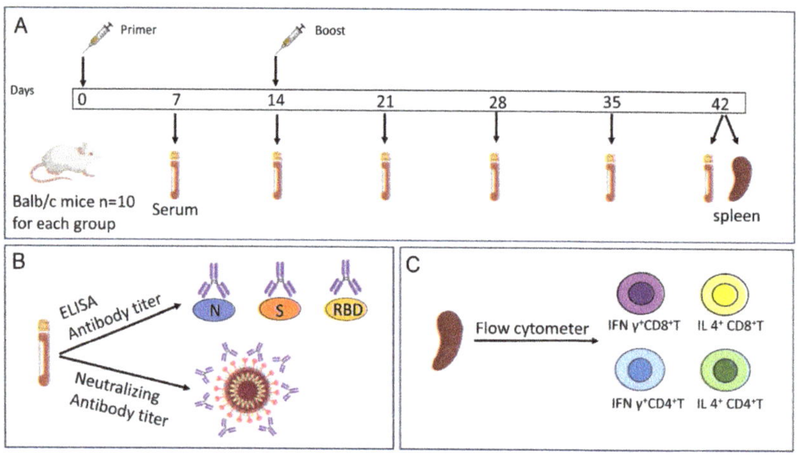

Figure 1. Animal immunization and sample analysis strategy. Balb/c mice were immunized at days 0 and 14. Serum was collected at 7, 14, 21, 28, 35, and 42 days, and spleens were isolated 42 days post-primer vaccination (**A**). Antibodies against S, RBD, N proteins, and neutralizing antibody in mouse serum were detected via indirect ELISA and microneutralization test methods (**B**). Spleens were analyzed using flow cytometry to measure the percentages of IFN γ+ and IL 4+ in CD4+/CD8+ T cells, respectively (**C**).

2.3. *Indirect ELISA Assay*

The indirect ELISA method was used to detect antibodies against S, RBD, and N proteins in mouse serum. The recombinant Delta strain SARS-CoV-2 N, S, or RBD proteins (SinoBiological, Beijing, China, N Cat: 40588-V07E29, S Cat: 40589-V08B16, RBD Cat: 40592-V08H90) were diluted to 1 μg/mL using PBS. Ninety-six-well ELISA plates were coated with antigens at 100 μL/well and placed in 4 °C refrigeration overnight. After washing the plate with PBS-T solution five times, 200 μL/well blocking solution (1% BSA in PBS) was added and blocking was undertaken for 1 h at room temperature. After discarding the blocking solution, mouse serum diluted with blocking solution was added, with 2-fold dilution starting at 800-fold, and samples were incubated at 37 °C for 1 h. After washing five times with PBS-T solution, HRP-labeled anti-mouse IgG antibody (Seracare/KPL, Milford, MA, USA) was added and diluted 10,000-fold with blocking solution. This was followed by washing the plate five more times. TMB substrate (Seracare/KPL, Milford, MA, USA) was added to the plate to produce a color change, and A450/630 was measured using a plate reader. Serum titer was defined as a dilution factor greater than 2.1 times the absorbance value of the negative well. An antibody titer less than 800 was assigned a value of 400 for the GMT calculation. For serum samples 42 days after immunization, tests for IgG antibody subtypes were also performed. The above steps were used for S protein coating and serum sample dilution, and HRP-labeled anti-mouse IgG antibodies were replaced with HRP-labeled anti-mouse IgG1/2a/2b/3 (Southern Biotech, Oxmoor Blvd, Birmingham, AL, USA) antibodies to detect the titer of different isotypes of antibodies.

2.4. Neutralizing Antibody Detection

Neutralizing antibody titers in mouse serum were determined using microneutralization tests, which were conducted in a biosafety level 3 (BSL-3) laboratory. Serum samples underwent inactivation in a water bath set at 56 °C for 30 min, then were diluted eight times with 199 media. Subsequently, 50 µL of the sample was placed into each of two wells in a pre-filled cell culture plate, which was then subjected to a two-fold dilution. Using 199 media, the SARS-CoV-2 Delta strain virus was diluted to 100 $CCID_{50}$/50 µL, and 50 µL was added per well to the cell culture plate for neutralization in a 37 °C, 5% CO_2 incubator for a duration of 120 min. Vero cells were diluted to a concentration of 1.2–2.0 $\times$ 10^5 cells/mL, and 100 µL was added to each well of the neutralized cell culture plate. The plate was then incubated for five days in a 37 °C, 5% CO_2 incubator. Cytopathies were visually inspected under a microscope, and neutralizing antibody titers were calculated based on the dilution at which half the cytopathies were presented. When calculating the geometric mean titers (GMTs), any antibody titers under 8 were given a value of 4. Additionally, 42 days post-immunization, neutralizing titers were also conducted against the prototype, Beta, and Gamma strains of SARS-CoV-2 using serum samples.

2.5. Flow Cytometry Analysis

After 42 days from the first immunization, mouse spleens were removed for antigen-specific T cell testing. Mouse spleens were suspended in PBS for trituration into a homogenate, incubated with Lysing buffer to lyse red blood cells, and subsequently filtered through a 40 µm cell strainer to obtain single splenocytes. Approximately 1 million splenocytes were seeded into each well, followed by the addition of a SARS-CoV-2 S peptide pool (SinoBiological, Beijing, China) and cultured at 37 °C with 5% CO_2 for 4 h. T cell surface markers were stained using anti-mouse CD3 (BV510), anti-mouse CD4 (FITC), anti-mouse CD8 (Brilliant Violet 785), and anti-mouse CD25 (Brilliant Violet 421). Following washing, cells were treated with Fixation and Permeabilization Solution. After incubation for 30 min at 4 °C in the dark and washing with Permeabilization washing buffer, anti-mouse IFN γ (PE) and anti-mouse IL4 (PE-CF594) were added for cytokine detection. The percentage of antigen-specific cytokine secretion T cells were analyzed using flow cytometry (NovoCyte D3130, AECA, Santa Clara, CA, USA).

2.6. Statistical Analyses

GraphPad Prism 8.0.1 was applied for statistical analysis. Binding antibodies against SARS-CoV-2 S, RBD, and N proteins and neutralization antibodies are presented as geometric mean titers (GMTs). Antigen-specific cytokine secretion T cells are presented as mean percentages. Significance was considered when p values were < 0.05.

3. Results

3.1. Detection of Antibodies against S, RBD, and N Proteins in Mouse Serum

Recombinant SARS-CoV-2 S, RBD, N proteins were used to detect the titer of bound antibodies in the serum of immunized mice. S protein antibodies were detected in mice with and without adjuvant vaccine groups seven days after the first immunization (Figure 2A). The 7-day and 14-day antibody titers of the Delta + MF59-like vaccine group were the highest, while the antibody titers in the Delta + CpG vaccine group were lower, which was similar to that in the unadjuvanted Delta vaccine group. After the booster immunization on the 14th day, the antibody titers of mice in all vaccine groups increased significantly, and the Delta + Alum vaccine group had the highest antibody titers. After the antibody titer in the Delta + CpG vaccine group reached the highest level on day 28, the antibody titer decreased significantly on day 35 and day 42. However, the antibody titers of the Delta + MF59-like and Delta + Alum vaccine groups decreased slightly after 35 days, suggesting that the two adjuvanted vaccines had better antibody persistence. The results of the RBD antibody test (Figure 2B) were similar to those of the S protein antibody, with the highest antibody titers in the Delta + MF59-like vaccine group at 7 and 14 days after the initial immunization, and the highest antibody titers in the Delta + Alum vaccine group after booster immunization. Unlike S protein antibodies, RBD-specific antibodies had better persistence, and there was no significant decrease in RBD antibody titers in each vaccine group during the 42-day observation period. At 7 and 14 days after the initial immunization, none of the groups exhibited N-specific antibodies (Figure 2C), and after booster immunization, the Delta + Alum vaccine group had the highest antibody titers, while the unadjuvanted Delta vaccine group still did not produce N-specific antibodies. This indicates that the Delta strain SARS-CoV-2 virus has poor N-protein immunogenicity, which is also consistent with our previous research results on the prototype virus [11].

3.2. Detection of IgG Subtype Antibodies against S Proteins in Mouse Serum

Next, we performed a titer analysis of the IgG subtype in serum collected over 42 days (Figure 2D). The IgG1 subtype titer was the highest in each group. Compared with the unadjuvanted Delta vaccine group, the enhancement of antibody subtypes was different between the three adjuvant vaccine groups: the antibody titers of IgG2b and IgG3 subtypes were significantly increased in the Delta + CpG vaccine group, the IgG2a, IgG2b, and IgG3 subtype antibody titers were significantly increased in the Delta + MF59-like vaccine group, and the IgG1 and IgG2a subtype antibody titers were significantly increased in the Delta + Alum vaccine group. The ratio of IgG2a/IgG1 antibody titers to the unadjuvanted Delta vaccine group was calculated, and the results are shown in Figure 2E. The Delta + CpG vaccine group is biased to IgG2a antibody response (Th1), the Delta + MF59-like vaccine group has a similar antibody response bias to the unadjuvanted Delta vaccine group, and the Delta + Alum vaccine group is more biased towards IgG1 antibody response (Th2).

3.3. Neutralizing Antibody Titers

The micro-neutralization method was used to detect the titer of Delta virus neutralizing antibody in the serum of immunized mice. As shown in Figure 3, neutralizing antibodies were detected in some mice in the Delta + MF59-like and Delta + Alum vaccine groups seven days after the first immunization. By day 14, neutralizing antibodies were detectable in most mice in the two groups, and the neutralizing antibody levels were slightly higher in the Delta + MF59-like vaccine group. After the booster immunization on day 14, neutralizing antibodies were detectable in mice in all vaccine groups. The neutralizing antibody levels in the Delta + MF59-like and Delta + Alum vaccine groups were higher than those in the unadjuvanted Delta and Delta + CpG vaccine groups, and the neutralizing antibody titers in the Delta + Alum vaccine group were the highest. During the 42-day observation period, there was no significant decrease in neutralizing antibody titers in each vaccine group. These results are consistent with RBD-binding antibody test results.

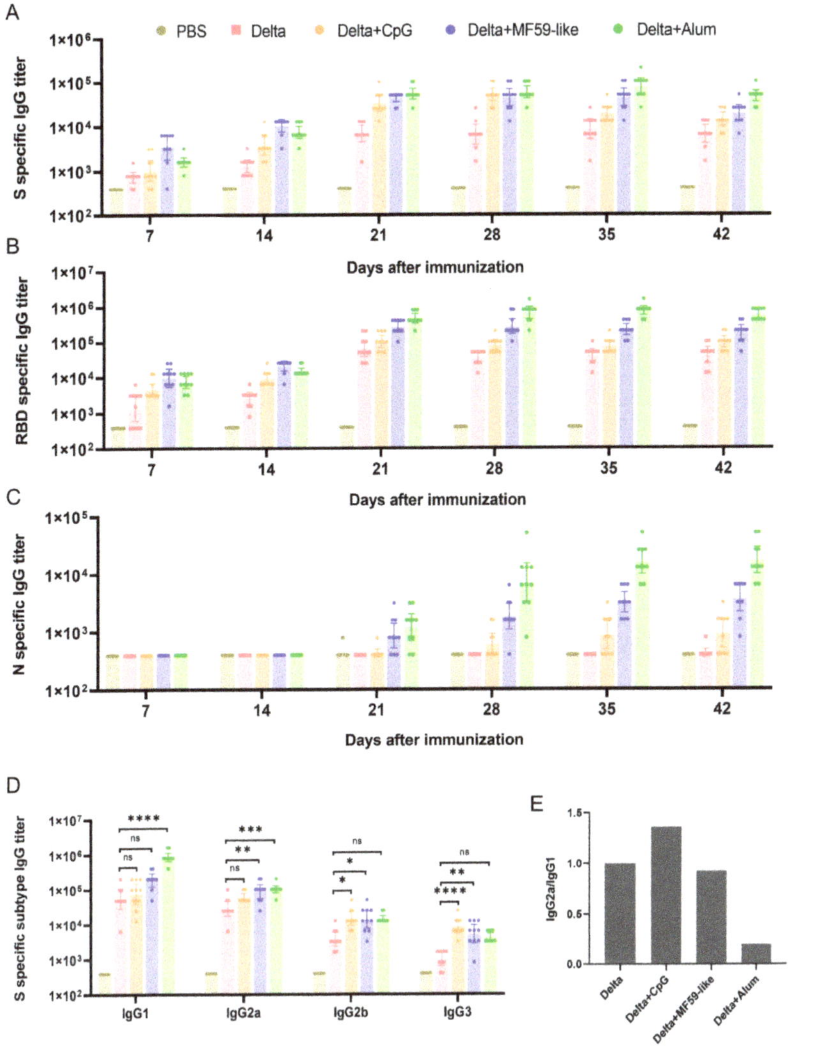

Figure 2. Serum binding antibody titers after immunization. Five groups of mice (n = 10) were immunized with Delta inactivated vaccine with different adjuvants at 14-day intervals. The changes in antibody titers against the S protein (**A**), RBD (**B**), and N protein (**C**) in serum at 7, 14, 21, 28, 35, and 42 days after the first immunization were measured. IgG isotype analysis against S protein was performed on serum 42 days after immunization, antibody titers against IgG1, IgG2a, IgG2b, and IgG3 were detected (**D**), and the relative values of IgG2a/IgG1 and the unadjuvanted Delta vaccine group were calculated (**E**). The data points represent the values detected from each mouse, the reference column height is the geometric mean of the titer (GMT), and the error bars indicate the 95% confidence interval. Statistical analyses were conducted using one-way ANOVA with Dunnett's multiple comparisons test. ns, $p > 0.05$; *, $p < 0.05$; **, $p < 0.01$; ***, $p < 0.001$; ****, $p < 0.0001$.

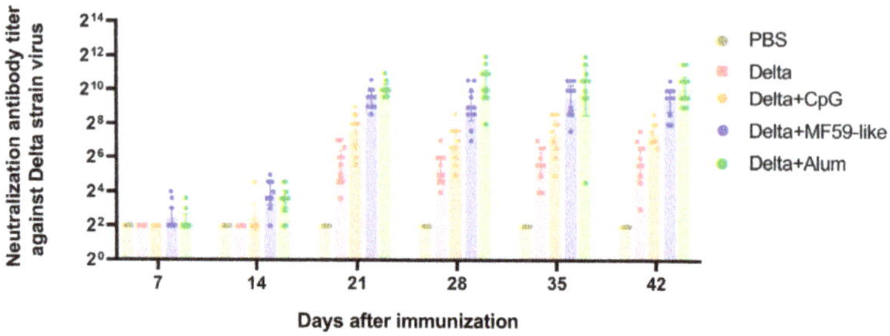

Figure 3. The serum neutralization titer was detected using a micro-neutralization assay. The live virus of the Delta strain was used for micro-neutralization experiments, and the serum neutralizing antibody titers of mice 7, 14, 21, 28, 35, and 42 days after the first immunization of each group were detected. The data points represent the values detected from each mouse, the reference column height is the geometric mean of the titer (GMT), and the error bars indicate the 95% confidence interval.

Next, we measured serum neutralizing antibody levels against the prototype, Beta, and Gamma strains of the SARS-CoV-2 virus at 42 days post-immunization (Table 2). The Delta + Alum vaccine group with the highest neutralizing antibody titer against the Delta strain also had the highest neutralizing antibody level against the other three strains. The Delta + MF59-like vaccine group had a slightly lower neutralizing antibody level, and the Delta + CpG vaccine group had the lowest neutralizing antibody level among the three adjuvanted vaccine groups, but it was still higher than that of the unadjuvanted Delta vaccine group.

Table 2. Neutralization antibodies to the Delta, prototype, Beta, and Gamma strains of the live SARS-CoV-2 virus (GMT and 95% confidence interval, $n = 10$ for each group).

Vaccines \ Virus	Delta	Prototype	Beta	Gamma
Delta	53.85 (27.85~104.1)	26.61 (11.81~59.96)	7.87 (4.42~14.01)	6.06 (3.56~10.33)
Delta + CpG	178.0 (129.1~245.3)	56.08 (32.68~96.23)	22.9 (12.33~42.51)	22.25 (11.96~41.39)
Delta + MF59-like	672.1 (431.7~1046)	250.4 (125.3~500.3)	112.2 (68.29~184.2)	103.4 (55.21~193.8)
Delta + Alum	1150 (714.9~1851)	1002 (670.9~1495)	209.3 (106.5~411.3)	237.6 (153.9~367)

3.4. T Cell Immune Responses Elicited by Inactivated Delta Vaccines with Different Adjuvants

Two doses of inactivated Delta vaccine with different adjuvants were administered to BALB/c mice at 14-day intervals. At 42 days following the initial vaccination, mice were euthanized and splenocytes were isolated. The isolated splenocytes were then stimulated with a SARS-CoV-2 S peptide pool before the detection of T cell immune responses. Compared with the PBS group, a significant increase in the proportion of IFN γ+ CD4+ T and IL4+ CD4+ T cells was observed in mice from the Delta + Alum vaccine group. Slight increases in the proportions of IFN γ+ CD4+ T and IL4+ CD4+ T cells were noted in the Delta + CpG and Delta + MF59-like vaccine groups, although these changes were not statistically significant (Figure 4B,C). This trend correlates with the highest antibody titer levels observed at 42 days in mice from the Delta + Alum vaccine group. Increases in the proportions of IFN γ+ CD8+ T and IL4+ CD8+ T were observed for the Delta + CpG, Delta + MF59-like, and Delta + Alum vaccine groups compared with those from the PBS group, and the proportion of IFN γ+ CD8+ T cells in the Delta + CpG group was significantly increased (Figure 4E,F).

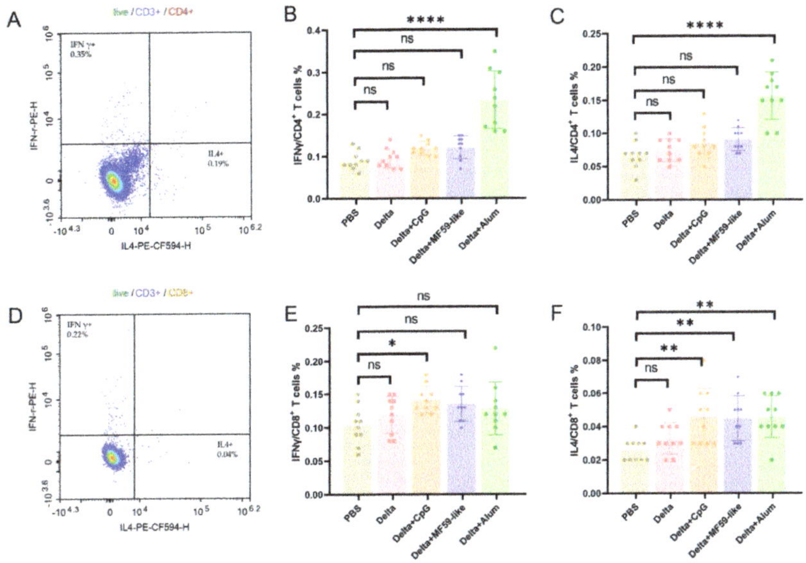

Figure 4. T cell immune responses in the vaccinated BALB/c mouse spleens. BALB/c mice received two doses of the inactivated Delta vaccine with various adjuvants at 14-day intervals. Splenocytes, isolated 42 days post-initial vaccination and stimulated with a SARS-CoV-2 S peptide pool, were analyzed using flow cytometry. The gating strategy of IFNγ/IL4 positive cells in CD3+CD4+/CD3+CD8+ T cells are presented in panel (**A**) and (**D**). The results of percentages of IFN γ+ in CD4+ T cells (**B**); IL 4+ in CD4+ T cells (**C**); IFN γ+ in CD8+ T cells (**E**) and IL 4+ in CD8+ T cells (**F**), respectively. Data are presented as mean ± SEM (n = 10). Statistical analyses were conducted using one−way ANOVA with Dunnett's multiple comparisons test. ns, $p > 0.05$; *, $p < 0.05$; **, $p < 0.01$; ****, $p < 0.0001$.

4. Discussion

Since the emergence of the COVID-19 pandemic, various vaccines including mRNA vaccines, adenovirus vector vaccines, recombinant vaccines, and inactivated vaccines have been utilized for the prevention and control of SARS-CoV-2, effectively reducing the rates of infection and severe cases. The mRNA vaccines, developed using new technology, were the first to be globally approved for the market. These vaccines were found to induce high titers of neutralizing antibodies and CD4/CD8 T cell responses in mice. In non-human primate models, mRNA-vaccinated groups exhibited significant viral clearance capabilities upon virus challenge [22,23]. Clinical studies indicate that mRNA vaccines have achieved over 90% efficacy in human populations [24,25]. Inactivated SARS-CoV-2 vaccines have also demonstrated protective effects in animal models [11]. Despite inducing lower levels of neutralizing antibodies in human populations and exhibiting lower efficiency in preventing symptomatic COVID-19 cases (ranging between 50.7% and 83.5%), they possess superior safety profiles [16,17,21,26]. Furthermore, the technology used to produce inactivated vaccines is mature and easily scalable for mass production. These vaccines can also be refrigerated and transported at temperatures of 2–8 °C without the need for freezing, providing significant advantages for global distribution, especially in developing countries. Thus, they remain a crucial tool against the pandemic threat of the virus [27].

The emergence of new COVID-19 variants has reduced the protective effects of various vaccines. The efficacy of mRNA vaccines against the Beta/Gamma variants ranges from 76 to 100%, against the Delta variant from 47.3 to 88%, and against the Omicron variant effectiveness has dropped to 23.7 from 47% [28–33]. The effectiveness of inactivated vaccines against the variants has also significantly decreased, with efficacy against the Delta variant ranging from 30.4 to 74.5% and against the Omicron variant at 37.9% [15,33,34]. The development of universal vaccines is an ideal approach to address emerging variants. Broad-spectrum vaccines prepared using different technologies have demonstrated protection against various VOCs in animal experiments, but no universal vaccines have been marketed so far [35–38]. Vaccines that include prevalent SARS-CoV-2 variants are a more feasible solution, and the WHO encourages the development of such vaccines on various platforms [39]. Vaccines with high immunogenicity that are capable of resisting SARS-CoV-2 variants can play a vital role in the prevention and control of the COVID-19 pandemic. This study involved preparing inactivated Delta variant vaccines containing three types of human adjuvants, CpG, MF59-like, and Alum, focusing on the immune response in mice and evaluating the role of adjuvants in order to screen for vaccines with better immunogenicity.

In this study, the effect of humoral immunity was first evaluated. After two immunizations on days 0 and 14, S- and RBD-specific IgG antibodies were rapidly produced in mice. The Delta + MF59-like vaccine group had the highest antibody titers on days 7 and 14 after the first immunization, suggesting that MF59-like can quickly induce antibody production. After the booster immunization on day 14, the Delta + Alum vaccine group had the highest antibody titers. The Delta + CpG group had the lowest titers among the three adjuvant vaccine groups, but these were still higher than the control group without an adjuvant. This is consistent with the results of Jie et al. for the prototype strain of the inactivated COVID-19 vaccine with different adjuvants [40]. As the most abundant protein in the SARS-CoV2 virus, N has an independent role in vaccine-induced protection [41]. Therefore, the N protein antibody was also evaluated in this study. After the second immunization, the three adjuvant vaccine groups produced low levels of N-protein-specific antibodies, suggesting that, like the prototype strain, the immunogenicity of the N protein of the Delta strain SARS-CoV-2 virus is also low [11]. The results of S-specific IgG antibody subtypes showed that regardless of the addition of adjuvants, the level of IgG1 antibodies was the highest. The titers of IgG1, IgG2a, IgG2b, and IgG3 subtype antibodies in the three adjuvant vaccine groups were all higher than those of the vaccine without adjuvants to varying degrees. Considering that the level of IgG1 antibodies is a marker of Th2 response, and IgG2a is a marker of Th1 response [42], we calculated the ratio of IgG2a/IgG1 of different vaccines and compared it with the group without adjuvants. The Delta + CpG group was more biased towards the Th1 response, the Delta + MF59-like vaccine group response was similar to the group without adjuvants, and the Delta + Alum vaccine group was more biased towards the Th2 response. Since balanced immune responses are more conducive to virus clearance, MF59-like adjuvant shows a promising ability to induce a more intense antiviral response in the host [43].

Preclinical studies in mice and non-human primates, as well as vaccine efficacy studies, suggest that neutralizing antibody levels are potential indicators for protection against SARS-CoV-2 [11,22,23]. In this study, after immunization in mice, the three adjuvant vaccine groups had higher levels of neutralizing antibodies than the group without an adjuvant, demonstrating an immunoenhancing effect. Similar to the results of S- and RBD-specific IgG antibody detection, the Delta + MF59-like vaccine group had the highest antibody titers on days 7 and 14 after the first immunization, and after the booster immunization on day 14, the Delta + Alum vaccine group had the highest antibody titers. Moreover, mice in the Delta + MF59-like and Delta + Alum vaccine groups not only produced high titers of neutralizing antibodies against the homologous Delta strain virus, but also showed good cross-neutralizing activity against the prototype, Beta, and Gamma strains of the SARS-CoV-2 virus, indicating their potential cross-protective effect.

Studies on patients infected with COVID-19 reveal that T-cell immunity is instrumental in combating SARS-CoV-2 infection. A decrease in T-cell counts has been linked to the emergence of severe conditions like pneumonia. On the other hand, adequate T-cell counts are associated with the rapid elimination of the virus and successful patient recovery [44,45]. Therefore, this study evaluated the S-protein-specific T cell immune response. Our results show that the antigen-specific CD4+ and CD8+ T cells in the three adjuvant vaccine groups were all higher than in the control group; in particular, the level of IFN γ+ CD8+ T cells in the Delta + CpG group was significantly increased. An advantage of the T-cell immune response is its ability to recognize conservative regions that exist in the S protein across various strains of the virus, suggesting that specific mutations of SARS-CoV-2 variants may not result in cellular immune evasion. Unlike COVID-19 vaccines targeting the S protein produced with other technologies, inactivated COVID-19 vaccines contain conservative antigens like the N protein. As antigenic changes in these conserved internal structural proteins are seldom observed in VOCs, T-cell immunity against these proteins is unlikely to be impacted by antibody escape mutations in VOCs. Therefore, inactivated COVID-19 vaccines are likely to remain effective against existing and future variants of SARS-CoV-2 [46–48].

There are some limitations in this study that need to be addressed. Firstly, although a large amount of previous research has shown a correlation between the level of neutralizing antibodies produced after vaccination and protective efficacy, this study did not conduct viral challenge research to directly evaluate the protective efficacy of different adjuvant vaccines. Secondly, due to some uncontrollable factors, we were unable to obtain mRNA vaccines for control experiments or the Omicron strain for cross-protection evaluation during the conduct of this study. This limits our ability to conduct a broader and deeper evaluation of cross-protection efficacy, which is crucial for assessing the effectiveness and efficiency of different vaccines against new virus variants. We are conducting research to address these issues.

As SARS-CoV-2 continues to mutate, the threat it poses in the future cannot be ignored. Our research has compared the immune effects of inactivated vaccines with different adjuvants in mice, providing a reference for the development of safer and more effective inactivated vaccines.

Author Contributions: Investigation, all authors. Methodology, K.X., J.L. and X.L.; software, X.G. and K.W.; validation, K.W., J.W. and Z.Q.; formal analysis, Y.Q.; resources, C.L.; data curation, K.W., K.X. and Y.Q.; writing—original draft preparation, K.X.; writing—review and editing, X.L. and C.L.; visualization, K.X.; supervision, Y.Q.; project administration, C.L.; funding acquisition, C.L. All authors have read and agreed to the published version of the manuscript.

Funding: This research was funded by the Beijing Natural Science Foundation (Grant no. L234053) and the Emergency Key Program of Guangzhou Laboratory (Grant no. EKPG21-30-1).

Institutional Review Board Statement: This study was approved by the Institutional Animal Care and Use Committee at National Institutes for Food and Drug Control, China, and study protocols were conducted in accordance with regulations regarding the management of laboratory animals (National Science and Technology Commission no. 2 of 31 October 1988) and "guidance notes on the treatment of experimental animals" (Chinese version (2006) no. 398). All institutional guidelines for animal care and use were strictly followed throughout this study.

Informed Consent Statement: Not applicable.

Data Availability Statement: Data are contained within the article.

Conflicts of Interest: The authors declare no conflicts of interest.

References

1. Zhu, N.; Zhang, D.; Wang, W.; Li, X.; Yang, B.; Song, J.; Zhao, X.; Huang, B.; Shi, W.; Lu, R.; et al. A Novel Coronavirus from Patients with Pneumonia in China, 2019. *N. Engl. J. Med.* **2020**, *382*, 727–733. [CrossRef] [PubMed]
2. Zhou, P.; Yang, X.L.; Wang, X.G.; Hu, B.; Zhang, L.; Zhang, W.; Si, H.R.; Zhu, Y.; Li, B.; Huang, C.L.; et al. A pneumonia outbreak associated with a new coronavirus of probable bat origin. *Nature* **2020**, *579*, 270–273. [CrossRef] [PubMed]
3. Huang, C.; Wang, Y.; Li, X.; Ren, L.; Zhao, J.; Hu, Y.; Zhang, L.; Fan, G.; Xu, J.; Gu, X.; et al. Clinical features of patients infected with 2019 novel coronavirus in Wuhan, China. *Lancet* **2020**, *395*, 497–506. [CrossRef] [PubMed]
4. Santacroce, L.; Charitos, I.A.; Carretta, D.M.; De Nitto, E.; Lovero, R. The human coronaviruses (HCoVs) and the molecular mechanisms of SARS-CoV-2 infection. *J. Mol. Med.* **2021**, *99*, 93–106. [CrossRef] [PubMed]

5. Gordon, D.E.; Jang, G.M.; Bouhaddou, M.; Xu, J.; Obernier, K.; White, K.M.; O'Meara, M.J.; Rezelj, V.V.; Guo, J.Z.; Swaney, D.L.; et al. A SARS-CoV-2 protein interaction map reveals targets for drug repurposing. *Nature* **2020**, *583*, 459–468. [CrossRef] [PubMed]
6. Cai, Y.; Zhang, J.; Xiao, T.; Peng, H.; Sterling, S.M.; Walsh, R.M., Jr.; Rawson, S.; Rits-Volloch, S.; Chen, B. Distinct conformational states of SARS-CoV-2 spike protein. *Science* **2020**, *369*, 1586–1592. [CrossRef] [PubMed]
7. Walls, A.C.; Park, Y.J.; Tortorici, M.A.; Wall, A.; McGuire, A.T.; Veesler, D. Structure, Function, and Antigenicity of the SARS-CoV-2 Spike Glycoprotein. *Cell* **2020**, *181*, 281–292.e286. [CrossRef]
8. Hoffmann, M.; Kleine-Weber, H.; Schroeder, S.; Kruger, N.; Herrler, T.; Erichsen, S.; Schiergens, T.S.; Herrler, G.; Wu, N.H.; Nitsche, A.; et al. SARS-CoV-2 Cell Entry Depends on ACE2 and TMPRSS2 and Is Blocked by a Clinically Proven Protease Inhibitor. *Cell* **2020**, *181*, 271–280.e278. [CrossRef]
9. Xu, K.; Sun, H.; Wang, K.; Quan, Y.; Qiao, Z.; Hu, Y.; Li, C. The Quantification of Spike Proteins in the Inactivated SARS-CoV-2 Vaccines of the Prototype, Delta, and Omicron Variants by LC-MS. *Vaccines* **2023**, *11*, 1002. [CrossRef]
10. Chi, W.Y.; Li, Y.D.; Huang, H.C.; Chan, T.E.H.; Chow, S.Y.; Su, J.H.; Ferrall, L.; Hung, C.F.; Wu, T.C. COVID-19 vaccine update: Vaccine effectiveness, SARS-CoV-2 variants, boosters, adverse effects, and immune correlates of protection. *J. Biomed. Sci.* **2022**, *29*, 82. [CrossRef]
11. Gao, Q.; Bao, L.; Mao, H.; Wang, L.; Xu, K.; Yang, M.; Li, Y.; Zhu, L.; Wang, N.; Lv, Z.; et al. Development of an inactivated vaccine candidate for SARS-CoV-2. *Science* **2020**, *369*, 77–81. [CrossRef] [PubMed]
12. Hu, B.; Guo, H.; Zhou, P.; Shi, Z.L. Author Correction: Characteristics of SARS-CoV-2 and COVID-19. *Nat. Rev. Microbiol.* **2022**, *20*, 315. [CrossRef] [PubMed]
13. Li, M.; Wang, H.; Tian, L.; Pang, Z.; Yang, Q.; Huang, T.; Fan, J.; Song, L.; Tong, Y.; Fan, H. COVID-19 vaccine development: Milestones, lessons and prospects. *Signal Transduct. Target. Ther.* **2022**, *7*, 146. [CrossRef] [PubMed]
14. Xia, S.; Zhang, Y.; Wang, Y.; Wang, H.; Yang, Y.; Gao, G.F.; Tan, W.; Wu, G.; Xu, M.; Lou, Z.; et al. Safety and immunogenicity of an inactivated SARS-CoV-2 vaccine, BBIBP-CorV: A randomised, double-blind, placebo-controlled, phase 1/2 trial. *Lancet Infect. Dis.* **2021**, *21*, 39–51. [CrossRef] [PubMed]
15. Jin, L.; Li, Z.; Zhang, X.; Li, J.; Zhu, F. CoronaVac: A review of efficacy, safety, and immunogenicity of the inactivated vaccine against SARS-CoV-2. *Hum. Vaccin. Immunother.* **2022**, *18*, 2096970. [CrossRef] [PubMed]
16. Tanriover, M.D.; Doganay, H.L.; Akova, M.; Guner, H.R.; Azap, A.; Akhan, S.; Kose, S.; Erdinc, F.S.; Akalin, E.H.; Tabak, O.F.; et al. Efficacy and safety of an inactivated whole-virion SARS-CoV-2 vaccine (CoronaVac): Interim results of a double-blind, randomised, placebo-controlled, phase 3 trial in Turkey. *Lancet* **2021**, *398*, 213–222. [CrossRef] [PubMed]
17. Ella, R.; Reddy, S.; Blackwelder, W.; Potdar, V.; Yadav, P.; Sarangi, V.; Aileni, V.K.; Kanungo, S.; Rai, S.; Reddy, P.; et al. Efficacy, safety, and lot-to-lot immunogenicity of an inactivated SARS-CoV-2 vaccine (BBV152): Interim results of a randomised, double-blind, controlled, phase 3 trial. *Lancet* **2021**, *398*, 2173–2184. [CrossRef]
18. Hu, L.; Sun, J.; Wang, Y.; Tan, D.; Cao, Z.; Gao, L.; Guan, Y.; Jia, X.; Mao, J. A Review of Inactivated COVID-19 Vaccine Development in China: Focusing on Safety and Efficacy in Special Populations. *Vaccines* **2023**, *11*, 1045. [CrossRef]
19. Shi, S.; Zhu, H.; Xia, X.; Liang, Z.; Ma, X.; Sun, B. Vaccine adjuvants: Understanding the structure and mechanism of adjuvanticity. *Vaccine* **2019**, *37*, 3167–3178. [CrossRef]
20. He, P.; Zou, Y.; Hu, Z. Advances in aluminum hydroxide-based adjuvant research and its mechanism. *Hum. Vaccin. Immunother.* **2015**, *11*, 477–488. [CrossRef]
21. Ella, R.; Vadrevu, K.M.; Jogdand, H.; Prasad, S.; Reddy, S.; Sarangi, V.; Ganneru, B.; Sapkal, G.; Yadav, P.; Abraham, P.; et al. Safety and immunogenicity of an inactivated SARS-CoV-2 vaccine, BBV152: A double-blind, randomised, phase 1 trial. *Lancet Infect. Dis.* **2021**, *21*, 637–646. [CrossRef] [PubMed]
22. Corbett, K.S.; Flynn, B.; Foulds, K.E.; Francica, J.R.; Boyoglu-Barnum, S.; Werner, A.P.; Flach, B.; O'Connell, S.; Bock, K.W.; Minai, M.; et al. Evaluation of the mRNA-1273 Vaccine against SARS-CoV-2 in Nonhuman Primates. *N. Engl. J. Med.* **2020**, *383*, 1544–1555. [CrossRef]
23. Vogel, A.B.; Kanevsky, I.; Che, Y.; Swanson, K.A.; Muik, A.; Vormehr, M.; Kranz, L.M.; Walzer, K.C.; Hein, S.; Guler, A.; et al. BNT162b vaccines protect rhesus macaques from SARS-CoV-2. *Nature* **2021**, *592*, 283–289. [CrossRef]
24. Baden, L.R.; El Sahly, H.M.; Essink, B.; Kotloff, K.; Frey, S.; Novak, R.; Diemert, D.; Spector, S.A.; Rouphael, N.; Creech, C.B.; et al. Efficacy and Safety of the mRNA-1273 SARS-CoV-2 Vaccine. *N. Engl. J. Med.* **2021**, *384*, 403–416. [CrossRef]
25. Polack, F.P.; Thomas, S.J.; Kitchin, N.; Absalon, J.; Gurtman, A.; Lockhart, S.; Perez, J.L.; Perez Marc, G.; Moreira, E.D.; Zerbini, C.; et al. Safety and Efficacy of the BNT162b2 mRNA Covid-19 Vaccine. *N. Engl. J. Med.* **2020**, *383*, 2603–2615. [CrossRef] [PubMed]
26. Fadlyana, E.; Rusmil, K.; Tarigan, R.; Rahmadi, A.R.; Prodjosoewojo, S.; Sofiatin, Y.; Khrisna, C.V.; Sari, R.M.; Setyaningsih, L.; Surachman, F.; et al. A phase III, observer-blind, randomized, placebo-controlled study of the efficacy, safety, and immunogenicity of SARS-CoV-2 inactivated vaccine in healthy adults aged 18-59 years: An interim analysis in Indonesia. *Vaccine* **2021**, *39*, 6520–6528. [CrossRef] [PubMed]
27. Han, X.; Xu, P.; Ye, Q. Analysis of COVID-19 vaccines: Types, thoughts, and application. *J. Clin. Lab. Anal.* **2021**, *35*, e23937. [CrossRef] [PubMed]
28. Abu-Raddad, L.J.; Chemaitelly, H.; Butt, A.A. Effectiveness of the BNT162b2 COVID-19 Vaccine against the B.1.1.7 and B.1.351 Variants. *N. Engl. J. Med.* **2021**, *385*, 187–189. [CrossRef]

29. Charmet, T.; Schaeffer, L.; Grant, R.; Galmiche, S.; Cheny, O.; Von Platen, C.; Maurizot, A.; Rogoff, A.; Omar, F.; David, C.; et al. Impact of original, B.1.1.7, and B.1.351/P.1 SARS-CoV-2 lineages on vaccine effectiveness of two doses of COVID-19 mRNA vaccines: Results from a nationwide case-control study in France. *Lancet Reg. Health Eur.* **2021**, *8*, 100171. [CrossRef]
30. Sheikh, A.; McMenamin, J.; Taylor, B.; Robertson, C. SARS-CoV-2 Delta VOC in Scotland: Demographics, risk of hospital admission, and vaccine effectiveness. *Lancet* **2021**, *397*, 2461–2462. [CrossRef]
31. Lopez Bernal, J.; Andrews, N.; Gower, C.; Gallagher, E.; Simmons, R.; Thelwall, S.; Stowe, J.; Tessier, E.; Groves, N.; Dabrera, G.; et al. Effectiveness of Covid-19 Vaccines against the B.1.617.2 (Delta) Variant. *N. Engl. J. Med.* **2021**, *385*, 585–594. [CrossRef] [PubMed]
32. Chemaitelly, H.; Yassine, H.M.; Benslimane, F.M.; Al Khatib, H.A.; Tang, P.; Hasan, M.R.; Malek, J.A.; Coyle, P.; Ayoub, H.H.; Al Kanaani, Z.; et al. mRNA-1273 COVID-19 vaccine effectiveness against the B.1.1.7 and B.1.351 variants and severe COVID-19 disease in Qatar. *Nat. Med.* **2021**, *27*, 1614–1621. [CrossRef] [PubMed]
33. Zhou, Z.; Zhu, Y.; Chu, M. Role of COVID-19 Vaccines in SARS-CoV-2 Variants. *Front. Immunol.* **2022**, *13*, 898192. [CrossRef] [PubMed]
34. Jara, A.; Undurraga, E.A.; Zubizarreta, J.R.; Gonzalez, C.; Acevedo, J.; Pizarro, A.; Vergara, V.; Soto-Marchant, M.; Gilabert, R.; Flores, J.C.; et al. Effectiveness of CoronaVac in children 3-5 years of age during the SARS-CoV-2 Omicron outbreak in Chile. *Nat. Med.* **2022**, *28*, 1377–1380. [CrossRef] [PubMed]
35. Su, S.; Li, W.; Jiang, S. Developing pan-beta-coronavirus vaccines against emerging SARS-CoV-2 variants of concern. *Trends Immunol.* **2022**, *43*, 170–172. [CrossRef] [PubMed]
36. Zhou, J.; Liu, Z.; Zhang, G.; Xu, W.; Xing, L.; Lu, L.; Wang, Q.; Jiang, S. Development of variant-proof severe acute respiratory syndrome coronavirus 2, pan-sarbecovirus, and pan-beta-coronavirus vaccines. *J. Med. Virol.* **2023**, *95*, e28172. [CrossRef] [PubMed]
37. Joyce, M.G.; Chen, W.H.; Sankhala, R.S.; Hajduczki, A.; Thomas, P.V.; Choe, M.; Martinez, E.J.; Chang, W.C.; Peterson, C.E.; Morrison, E.B.; et al. SARS-CoV-2 ferritin nanoparticle vaccines elicit broad SARS coronavirus immunogenicity. *Cell Rep.* **2021**, *37*, 110143. [CrossRef]
38. Wang, C.Y.; Peng, W.J.; Kuo, B.S.; Ho, Y.H.; Wang, M.S.; Yang, Y.T.; Chang, P.Y.; Shen, Y.H.; Hwang, K.P. Toward a pan-SARS-CoV-2 vaccine targeting conserved epitopes on spike and non-spike proteins for potent, broad and durable immune responses. *PLoS Pathog.* **2023**, *19*, e1010870. [CrossRef]
39. WHO. Statement on the Antigen Composition of COVID-19 Vaccines. 2023. Available online: https://www.who.int/news/item/18-05-2023-statement-on-the-antigen-composition-of-covid-19-vaccines (accessed on 18 May 2023).
40. Yang, J.; Li, B.; Yang, D.; Wu, J.; Yang, A.; Wang, W.; Lin, F.; Wan, X.; Li, Y.; Chen, Z.; et al. The immunogenicity of Alum+CpG adjuvant SARS-CoV-2 inactivated vaccine in mice. *Vaccine* **2023**, *41*, 6064–6071. [CrossRef]
41. Yang, H.; Xie, Y.; Lu, S.; Sun, Y.; Wang, K.; Li, S.; Wang, J.; Liao, G.; Li, C. Independent Protection and Influence of the Spike-Specific Antibody Response of SARS-CoV-2 Nucleocapsid Protein (N) in Whole-Virion Vaccines. *Vaccines* **2023**, *11*, 1681. [CrossRef]
42. Haghighi, M.; Khorasani, A.; Karimi, P.; Keshavarz, R.; Mahdavi, M. Different Formulations of Inactivated SARS-CoV-2 Vaccine Candidates in Human Compatible Adjuvants: Potency Studies in Mice Showed Different Platforms of Immune Responses. *Viral Immunol.* **2022**, *35*, 663–672. [CrossRef] [PubMed]
43. Bai, S.; Kang, Y.; Chen, W.; Xie, H.; Zhang, L.; Lv, M.; Wang, J.; Wu, J.; Zhao, W. Comparison of Immunogenicity of Alum and MF59-Like Adjuvant Inactivated SARS-CoV-2 Vaccines Against SARS-CoV-2 Variants in Elderly Mice. *Viral Immunol.* **2023**, *36*, 526–533. [CrossRef] [PubMed]
44. Grifoni, A.; Weiskopf, D.; Ramirez, S.I.; Mateus, J.; Dan, J.M.; Moderbacher, C.R.; Rawlings, S.A.; Sutherland, A.; Premkumar, L.; Jadi, R.S.; et al. Targets of T Cell Responses to SARS-CoV-2 Coronavirus in Humans with COVID-19 Disease and Unexposed Individuals. *Cell* **2020**, *181*, 1489–1501.e1415. [CrossRef] [PubMed]
45. Moss, P. The T cell immune response against SARS-CoV-2. *Nat. Immunol.* **2022**, *23*, 186–193. [CrossRef]
46. Huang, R.; Ying, L.; Wang, J.; Xia, J.; Zhang, Y.; Mao, H.; Zhang, R.; Zang, R.; Le, Z.; Shu, Q.; et al. Non-spike and spike-specific memory T cell responses after the third dose of inactivated COVID-19 vaccine. *Front. Immunol.* **2023**, *14*, 1139620. [CrossRef]
47. Li, Y.; Wang, X.; Jin, J.; Ma, Z.; Liu, Y.; Zhang, X.; Su, B. T-cell responses to SARS-CoV-2 Omicron spike epitopes with mutations after the third booster dose of an inactivated vaccine. *J. Med. Virol.* **2022**, *94*, 3998–4004. [CrossRef]
48. Schultz, B.M.; Melo-Gonzalez, F.; Duarte, L.F.; Galvez, N.M.S.; Pacheco, G.A.; Soto, J.A.; Berrios-Rojas, R.V.; Gonzalez, L.A.; Moreno-Tapia, D.; Rivera-Perez, D.; et al. A Booster Dose of CoronaVac Increases Neutralizing Antibodies and T Cells that Recognize Delta and Omicron Variants of Concern. *mBio* **2022**, *13*, e0142322. [CrossRef]

vaccines

MDPI

Article

Heterologous Vector—mRNA Based SARS-CoV-2 Vaccination Strategy Appears Superior to a Homologous Vector—Based Vaccination Scheme in German Healthcare Workers Regarding Humoral SARS-CoV-2 Response Indicating a High Boosting Effect by mRNA Vaccines

Catharina Gerhards [1,*], Margot Thiaucourt [1], Michael Hetjens [2], Verena Haselmann [1], Michael Neumaier [1] and Maximilian Kittel [1]

1 Institute for Clinical Chemistry, Medical Faculty Mannheim, University of Heidelberg, Theodor Kutzer Ufer 1-3, 68167 Mannheim, Germany
2 Department of Biomedical Informatics, Center for Preventive Medicine and Digital Health Baden-Württemberg, Medical Faculty Mannheim, Heidelberg University, 68167 Mannheim, Germany
* Correspondence: catharina.gerhards@umm.de

Abstract: Background: Longitudinal humoral SARS-CoV-2 (severe acute respiratory syndrome coronavirus type 2) immunity for up to 15 months due to vaccination, the efficacy of vaccination strategies (homologous, vector–vector versus heterologous, vector–mRNA), the influence of vaccination side effects, and the infection rate in German healthcare workers need to be investigated. Methods: In this study, 103 individuals vaccinated against SARS-CoV-2 were enrolled to examine their anti-SARS-CoV-2 anti-N- and anti-RBD/S1-Ig levels. A total of 415 blood samples in lithium heparin tubes were prospectively obtained, and a structured survey regarding medical history, type of vaccine, and vaccination reactions was conducted. Results: All participants demonstrated a humoral immune response, among whom no values decreased below the positivity cutoff. Five to six months after the third vaccination, three participants showed anti-RBD/S1 antibodies of less than 1000 U/mL. We observed higher levels for heterologous mRNA-/vector-based combinations compared to pure vector-based vaccination after the second vaccination, which is harmonized after a third vaccination with the mRNA-vaccine only in both cohorts. The incidence of vaccine breakthrough in a highly exposed cohort was 60.3%. Conclusion: Sustained long-term humoral immunity was observed, indicating the superiority of a heterologous mRNA-/vector-based combination compared to pure vector-based vaccination. There was longevity of anti-RBD/S1 antibodies of at least 4 and up to 7 months without external stimulus. Regarding vaccination reactogenity, the occurrence of local symptoms as pain at the injection site was increased after the first mRNA application compared to the vector–vector cohort with a general decrease in adverse events at later vaccination time points. Overall, a correlation between the humoral vaccination response and vaccination side effects was not observed. Despite the high prevalence of vaccine breakthroughs, these only occurred in the later course of the study when more infectious variants, which are, however, associated with milder courses, were present. These results provide insights into vaccine-related serologic responses, and the study should be expanded using additional vaccine doses and novel variants in the future.

Keywords: antibody dynamics; antibody kinetics; anti-SARS-CoV-2 antibodies; longitudinal assessment; serological immune response; humoral response post-vaccination; vaccination breakthrough; side effects; vaccination strategy

1. Introduction

The key strategy over the past year to overcome the global pandemic caused by severe acute respiratory syndrome coronavirus 2 (SARS-CoV-2) has been to provide vaccinations to stimulate the immune system prior to possible exposure to infection, thereby minimizing the extent of clinical manifestations [1]. Vaccination agents available in Germany are messenger RNA (mRNA)-based products such as BNT162b2 (Comirnaty, Pfizer-BioNTech, Mainz, Germany) and mRNA-1273 (Spikevax, Moderna) and vector-based products such as ChAdOx1 nCoV-19 (AstraZeneca, Cambridge, UK) and Ad16.COV.2.S (Janssen Pharmaceutical K.K.) [1–3]. Different vaccines are recommended depending on the medical history and age of the patient. In the course of the first immunization phase with the adeno-vector-based vaccine ChAdOx1 nCoV-19 (AstraZeneca), venous sinus thrombosis occurred, particularly in younger patients. Thus, vaccination guidelines were adjusted for patients younger than 30 years. The adaptations included a second vaccination with an mRNA-based vaccine, resulting in a heterologous vaccine combination [4]. Consequently, the population was divided into two cohorts: those who had received either homologous or heterologous vaccines for the first two vaccinations. Regarding the mentioned vaccination strategies, a homologous vaccination means that the same vaccine is administered twice (vector-based only), and a heterologous vaccination combines two different types of vaccines (mRNA- and vector-based). Due to the described adaption of vaccination guidelines, a combination of miscellaneous vaccines could have been used for the second and third vaccination in Germany. Therefore, different immune responses were reported in the described cohorts and should be investigated on the basis of humoral methods over a long time period. In general, the immune response can be assessed via both humoral and cellular systems. Thus, when antibody titers decrease, immunity can still be maintained at the cellular level. Regarding humoral SARS-CoV-2 immunity, the serological response can be assessed via antibodies directed against the nucleocapsid (N) protein and the spike (S) subunit 1 receptor-binding domain (RBD) of SARS-CoV-2 (RBD/S1). A disease-specific formation of antibodies can be detected by anti-N-abs. In contrast, both vaccination and COVID-19 can stimulate the production of anti-SARS-CoV-2-RBD/S1 abs. There are various epitopes against which antibodies can be formed, such as the receptor-binding domain (RBD) or the S1/S2 domain of the spike glycoprotein [5]. As described in previous studies, protective immunity mediated by antibodies with virus-neutralizing capacity correlates with anti-RBD/S1 antibodies [6–8]. Thus, anti-RBD/S1-pan-Ig serves as a surrogate parameter to monitor humoral SARS-CoV-2 immunity. For this reason, we focused only on antibody detection as an inexpensive and easy-to-implement assay in routine care [1–3].

A previous study including health workers who had been vaccinated with BNT162b2 (Comirnaty, Pfizer-BioNTech) revealed an initial increase in anti-RBD/S1 antibodies 2 weeks after the first dose. The maximum observed period was 5 weeks after study inclusion, leading to investigation via longitudinal follow-up. A further study examined a slightly longer period, up to 13 weeks after the first vaccination, and showed an increased anti-RBD/S1 antibody titer starting 3 weeks after the initial dose, which was exceeded after the second vaccination. Moreover, the researchers described a half-time decline in the anti-RBD/S1 antibody titer in a homologous vaccinated cohort 13 weeks after the initial dose, but did not assess the time for twofold decline using the heterologous strategy [9]. The dynamics and robustness of anti-RBD/S1 antibodies need to be investigated for homologous and heterologous vaccination strategies over a longer period, including the effects of the third vaccination.

Evaluating the influence of various vaccination combinations, some studies have reported differences in the response of the humoral immune system [10,11]. Results have shown higher humoral reactogenicity from heterologous vaccination with the inactivated CoronaVac vaccine (Sinovac Biotech, Beijing, China) and the vector-based ChAdOx1 nCoV-19 vaccine (AstraZeneca), compared to the homogeneous CoronaVac strategy. These results should be extended to include combinations of mRNA- and vector-based vaccines. Additionally, by including a third vaccination in the analysis, more possible combinations could be investigated. In this regard, a previous study examined the efficacy against pathogens of booster strategies with the CoronaVac, ChAdOx1 nCoV-19, and NT162b2 vaccines, suggesting the superiority of heterologous combinations [9,10].

Therefore, long-term observations of more than 1 year, including a period without stimulation of up to 5.7 months after the third vaccination, need to be examined in order to investigate the robustness of antibody titers. A comparison of heterologous mRNA-/vector-based and homologous vector-based vaccination regimes is a central component of our study, as they affect a large number of highly exposed health care workers and the general public and, therefore, urgently need to be addressed in the long term. This refers to differences in antibody response, vaccination side effects, and the prevalence or timing of vaccination breakthroughs. Thus, in this study, we present the results of a longitudinal evaluation of antibody dynamics up to 15 months after the initial dose of the SARS-CoV-2 vaccine in healthy German workers. The aims of this study were as follows: (i) to examine anti-SARS-CoV-2 antibody titers, at six appointments to reveal vaccination-related humoral immune response in dependence of the vaccination strategy andantibody robustness 5 months after the third vaccination without external stimulation; (ii) to investigate the correlation between antibody titers and clinical vaccination reactions; and (iii) to illuminate the infection rates after vaccination, in the context of circulating variants.

2. Materials and Methods

2.1. Participant Recruitment and Sample Collection

Employees of the University Medical Center Mannheim, Germany, were invited to participate in the Immunitor-III study during the SARS-CoV-2 vaccine campaign. From February 2021 to June 2022, SARS-CoV-2 vaccinated individuals aged 18 years or older were enrolled in the study and underwent longitudinal blood sampling at six appointments. A secure web platform, Research Electronic Data Capture (REDCap), was utilized for the acquisition of standardized surveys addressing medical and personal data of participants at study entrance, and clinical data related to vaccination and SARS-CoV-2 status in the course of the study. Data collection was performed in pseudonymized form. The study was approved by the Institutional Review Board (2020-556N), informed consent was obtained from all participants, and the study was conducted in accordance with the Declaration of Helsinki.

For long-term surveillance of the humoral SARS-CoV-2 response and to carry out further analysis, blood samples were collected in 7.5 mL lithium heparin tubes (S-Monovette, Sarstedt AG & Co., Nümbrecht, Germany) at 6 time intervals: time of study enrollment (before or on the date of the first vaccination), 2 months after the first vaccination and before the second vaccination, 1.5 months after the second vaccination (6 weeks), up to 6 months after the second vaccination (24 weeks), 1.5 months after the third vaccination (7 weeks), and 6 months after the third vaccination (19 weeks) (for study protocol, see Figures 1 and 2). Samples were centrifuged at 2000$\times$ g for 10 min at 18 °C, and plasma was aliquoted and stored at -80 °C until analysis.

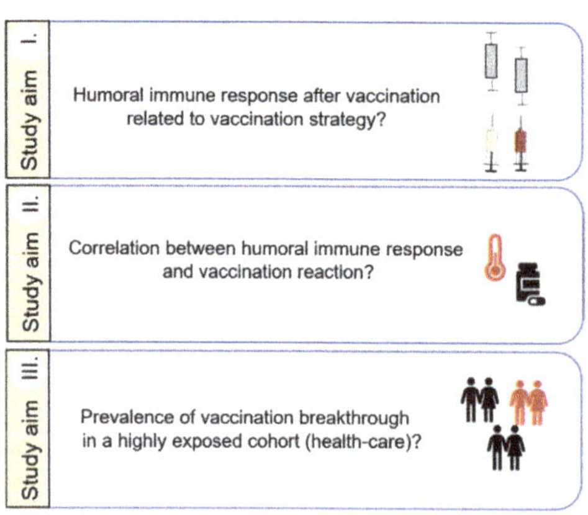

Figure 1. Study aims.

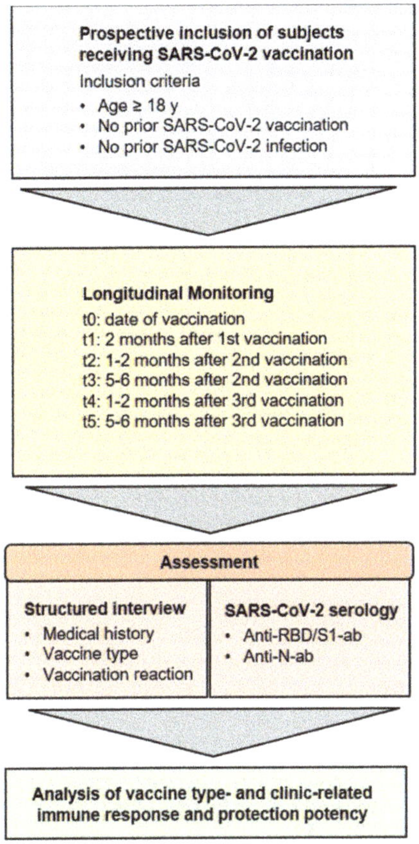

Figure 2. Inclusion criteria and analysis. Study concept and research objectives. Adults without a previous SARS-CoV-2 infection or vaccination were enrolled between February 2021 and June 2022 and were longitudinally assessed for clinical parameters and humoral immune responses.

2.2. Anti-SARS-CoV-2 Antibody Detection

The detection of SARS-CoV-2 anti-N and anti-RBD/S1 antibodies was performed using the same platform as that described in previous work [12]. For the detection of anti-SARS-CoV-2 anti-N Pan-Ig, a qualitative CE- and FDA-approved electrochemiluminescence immunoassay (ECLIA) with a positivity cut-off index (COI) $\geq$ 1.0 was used. The quantitative Elecsys® Anti-SARS-CoV-2 S assay (Roche, Mannheim, Germany), similarly CE- and FDA-approved, was utilized to assess total anti-SARS-CoV-2 anti-RBD/S1 Ig, evaluated as reactive from a titer of 0.8 U/mL. After internal verification in line with DIN EN ISO 15189, the analyses were performed according to the manufacturer's instructions at an accredited laboratory.

2.3. General Data Analysis

All clinical data were systematically recorded by the REDCap platform (Vanderbilt University, Nashville, TN, USA). Statistical analyses were performed in Microsoft Excel 2019 (Microsoft, Redmond, WA, USA) and RStudio (version 4.1.2; RStudio, Boston, MA, USA) [9]. For all statistical analyses, a p-value < 0.05 was considered statistically significant and was indicated by an asterisk (*), and $p < 0.01$ was indicated by double asterisks (**). Non-normally distributed continuous variables were compared by the Kruskal–Wallis rank sum test, and for categorical variables, Fisher's exact test was performed. Graphs were plotted using GraphPad Prism 7.05 (Dotmatics, Boston, MA, USA), RStudio (version 4.1.2; RStudio, Boston, MA, USA) [13], and Microsoft PowerPoint 2019 (Microsoft, Redmond, WA, USA).

3. Results

3.1. Demographics and Vaccination Strategies

A total of 103 participants were enrolled in the study, among whom 51 were included in longitudinal follow-up up to 13.5 ± 0.6 months after recruitment. During the 462 study days, 415 blood samples were collected in lithium heparin tubes and analyzed. Demographic data of the study cohort, including medical history and pre-existing medical conditions, are given in Table 1. Among the 103 participants, some individuals did not complete the structured questionnaire online or omitted particular questions, resulting in potential variation of the maximum participant size in the anamnestic data. Among the participants, 75.7% were female (f) (f = 78/103), the mean age was 39.64 ± 14.81 years, and the mean body mass index (BMI) was 25.32 (BMI range 15–50). At the initial interview, 24.7% were smokers, the majority of whom (7/20) smoked 11 to 20 cigarettes per day. At t_0, one participant reported SARS-CoV-2 positivity in the past (1.2%). A total of 97% were health care workers, among whom 67 out of 82 reported being a medical professional. In terms of pre-existing diseases, 12.3% suffered from pulmonary disease, 7.4% from vascular disease, 3.7% from autoimmune disease, 2.5% from cancer in remission, and 1.2% from diabetes. Other diseases were also reported, the most frequent of which was hypothyroidism (6/17). A total of 40.7% took daily medication, and none were on immunosuppressive therapy. In this context, 9 out of 37 participants who presented allergies took antiallergic medication, and none of them were currently undergoing hyposensitization. Among the 50 participants who knew their blood type and rhesus factor, the majority indicated that it was 0+ (21%), followed by A+ (18.5%).

Table 1. Patient demographics and clinical history.

Variable	Overall
	n = 103
Demographics	
Sex F/M (%)	78/25 (76.7/24.3)
Age (mean (SD))	39.64 (14.81)
BMI (mean (SD))	25.32 (5.10)
Smoking	20/81 (24.7%)
1–5 cigarettes/day	4 (4.9%)
6–10 cigarettes/day	6 (7.4%)
11–20 cigarettes/day	7 (8.6%)
31–40 cigarettes/day	3 (3.7%)
Pre-existing disease	
Pulmonary disease	10/81 (12.3%)
Vascular disease	6/81 (7.4%)
Autoimmune disease	3/81 (3.7%)
Cancer (in remission)	2/81 (2.5%)
Diabetes	1/81 (1.2%)
Other disease	17/81 (21.0%)
Hypothyroidism	6/81 (7.4%)
Allergy	37/81 (45.7%)
Medication	
Any medication	33/81 (40.7%)
Immunosuppression	0/81 (0%)
Antiallergic medication	9/81 (11.1%)

Table 1. *Cont.*

Variable	Overall
	n = 103
Blood type	
Unknown	31 (38.3%)
0+	17 (21.0%)
A+	15 (18.5%)
B+	3 (3.7%)
AB+	4 (4.9%)
0-	4 (4.9%)
A-	6 (7.4%)
AB-	1 (1.2%)
SARS-CoV-2 anamnesis	
Contact (1st vaccination)	13/82 (15.9%)
Previous infection	2/103 (1.9%)
SARS-CoV-2 vaccination	
First dose	
ChAdOx1 nCoV-19 (AstraZeneca)	85 (94.4%)
BNT162b2 (Pfizer-BioNTech)	4 (4.4%)
mRNA-1273 (Moderna)	1 (1.1%)
Second dose	
ChAdOx1 nCoV-19 (AstraZeneca)	27 (32.1%)
BNT162b2 (Pfizer-BioNTech)	55 (65.5%)
mRNA-1273 (Moderna)	2 (2.4%)
Third dose	
ChAdOx1 nCoV-19 (AstraZeneca)	0 (0%)
BNT162b2 (Pfizer-BioNTech)	59 (84.5%)
mRNA-1273 (Moderna)	11 (15.5%)
Vaccination strategy	
First and second dose	
Homologous (AstraZeneca)	26 (31.0%)
Heterologous (AstraZeneca/ BioNTech)	53 (63.1%)
Homologous (BioNTech)	4 (4.8%)
Homologous (Moderna)	1 (1.2%)
Second and third dose	
Homologous (mRNA vaccines)	47 (66.2%)
Heterologous (mRNA/vector)	24 (33.8%)

For the initial vaccination, 94.4% received the vector-based ChAdOx1 nCoV-19 (AstraZeneca), 4.4% received the mRNA-based NT162b2 (Comirnaty, Pfizer-BioNTech), and 1.1% received the mRNA-1273 (Spikevax, Moderna). For the second vaccination, the majority received the NT162b2 vaccine (65.5%), followed by ChA-dOx1 (32.1%) and mRNA-1273 (2.4%). Booster vaccination was performed exclusively with mRNA vaccines, predominantly NT162b2 (84.5%) (15.5% mRNA-1273).

From these data, different vaccination strategies could be deduced, i.e., homologous vector-based or heterologous vector- and mRNA-based for the first and second vaccinations. However, for the third vaccination, only mRNA vaccines were used, resulting in a complete absence of purely vector-vaccinated participants at the t_4 appointment. Regarding vaccine breakthroughs, during the spread of the Delta variant (B.1.617.2), one participant had a positive SARS-CoV-2 qPCR result (1.2%, n = 83). In spring 2022, during the expansion of the Omicron variant (B1.1.529), 31 of 84 participants (36.9%) experienced a vaccine breakthrough. Following the last appointment, vaccination breakthroughs were followed up, and SARS-CoV-2 positivity was reported in 44 of 73 participants (60.3%). All characteristics are summarized in more detail in Table 1.

3.2. Overall Antibody Trends

Initial blood sampling was performed before or on the same day as the first vaccination in order to establish a baseline value. Two participants were found to be seropositive for anti-SARS-CoV-2-anti-N and anti-SARS-CoV-2 anti-RBD/S1 antibodies; one had reported the COVID-19 disease, and one had a subclinical course. In addition, another individual who showed slightly positive anti-SARS-CoV-2 anti-RBD/S1 antibodies reported having a previous vaccination. From t_1, an average of 2.3 months after the first vaccination but before the second vaccination, 98.8% of the participants (81/82) demonstrated anti-RBD/S1 positivity. In one seronegative case, however, there were only 5 days between vaccination and blood sampling. The median t_1 titer was 71 U/mL (IQR 42, 112), a significant increase from t_0 ($p < 0.001$). Furthermore, we observed an enormous increase at t_2, 27 ± 9 days after the second vaccination, to a median of 7201 U/mL (IQR 1793, 10840) ($p < 0.001$). Antibodies significantly decreased to 784 U/mL (IQR 322, 1414) approximately 5.6 ± 0.7 months after the second vaccination ($p < 0.001$). Ten participants had already received their third vaccination prior to this appointment (4–38 days prior). This was the first appointment where anti-SARS-CoV-2-anti-N-ab positivity was observed that was not known at study inclusion, with a value of 12.9 COI. Three additional participants were tested for anti-SARS-CoV-2 anti-N-ab at t_4 and 10 more at t_5, 13.3 ± 0.5 months after study inclusion. The anti-SARS-CoV-2 anti-N-ab of the participant who was initially positive decreased from 101.8 to 51.2 COI, during the course of the study. Regarding all N-ab positive subjects a medium N-ab titer of 29.5 ± 27.5 COI was observed at t_5 ($n = 14$). On average, there were 49.4 ± 33.1 days between the positive qPCR result and the sample collection. A total of 3 out of 15 participants did not have a positive qPCR or antigen test result, of whom one reported a high-risk contact, resulting in cold symptoms, and two had an inapparent course. Since most participants ($n = 12/15$) had a positive qPCR result between t_4 and t_5, no longitudinal assessment of N-ab is available. For two participants, a percentage increase to 195.2 ± 29% was observed between t_4 and t_5. One SARS-CoV-2 qPCR-positive participant did not attend the t_5 appointment. The overall and individual development of N-ab is shown in Figure 3 and Supplemental Table S1.

The third vaccination resulted in a further significant increase in anti-SARS-CoV-2 anti-RBD/S1-ab to 11,490 U/mL (IQR 9239, 19,080) after 1.6 months. Finally, 4.6 months after the third vaccination, the titer decreased to a median of 6557 U/mL (IQR 3304, 11,506). During an interval without artificial stimulus, anti-SARS-CoV-2 anti-RBD/S1-ab decreased to 20 ± 17% between t_2 and t_3 ($n = 58$, 5.4 months after the second dose) and to 39 ± 15% between t_4 and t_5 ($n = 28$, 4.4 months after the third dose). Overall antibody dynamics are presented in Figure 4 and Table 2.

Table 2. SARS-CoV-2 antibodies.

Variable	Overall
t_0 anti-N abs (median (IQR))	0.09 (0.09, 0.09)
t_1 anti-N abs (mean (SD))	0.94 (7.70)
t_2 anti-N abs (median (IQR))	0.09 (0.09, 0.09)
t_3 anti-N abs (median (IQR))	0.07 (0.07, 0.09)
t_4 anti-N abs (median (IQR))	0.08 (0.08, 0.09)
t_5 anti-N abs (median (IQR))	0.06 (0.06, 1.50)
t_0 anti-RBD/S1 (median (IQR))	0.40 (0.40, 0.40)
t_1 anti-RBD/S1 (median (IQR))	71.43 (41.76, 111.68)
t_2 anti-RBD/S1 (median (IQR))	7201.00 (1793.00, 10,839.50)
t_3 anti-RBD/S1 (median (IQR))	791.70 (376.95, 1619.00)
t_4 anti-RBD/S1 (median (IQR))	11,490.00 (9239.00, 19,079.50)
t_5 anti-RBD/S1 (median (IQR))	6557.00 (3304.25, 11,506.25)

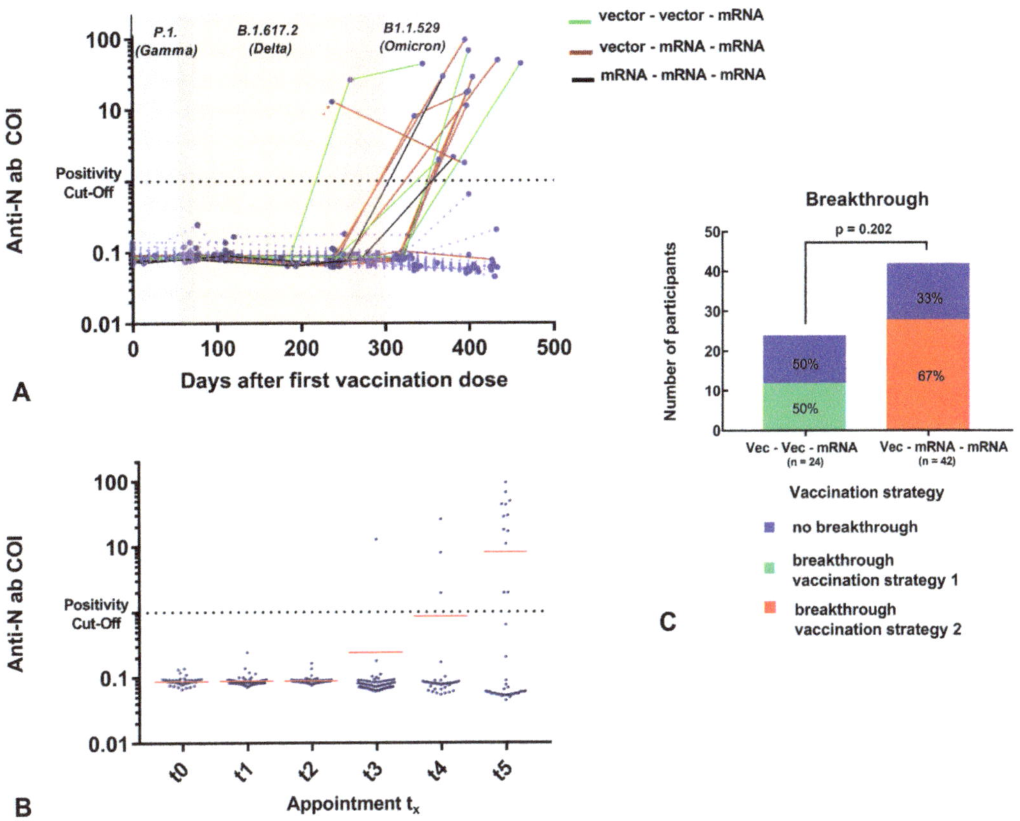

Figure 3. (**A**) Individual anti-N Ig titer (COI; *y*-axis) related to days after first vaccination (days; *x*-axis). Two values of one subject (titer of t_0 = 0 days and t_3 =238 days) were not connected due to missing data of t_1 and t_2, which could imply an earlier ab increase. The variants circulating in the respective period with a chronological overlap are highlighted in color (gray = P1. (Gamma), apricot = B.1.617.2 (Delta), light pink = B1.1.529 (Omicron)). Subjects with vaccination breakthrough before t_5 are illustrated: Vector–vector–mRNA cohort: green, vector–mRNA–mRNA cohort: red, mRNA–mRNA –mRNA: black (individual cases). Most of the subjects experienced a breakthrough after the observation period (>t_5). (**B**) Total anti-N Ig titer (COI; *y*-axis) at different time points (t_0: date of vaccination; t_1: 2 months after first vaccination; t_2: 1–2 months after second vaccination; t_3: 5–6 months after second vaccination; t_4: 1–2 months after third vaccination; t_5: 5–6 months after third vaccination (*x*-axis) for 101 participants (at time of study inclusion). Subjects with positive N-ab results at t_0 were excluded from this analysis due to infection prior first vaccination. Mean anti-N-Ig titer is illustrated in red. (**C**): The percentage distribution of vaccination breakthroughs depending on the vaccination strategy is illustrated. Most of the participants were SARS-CoV-2-infected after t_5 appoinment.

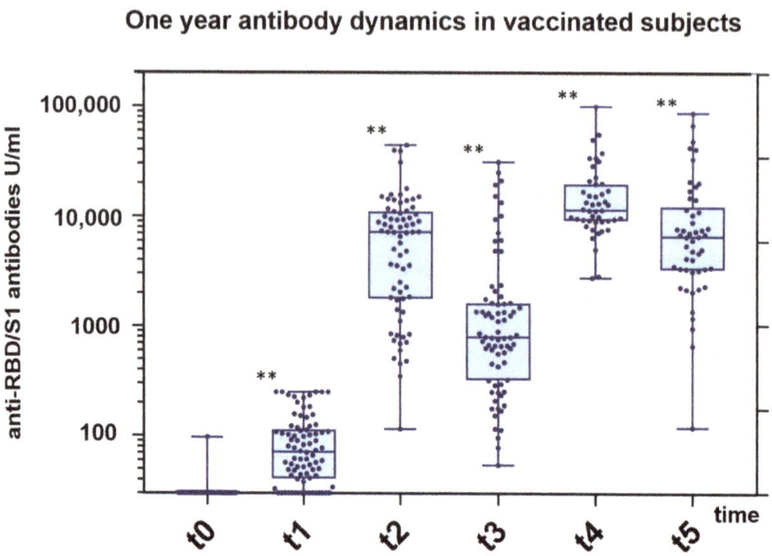

Figure 4. Total anti-RBD/S1 Ig titer related to vaccination dose. Boxplots for anti-RBD/S1 (U/mL; y-axis) at different time points (t_0: date of vaccination; t_1: 2 months after first vaccination; t_2: 1–2 months after second vaccination; t_3: 5–6 months after second vaccination; t_4: 1–2 months after third vaccination; t_5: 5–6 months after third vaccination (x-axis) for 100 participants (at time of study inclusion). In plots, interquartile ranges are shown in boxes with a line representing the median and whiskers representing the minimum and maximum data values. Dots indicate individual values. Antibody titers were compared to previous time points by the Kruskal–Wallis rank sum test. p-values < 0.05 were considered to be significant; ** p-values < 0.001.

3.3. *Anti-RBD/S1 Antibody Dynamics Correlated with Vaccination Strategy*

As described, diverse subcohorts with different vaccine combinations were considered. In terms of follow-ups after the second vaccination, we compared two cohorts, hom1 and het1. Participants who received only mRNA-1273 (Spikevax, Moderna) or mRNA-based NT162b2 (Comirnaty, Pfizer-BioNTech) were excluded from the comparison due to the small number of cases in the study population. Thus, since all participants in this comparison received ChAdOx1 (AstraZeneca) as their first vaccine, those who received mRNA-based NT162b2 (Comirnaty, Pfizer-BioNTech) for the second vaccination were assigned to het1, and those who received ChAdOx1 (AstraZeneca) were assigned to hom1. The baseline values of both cohorts were comparable due to the use of identical vaccines ($p = 0.163$). In contrast, significantly higher titers were detected in participants who received heterologous vaccinations, with a median of 9378.50 U/mL (7180.00, 14,103.00), compared to 826.30 U/mL (600.90, 1711.00) ($p < 0.001$) at t_2. Up to 6.9 months after the second vaccination, the heterologous cohort still showed a tendency towards higher anti-RBD/S1-abs compared to the homologous vector-vaccinated group (1999 $\pm$ 3827 U/mL versus 495 $\pm$ 1034 U/mL, $p = 0.102$). After the third vaccination, the participants who previously received heterologous ChAdOx1/mRNA-based NT162b2 were categorized as hom2 and those who received homologous ChAdOx1 as het2, as the booster vaccination was administered with mRNA-based vaccines only. Convergence of the median titers of both cohorts was observed, with a median value of 12,852 (9,308, 17,567.25) in hom2 and 10,582 (8,592.25, 20,863) in het2 ($p = 0.714$). All cohort-related titers are illustrated in Figure 5A. Moreover, Figure 5B shows the individual anti-RBD/S1 abs of subjects infected during the observation phase (before t_5) related to the vaccination strategy. A reference to the second and third vaccination (the first vaccination was day 0) and the circulating SARS-CoV-2 variants are presented. Moreover, we have indicated the approximate time of the first inapparent course (no positive test) and the first positive qPCR result, in order to allow an estimation of when the vaccination effect is supplemented by an infection stimulus.

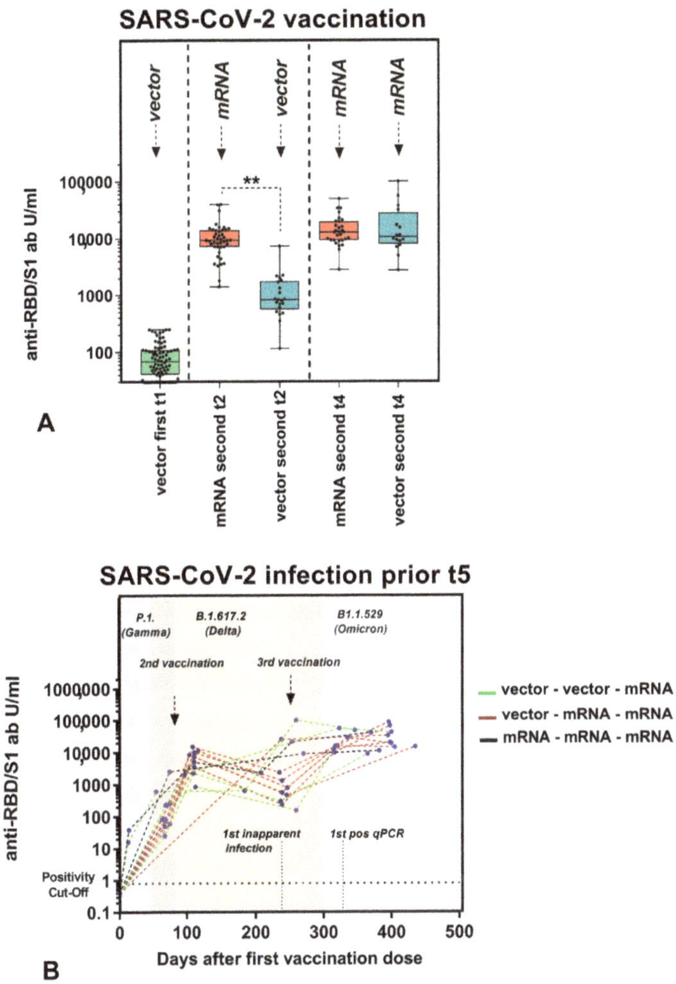

Figure 5. (**A**) Influence of vaccination strategy on antibody production. Dot plots illustrating anti-RBD/S1 antibodies (U/mL; *x*-axis) related to the vaccination schedule. All participants received ChAdOx1 nCoV-19 as the first vaccination. The homologous cohort received vector-based ChAdOx1 nCoV-19, and the heterologous cohort received mRNA-based NT162b2 as the second dose. This was reversed with the third vaccination, as all participants were vaccinated with NT162b2 or mRNA-1273. Antibody titers were compared between cohorts after the second and third vaccinations by the Kruskal–Wallis rank sum test. *p*-values 0.05 were considered to be significant; ** *p*-values < 0.001. (**B**) Individual anti-RBD/S1 Ig titer (U/mL; *y*-axis) related to days after the first vaccination (days; *x*-axis). The variants circulating in the respective period with a chronological overlap are highlighted in color (gray = P1. (Gamma), apricot = B.1.617.2 (Delta), light pink = B1.1.529 (Omicron)). Approximation of second/third vaccination, first inapparent infection and first positive qPCR result is illustrated. Vector–vector–mRNA cohort: green, vector–mRNA–mRNA cohort: red, mRNA –mRNA–mRNA: black (individual cases). Most of the subjects experienced a breakthrough after the observation period (t_5).

3.4. Anti-RBD/S1 Antibody Dynamics Correlated with Vaccination Reaction

We considered the humoral response to vaccination separately for participants with and without early vaccination reactions. With regard to the first vaccination, only 9% of participants (n = 7) showed no reaction, and among those who were symptomatic, 85% felt groggy (n = 63), 80% reported localized pain (n = 59), and 77% experienced a headache (n = 57). Anti-SARS-CoV-2 anti-RBD-ab did not differ in the cohorts, with values of 39.70 U/mL (21.84, 89.92) for the asymptomatic cohort and 72.60 U/mL (43.52, 119.35) for the symptomatic cohort (p = 0.160). Only 5 subjects presented with only local symptoms, while 69 presented with systemic symptoms, and no difference was observed in anti-SARS-CoV-2 anti-RBD-ab titers (systemic: 72.51 U/mL (43.12, 115.65); local: 99.45 U/mL (70.43, 121.80); p = 0.575). Regarding the second vaccination, 7 participants reported no symptoms and 57 described symptoms, among which itching was the most common (95%), followed by swelling and erythema (86%) and pain at the site of injection (77%). The anti-SARS-CoV-2 anti-RBD-ab titer, 1.4 months after the second vaccination, was not affected by the appearance of symptoms (6175 U/mL (1889.75, 8967) versus 7158 U/mL (1710, 10,790.75); p = 0.878). In addition, among symptomatic participants, those with systemic symptoms (n = 45) did not have higher titers than those with local symptoms (n = 12) (p = 0.422). After the third vaccination, 34/51 participants reported experiencing a reaction. However, we did not observe any significant effect on the antibody level (9880 U/mL (8828.50, 13,054.25) versus 12,780 U/mL (9001, 19,885); p = 0.183). With regard to the reactogenicity depending on the vaccination strategy, a higher incidence of local pain was found in the heterologous cohort after the second vaccination (p = 0.021). The vaccination reactions after the first, second, and third vaccination are illustrated in Figure 6. A comparison of homologous vector–vector and heterologous vector–mRNA vaccination strategy was compared after the second vaccination.

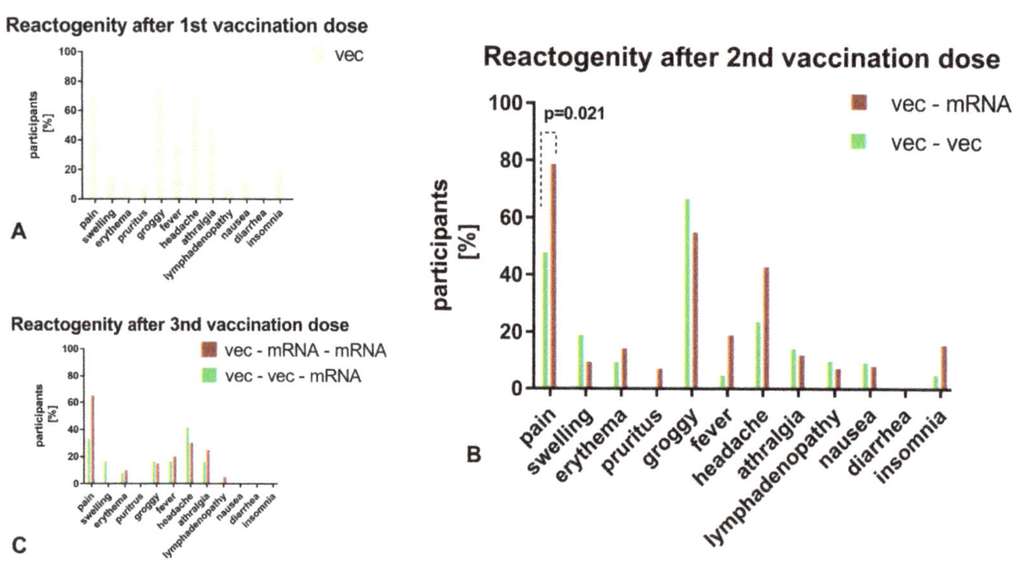

Figure 6. (**A**): For the first vaccination, only symptoms after the vector-based immunization are illustrated due to a small number of mRNA-vaccinated participants. (**B**,**C**): Depending on the vaccination scheme (homologous versus heterologous), the vaccination reactions are compared after the second vaccination. Only significant differences are indicated by a p-value. Vector–vector/+mRNA cohort: green, vector–mRNA/+mRNA cohort: red.

4. Discussion

COVID-19 vaccines play a crucial role in the fight against new variants of concern. They provide protection against severe illness and hospitalization, reducing the burden on the healthcare system [14]. By reducing the number of people who become infected with COVID-19, vaccines help to slow the spread of the virus, including new variants. It is possible that new variants could reduce the efficacy of existing vaccines [15]. In particular, the occurrence of unpredictable, potentially life-threatening adverse events has led to the unanticipated use of heterologous vaccination regimens [16].

In this study, we investigated the long-term humoral immune response to SARS-CoV-2 vaccination as a function of different vaccine combinations and individual clinical contexts. This is of current interest, as other studies have already shown that a combination can ensure unusually strong stimulation of the immune system [1–3,17].

Numerous publications have been published on heterologous vaccine regimens and their immunogenic activity. A strong humoral and cellular immunological response has been proven in animal experiments, indicating a high protective effect [18]. Some studies have described the humoral response to heterologous vaccination in the form of a prospective longitudinal study. Wanlapakorn et al. were able to demonstrate the superiority of a heterologous vaccination approach based on a vector-based vaccine combined with attenuated viral proteins [11,19]. Likewise, it was proven that heterologous vaccination regimens are superior in terms of the antibody titer level that develops in response to so-called booster vaccinations [20,21]. Interestingly, a comparison of our data with other publications indicates that the sequence of administration of the mRNA vaccine and the vaccine manufacturer are of secondary importance. However, this superiority of a heterologous combination refers only to subjects with first vector-based vaccination compared with double-vector vaccination and should also be compared with pure mRNA vaccinated subjects in follow-up studies. Thus, in contrast to the studies reported above, we distinguished between homologous and heterologous vaccines in terms of mRNA- and vector-based agents in the second vaccination. However, anti-SARS-CoV-2 anti-RBD-ab levels were significantly higher in our heterologous cohort, administered BNT162b2 as the second dose, than in the cohort administered only ChAdOx1nCov-19 ($p < 0.001$). Moreover, considering the anti-RBD/S-antibody levels up to 6.9 months without stimulus after the second vaccination, tendencies for a higher titer in the heterogeneous vaccine combination was observed ($p = 0.102$). This might indicate prolonged longevity of the antibodies, presumably due to a higher initial titer level. In the longitudinal assessment, anti-SARS-CoV-2 anti-RBD-ab titers were equalized after the booster vaccination in both cohorts. Since the titer converged after the third vaccination, when everyone had received at least one mRNA vaccination, there was some indication that this significant booster effect by an mRNA vaccine is possible after both the second and third vaccination. It should be evaluated whether, in resource-limited settings, a heterologous vaccine regimen with two different vaccines and two administration points would be preferable to a regimen with two homologous vaccines and one booster. This would not only alleviate the financial and workforce burden but also allow vaccination campaigns to be accelerated [22,23].

Supplementing the effect of greater antibody titers achieved by the heterologous vaccination regimen, respectively, a high boosting effect of vector-vaccinated subjects via the application of an mRNA vaccine, as described in this article, it is important to emphasize that other studies have already demonstrated that adenovirus vector-based SARS-CoV-2 vaccines lead to a more modest titer development compared to mRNA-based vaccines. For instance, in their comparison of different booster vaccination regimens, Atmar et al. were able to demonstrate a higher degree of immunogenicity in mRNA vaccines with respect to their neutralization capacity [24]. Differentiating from our study, it has to be stated that, in contrast to our approach, homologous basic immunization with different booster vaccines was compared. The comparison of the antibody formation effect in heterologous basic vaccination compared to other published studies indicates that this effect is significantly more prominent [25,26].

Due to public concerns that vaccination reactions are only associated with the development of protective antibodies, those were evaluated in the clinical context [27]. Evaluation [25,26] of this hypothesis indicated no significant correlation between the occurrence of a vaccination reaction and antibody development. Nevertheless, it should be pointed out that a high proportion of study participants (91%) reported a reaction after the first vaccination, but they had no history of reactions to other vaccinations. The most common symptom was a feeling of light headedness, followed by local pain at the injection site, and headache at administration. Notably, systemic reactions have been reported for vector-based vaccines and were confirmed in our study [28]. Regarding the second vaccination, the most predominant symptom was a local reaction with pain at the injection site, reported by 77% of participants.. Considering the reactogenicity in dependence of the vaccine administration, this might be explained due to a higher existence of local symptoms as pain at the injection site caused by mRNA vaccines. These data are in agreement with the results of other observational studies. A correlation between antibody formation and vaccination response was not demonstrated at any vaccination time point. Furthermore, in both our study and others, a decrease in adverse events at later vaccination time points was observed [29].

In addition to antibody development, within the cohort of middle-aged healthcare professionals, we further investigated the persistence of antibody titers in the absence of an external stimulus, such as vaccination or infection. Glöckner et al. described a significant decrease in anti-SARS-CoV-2 anti-RBD-ab at 13 weeks after the third vaccination [9]. In this study, for a mean observation period of 19 weeks after the second vaccination, a significant decrease in antibody titers with persistent seropositivity was observed in all 58 participants, excluding those with a confirmed breakthrough infection. Sugiyama et al. proposed an anti-SARS-CoV-2 anti-RBD-ab titer below 338 BAU/mL as a possible cutoff for protection against a severe variant infection [30]. In our long-term cohort, assessed up to 5.7 months after the third vaccination, 94% of participants showed titers higher than 1000 U/mL, and only one participant had a value less than 338 U/mL [31]. This indicates a long-lasting immunological response due to the heterologous vaccination strategy. The decline in antibody titers as well as neutralizing activity in plasma is a physiological phenomenon. Nevertheless, this is not immediately associated with higher susceptibility to severe infection since RBD-specific memory B cells are persistent [32].

According to previous studies [33,34], it is important to clarify that the analytical assays used in this study to assess humoral immunity are considered to be surrogate markers. Although several studies have addressed the correlation between virus neutralizing assays as the gold standard and anti-SARS-CoV-2 anti-RBD-ab assays, the authors suggest a value higher than 2000 BAU/mL as a possible good surrogate parameter for the prediction of immunity [35]. This suggests a limitation of serological assays due to the limited comparability among manufacturers. Therefore, we recommend using the same methodology for longitudinal monitoring of patients. Besides the lack of harmonization, the recommended cutoff values might be problematic in the context of emerging novel virus variants of concern [36]. Only if these limitations are rigorously addressed can they be used as a cost-effective surrogate marker for virus neutralization capacity [7,37]. T cell-mediated immune responses provide the other essential branch of the immune defenses in targeting intracellular pathogens [33,34]. For example, T-cell responses have been demonstrated to play a critical role in viral elimination during infection with influenza viruses or SARS-CoV. This cellular immune response is both stimulated by infection and by the vaccines that were evaluated in this study. It could be shown that the cellular and humoral immune response correlate predominantly well [38]. However, it must be mentioned that the cellular immune response is significantly more robust against viral mutations and is almost not affected by the exchange of single amino acids [39]. A very special case, justifying the much more complex investigation of cellular immune responses, is the examination of vaccinated subjects without detectable antibodies [40]. Since these constellations did not occur in the study collective, no cellular immunity investigation was carried out.

The occurrence of breakthrough infections during the 15-month observation period in this high-risk group due to elevated exposure is another aspect of the research addressed in the study. Nearly all breakthrough infections observed were attributable to the first Omicron variant (B.1.1.529/BA.1). It is not possible to draw a conclusion regarding the protective efficacy of vaccine combinations due to the relatively heterogeneous study population and the limited number of cases. However, despite the high number of breakthrough infections (60.3%; 44/73), no severe disease courses were reported. We did not observe a significant difference in the breakthrough rate between the "vector–mRNA–mRNA" and the "vector–vector–mRNA cohort". In contrast to only considering the vaccination breakthrough rates, it is common to evaluate the efficacy of vaccines based on hospitalization rates [41]. In this regard, further large-scale studies would be required to clarify whether the significantly higher humoral immune response to heterologous vaccine combinations could be associated with an increased efficacy of the virus neutralization capability. Furthermore, when considering the subcollective of breakthrough infections, it should be noted that the symptoms were very moderate, despite existing comorbidities as well as the absence of COVID-associated hospitalizations. These observations are in line with other published data, which state that a high antibody titer is not only associated with a probably higher protection against infection but also with a significantly reduced level of symptoms [42,43]. In addition, almost all of the SARS-CoV-2 infections occurred during the circulation of the Omicron variant. This may at least indicate that although variants with higher infectivity can lead to breakthroughs, infections with variants that cause more severe courses might be avoided in our highly exposed observation cohort.

However, several limitations of the data need to be critically considered. In order to make a stronger statement about the humoral immune response in relation to the vaccine combination, the inclusion of a control cohort with a homologous mRNA vaccine regimen would have been of great value. Nevertheless, the predominance of a heterologous mRNA/vector-based vaccine combination compared with vector-based vaccination alone can be proven in terms of antibody titers. Regrettably, individuals without an early vaccination response were significantly under-represented, limiting the power of the null hypothesis. One important aspect to emphasize is that the use of a combination of vaccines is an approach that has not been investigated by any manufacturer in a large-scale controlled registration program. The majority of the available scientific data is based on observational studies. Therefore, it cannot be ruled out that there may be adverse effects that have not yet been identified due to the lack of structured surveillance. In addition, a survey of individual COVID-19 symptoms of all vaccine breakthroughs could not be performed in comparable detail to the survey of vaccine reactions, because most experienced illnesses after the observation period. This resulted in more open responses indicating mild COVID-19 courses that could have been implemented more systematically if the study design had been longer. An additional limitation concerns the assays used to assess immunity. Even though the methods used in this study are well researched and have high congruence, virus neutralization assays are considered the gold standard for the assessment of the humoral immune capacity. Assays that assess cellular immunity would complement these methods and should be addressed in follow-up projects. Despite these limitations, the authors consider that important questions in the study were answered, and insights can be gained regarding the humoral antibody response after vaccinations, depending on the clinic and vaccine combination.

In conclusion, we were able to demonstrate persistent humoral antibody titers up to 15 months after the initial vaccination, suggesting that a heterologous mRNA/vector combination is preferable to a vector vaccination only. For this combination, the sequence of vaccines is not likely to be important, as we observed antibody titer similarity between the two vaccination regimens, indicating a high booster effect by mRNA vaccines as second or third dose. A correlation between the humoral vaccination response and early local or systemic reactions was not observed. Furthermore, the humoral response was long-lasting over an observation period of at least 4 months, but up to 7 months in the absence of an external stimulus. The data indicate evidence of protection against infection with a strict use of the assessed vaccination regimes. These results provide insight into vaccine-related serologic responses, and the study could be expanded in the future with respect to additional vaccine doses and new variants.

Supplementary Materials: The following supporting information can be downloaded at: https://www.mdpi.com/article/10.3390/vaccines11030701/s1, Table S1: Individual Breakthrough data.

Author Contributions: M.N. and V.H. designed the study. M.T., C.G. and M.K. were responsible for data collection and management. M.T. and C.G. enrolled patients and collected blood samples. M.H. structured the survey on the REDCap platform. C.G. was responsible for the statistical analysis and preparation of tables and figures. M.K. and C.G. were responsible for data interpretation. M.K. and C.G. wrote the manuscript draft. All authors have read and agreed to the published version of the manuscript.

Funding: This research received no external funding.

Institutional Review Board Statement: Institutional Review Board (2020-556N) approved the study protocol and the study was conducted in accordance with the Declaration of Helsinki.

Informed Consent Statement: Informed written consent was obtained from each subject involved in this study.

Data Availability Statement: The data presented in this study are available on request from the corresponding author.

Acknowledgments: We thank Sultan Günaydin and Ingrid Brechtel for managing the appointments, Awat Hama and Ameli Götz for enrolling participants, Ameli Götz for aliquoting the samples, and Ruth Leiblein for performing the analysis. We also thank all the people who participated in this study.

Conflicts of Interest: The authors have declared no conflict of interest.

Abbreviations

Ab	Antibody
BMI	Body mass index
CE	Communauté Européenne
COI	Cut-off index
COVID-19	Coronavirus disease 2019
ECLIA	Electrochemiluminescence immunoassay
FDA	Food and Drug Administration
Ig	Immunoglobulin
ISO	International Organization for Standardization
qPCR	Quantitative polymerase chain reaction
RBD	Receptor-binding domain
REDCap	Research Electronic Data Capture
mRNA	Messenger ribonucleic acid
S	Spike protein
SARS-CoV-2	Severe acute respiratory syndrome coronavirus 2

References

1. Baden, L.R.; El Sahly, H.M.; Essink, B.; Kotloff, K.; Frey, S.; Novak, R.; Diemert, D.; Spector, S.A.; Rouphael, N.; Creech, C.B.; et al. Efficacy and Safety of the mRNA-1273 SARS-CoV-2 Vaccine. *N. Engl. J. Med.* **2021**, *384*, 403–416. [CrossRef]
2. Walsh, E.E.; Frenck, R.W.; Falsey, A.R.; Kitchin, N.; Absalon, J.; Gurtman, A.; Lockhart, S.; Neuzil, K.; Mulligan, M.J.; Bailey, R.; et al. Safety and Immunogenicity of Two RNA-Based COVID-19 Vaccine Candidates. *N. Engl. J. Med.* **2020**, *383*, 2439–2450. [CrossRef]
3. Zhao, J.; Zhao, S.; Ou, J.; Zhang, J.; Lan, W.; Guan, W.; Wu, X.; Yan, Y.; Zhao, W.; Wu, J.; et al. COVID-19: Coronavirus Vaccine Development Updates. *Front. Immunol.* **2020**, *11*, 602256. [CrossRef] [PubMed]
4. Franceschi, A.M.; Petrover, D.R.; McMahon, T.M.; Libman, R.B.; Giliberto, L.; Clouston, S.A.P.; Castillo, M.; Kirsch, C. Retrospective review COVID-19 vaccine induced thrombotic thrombocytopenia and cerebral venous thrombosis-what can we learn from the immune response. *Clin. Imaging* **2022**, *90*, 63–70. [CrossRef] [PubMed]

5. Horn, M.P.; Jonsdottir, H.R.; Brigger, D.; Damonti, L.; Suter-Riniker, F.; Endrich, O.; Froehlich, T.K.; Fiedler, M.; Largiadèr, C.R.; Marschall, J.; et al. Serological testing for SARS-CoV-2 antibodies in clinical practice: A comparative diagnostic accuracy study. *Allergy* **2022**, *77*, 2090–2103. [CrossRef]
6. Fernandes-Matano, L.; Salas-Lais, A.G.; Grajales-Muñiz, C.; Hernández-Ávila, M.; Garfias-Becerra, Y.O.; Rodríguez-Sepúlveda, M.C.; Segura-Sánchez, C.; Montes-Herrera, D.; Mendoza-Sánchez, D.; Angeles-Martínez, J.; et al. Longevity and Neutralizing Capacity of IgG Antibodies against SARS-CoV-2 Generated by the Application of BNT162b2, AZD1222, Convidecia, Sputnik V, and CoronaVac Vaccines: A Cohort Study in the Mexican Population. *Microbiol. Spectr.* **2023**, *11*, e02376-22. [CrossRef] [PubMed]
7. Meyer, B.; Torriani, G.; Yerly, S.; Mazza, L.; Calame, A.; Arm-Vernez, I.; Zimmer, G.; Agoritsas, T.; Stirnemann, J.; Spechbach, H.; et al. Validation of a commercially available SARS-CoV-2 serological Immunoassay. *Clin. Microbiol. Infect.* **2020**, *26*, 1386–1394. [CrossRef]
8. Bonelli, F.; Sarasini, A.; Zierold, C.; Calleri, M.; Bonetti, A.; Vismara, C.; Blocki, F.A.; Pallavicini, L.; Chinali, A.; Campisi, D.; et al. Clinical and Analytical Performance of an Automated Serological Test That Identifies S1/S2-Neutralizing IgG in COVID-19 Patients Semiquantitatively. *J. Clin. Microbiol.* **2020**, *58*, e01224-20. [CrossRef]
9. Glöckner, S.; Hornung, F.; Baier, M.; Weis, S.; Pletz, M.W.; Deinhardt-Emmer, S.; Löffler, B.; the CoNAN Study Group. Robust Neutralizing Antibody Levels Detected after Either SARS-CoV-2 Vaccination or One Year after Infection. *Viruses* **2021**, *13*, 2003. [CrossRef]
10. Jara, A.; Undurraga, E.A.; Zubizarreta, J.R.; González, C.; Pizarro, A.; Acevedo, J.; Leo, K.; Paredes, F.; Bralic, T.; Vergara, V.; et al. Effectiveness of homologous and heterologous booster doses for an inactivated SARS-CoV-2 vaccine: A large-scale prospective cohort study. *Lancet Glob. Health* **2022**, *10*, e798–e806. [CrossRef]
11. Wanlapakorn, N.; Suntronwong, N.; Phowatthanasathian, H.; Yorsaeng, R.; Vichaiwattana, P.; Thongmee, T.; Auphimai, C.; Srimuan, D.; Thatsanatorn, T.; Assawakosri, S.; et al. Safety and immunogenicity of heterologous and homologous inactivated and adenoviral-vectored COVID-19 vaccine regimens in healthy adults: A prospective cohort study. *Hum. Vaccines Immunother.* **2022**, *18*, 2029111. [CrossRef]
12. Gerhards, C.; Thiaucourt, M.; Kittel, M.; Becker, C.; Ast, V.; Hetjens, M.; Neumaier, M.; Haselmann, V. Longitudinal assessment of anti-SARS-CoV-2 antibody dynamics and clinical features following convalescence from a COVID-19 infection. *Int. J. Infect. Dis.* **2021**, *107*, 221–227. [CrossRef] [PubMed]
13. R: A Language and Environment for Statistical Computing. Available online: https://www.R-project.org (accessed on 1 October 2021).
14. Mason, T.F.D.; Whitston, M.; Hodgson, J.; Watkinson, R.E.; Lau, Y.-S.; Abdulrazeg, O.; Sutton, M. Effects of BNT162b2 mRNA vaccine on COVID-19 infection and hospitalisation amongst older people: Matched case control study for England. *BMC Med.* **2021**, *19*, 275. [CrossRef] [PubMed]
15. Mostaghimi, D.; Valdez, C.N.; Larson, H.T.; Kalinich, C.C.; Iwasaki, A. Prevention of host-to-host transmission by SARS-CoV-2 vaccines. *Lancet Infect. Dis.* **2022**, *22*, e52–e58. [CrossRef] [PubMed]
16. Fadul, A.; ELMustafa, A.; Abdelmahmuod, E.; Abdulgayoom, M.; Ali, E.; Al-warqi, A.; Al-yahary, H. COVID-19 Vaccine-Induced Cerebral Sinus Thrombosis: Coincidence vs. Cause? *Cureus* **2022**, *14*, e26436. [CrossRef]
17. Corbett, K.S.; Edwards, D.K.; Leist, S.R.; Abiona, O.M.; Boyoglu-Barnum, S.; Gillespie, R.A.; Himansu, S.; Schäfer, A.; Ziwawo, C.T.; DiPiazza, A.T.; et al. SARS-CoV-2 mRNA vaccine design enabled by prototype pathogen preparedness. *Nature* **2020**, *586*, 567–571. [CrossRef]
18. Spencer, A.J.; McKay, P.F.; Belij-Rammerstorfer, S.; Ulaszewska, M.; Bissett, C.D.; Hu, K.; Samnuan, K.; Blakney, A.K.; Wright, D.; Sharpe, H.R.; et al. Heterologous vaccination regimens with self-amplifying RNA and adenoviral COVID vaccines induce robust immune responses in mice. *Nat. Commun.* **2021**, *12*, 2893. [CrossRef]
19. Sritipsukho, P.; Khawcharoenporn, T.; Siribumrungwong, B.; Damronglerd, P.; Suwantarat, N.; Satdhabudha, A.; Chaiyakulsil, C.; Sinlapamongkolkul, P.; Tangsathapornpong, A.; Bunjoungmanee, P.; et al. Comparing real-life effectiveness of various COVID-19 vaccine regimens during the delta variant-dominant pandemic: A test-negative case-control study. *Emerg. Microbes Infect.* **2022**, *11*, 585–592. [CrossRef]
20. Larkin, H.D. New Data on Heterologous COVID-19 Vaccine Combinations. *JAMA* **2022**, *328*, 916. [CrossRef]
21. Kim, D.-I.; Lee, S.J.; Park, S.; Kim, P.; Lee, S.M.; Lee, N.; Shum, D.; Kim, D.H.; Kim, E.H. Immunogenicity and Durability of Antibody Responses to Homologous and Heterologous Vaccinations with BNT162b2 and ChAdOx1 Vaccines for COVID-19. *Vaccines* **2022**, *10*, 1864. [CrossRef] [PubMed]
22. Duarte-Salles, T.; Prieto-Alhambra, D. Heterologous vaccine regimens against COVID-19. *Lancet* **2021**, *398*, 94–95. [CrossRef] [PubMed]
23. Serebro, D. COVID-19 Vaccine Financing and Purchasing in Africa: Wherefrom the Money? 2021. Available online: https://www.cabri-sbo.org/en/blog/2021/covid-19-vaccine-financing-and-purchasing-in-africa-wherefrom-the-money (accessed on 14 February 2023).
24. Atmar, R.L.; Lyke, K.E.; Deming, M.E.; Jackson, L.A.; Branche, A.R.; El Sahly, H.M.; Rostad, C.A.; Martin, J.M.; Johnston, C.; Rupp, R.E.; et al. Homologous and Heterologous COVID-19 Booster Vaccinations. *N. Engl. J. Med.* **2022**, *386*, 1046–1057. [CrossRef] [PubMed]

25. Wanlapakorn, N.; Suntronwong, N.; Phowatthanasathian, H.; Yorsaeng, R.; Thongmee, T.; Vichaiwattana, P.; Auphimai, C.; Wongsrisang, L.; Klinfueng, S.; Sudhinaraset, N.; et al. Immunogenicity of heterologous inactivated and adenoviral-vectored COVID-19 vaccine: Real-world data. *Vaccine* **2022**, *40*, 3203–3209. [CrossRef]
26. Adjobimey, T.; Meyer, J.; Sollberg, L.; Bawolt, M.; Berens, C.; Kovačević, P.; Trudić, A.; Parcina, M.; Hoerauf, A. Comparison of IgA, IgG, and Neutralizing Antibody Responses Following Immunization with Moderna, BioNTech, AstraZeneca, Sputnik-V, Johnson and Johnson, and Sinopharm's COVID-19 Vaccines. *Front. Immunol.* **2022**, *13*, 917905. [CrossRef] [PubMed]
27. Coggins, S.A.; Laing, E.D.; Olsen, C.H.; Goguet, E.; Moser, M.; Jackson-Thompson, B.M.; Samuels, E.C.; Pollett, S.D.; Tribble, D.R.; Davies, J.; et al. Adverse Effects and Antibody Titers in Response to the BNT162b2 mRNA COVID-19 Vaccine in a Prospective Study of Healthcare Workers. *Open Forum Infect. Dis.* **2022**, *9*, ofab575. [CrossRef] [PubMed]
28. Sánchez-Saez, F.; Peiró, S.; Cuenca, L.; Vanaclocha, H.; Limón, R.; Salas, D.; Burgos, J.S.; Sánchez-Payá, J.; Meneu, R.; Díez, J.; et al. Side effects during the week after first dose vaccination with four COVID-19 vaccines. Results of the ProVaVac Survey Study with 13,837 people in Spain. *Vaccine* **2022**, *40*, 5942–5949. [CrossRef] [PubMed]
29. Atmar, R.L.; Lyke, K.E.; Deming, M.E.; Jackson, L.A.; Branche, A.R.; El Sahly, H.M.; Rostad, C.A.; Martin, J.M.; Johnston, C.; Rupp, R.E.; et al. Heterologous SARS-CoV-2 Booster Vaccinations—Preliminary Report. *medRxiv* **2021**. [CrossRef]
30. Sugiyama, K.; Suzuki, C.; Aoyama, M.; Toyota, N.; Nakagawa, N.; Shozu, M.; Nakai, K.; Iwano, K. Long-term observation of antibody titers against SARS-CoV-2 following vaccination. *Public Health Pract.* **2022**, *4*, 100297. [CrossRef]
31. Irsara, C.; Egger, A.E.; Prokop, W.; Nairz, M.; Loacker, L.; Sahanic, S.; Pizzini, A.; Sonnweber, T.; Holzer, B.; Mayer, W.; et al. Clinical validation of the Siemens quantitative SARS-CoV-2 spike IgG assay (sCOVG) reveals improved sensitivity and a good correlation with virus neutralization titers. *Clin. Chem. Lab. Med. CCLM* **2021**, *59*, 1453–1462. [CrossRef]
32. Gaebler, C.; Wang, Z.; Lorenzi, J.C.C.; Muecksch, F.; Finkin, S.; Tokuyama, M.; Cho, A.; Jankovic, M.; Schaefer-Babajew, D.; Oliveira, T.Y.; et al. Evolution of antibody immunity to SARS-CoV-2. *Nature* **2021**, *591*, 639–644. [CrossRef]
33. Zhong, W.; Roberts, A.D.; Woodland, D.L. Antibody-Independent Antiviral Function of Memory CD4+ T Cells In Vivo Requires Regulatory Signals from CD8+ Effector T Cells. *J. Immunol.* **2001**, *167*, 1379–1386. [CrossRef]
34. Zhao, J.; Alshukairi, A.N.; Baharoon, S.A.; Ahmed, W.A.; Bokhari, A.A.; Nehdi, A.M.; Layqah, L.A.; Alghamdi, M.G.; Al Gethamy, M.M.; Dada, A.M.; et al. Recovery from the Middle East respiratory syndrome is associated with antibody and T cell responses. *Sci. Immunol.* **2017**, *2*, eaan5393. [CrossRef] [PubMed]
35. Meschi, S.; Matusali, G.; Colavita, F.; Lapa, D.; Bordi, L.; Puro, V.; Leoni, B.D.; Galli, C.; Capobianchi, M.R.; Castilletti, C. Predicting the protective humoral response to a SARS-CoV-2 mRNA vaccine. *Clin. Chem. Lab. Med. CCLM* **2021**, *59*, 2010–2018. [CrossRef]
36. Kittel, M.; Eichner, R.; Aida, S.; Bode, A.; Ast, V.; Kessler, A.; Neumaier, M.; Wölfel, R.; Haselmann, V. Results of a European-Wide External Quality Assessment (EQA) Scheme for Serological Detection of Anti-SARS-CoV-2 (CoVimm)—Pitfalls of Routine Application. *Viruses* **2022**, *14*, 1662. [CrossRef]
37. Roche Diagnostics Deutschland GmbH. ElecsysT Anti-SARS-CoV-2 S. Available online: https://assets.cwp.roche.com/f/94122/x/37b1d9536f/factsheet-elecsys-anti-sars-cov-2-s_v3.pdf (accessed on 21 February 2023).
38. Goel, R.R.; Painter, M.M.; Apostolidis, S.A.; Mathew, D.; Meng, W.; Rosenfeld, A.M.; Lundgreen, K.A.; Reynaldi, A.; Khoury, D.S.; Pattekar, A.; et al. mRNA vaccines induce durable immune memory to SARS-CoV-2 and variants of concern. *Science* **2021**, *374*, abm0829. [CrossRef] [PubMed]
39. Tarke, A.; Sidney, J.; Methot, N.; Yu, E.D.; Zhang, Y.; Dan, J.M.; Goodwin, B.; Rubiro, P.; Sutherland, A.; Wang, E.; et al. Impact of SARS-CoV-2 variants on the total CD4+ and CD8+ T cell reactivity in infected or vaccinated individuals. *Cell Rep. Med.* **2021**, *2*, 100355. [CrossRef] [PubMed]
40. Naranbhai, V.; Nathan, A.; Kaseke, C.; Berrios, C.; Khatri, A.; Choi, S.; Getz, M.A.; Tano-Menka, R.; Ofoman, O.; Gayton, A.; et al. T cell reactivity to the SARS-CoV-2 Omicron variant is preserved in most but not all prior infected and vaccinated individuals. *medRxiv* **2022**. [CrossRef]
41. Tartof, S.Y.; Slezak, J.M.; Fischer, H.; Hong, V.; Ackerson, B.K.; Ranasinghe, O.N.; Frankland, T.B.; Ogun, O.A.; Zamparo, J.M.; Gray, S.; et al. Effectiveness of mRNA BNT162b2 COVID-19 vaccine up to 6 months in a large integrated health system in the USA: A retrospective cohort study. *Lancet* **2021**, *398*, 1407–1416. [CrossRef]
42. Takita, M.; Yoshida, T.; Tsuchida, T.; Nakagama, Y.; Kido, Y.; Suzuki, S.; Imamura, M.; Kawahata, K.; Shimizu, G.; Yoshida, H.; et al. Low SARS-CoV-2 antibody titers may be associated with poor clinical outcomes for patients with severe COVID-19. *Sci. Rep.* **2022**, *12*, 9147. [CrossRef]
43. Amjadi, M.F.; O'Connell, S.E.; Armbrust, T.; Mergaert, A.M.; Narpala, S.R.; Halfmann, P.J.; Bashar, S.J.; Glover, C.R.; Heffron, A.S.; Taylor, A.; et al. Specific COVID-19 Symptoms Correlate with High Antibody Levels against SARS-CoV-2. *ImmunoHorizons* **2021**, *5*, 466–476. [CrossRef]

vaccines

MDPI

Brief Report

Measuring Variant-Specific Neutralizing Antibody Profiles after Bivalent SARS-CoV-2 Vaccinations Using a Multivariant Surrogate Virus Neutralization Microarray

David Niklas Springer [1], Eva Höltl [2], Katja Prüger [1], Elisabeth Puchhammer-Stöckl [1], Judith Helene Aberle [1], Karin Stiasny [1,*,†] and Lukas Weseslindtner [1,*,†]

[1] Center for Virology, Medical University of Vienna, A-1090 Vienna, Austria; katja.prueger@meduniwien.ac.at (K.P.); elisabeth.puchhammer@meduniwien.ac.at (E.P.-S.); judith.aberle@meduniwien.ac.at (J.H.A.)

[2] Center for Public Health, Medical University of Vienna, A-1090 Vienna, Austria; eva.a.hoeltl@meduniwien.ac.at

* Correspondence: karin.stiasny@meduniwien.ac.at (K.S.); lukas.weseslindtner@meduniwien.ac.at (L.W.)

† These authors contributed equally to this work.

Abstract: The capability of antibodies to neutralize different SARS-CoV-2 variants varies among individuals depending on the previous exposure to wild-type or Omicron-specific immunogens by mono- or bivalent vaccinations or infections. Such profiles of neutralizing antibodies (nAbs) usually have to be assessed via laborious live-virus neutralization tests (NTs). We therefore analyzed whether a novel multivariant surrogate-virus neutralization test (sVNT) (adapted from a commercial microarray) that quantifies the antibody-mediated inhibition between the receptor angiotensin-converting enzyme 2 (ACE2) and variant-specific receptor-binding domains (RBDs) can assess the neutralizing activity against the SARS-CoV-2 wild-type, and Delta Omicron BA.1, BA.2, and BA.5 subvariants after a booster with Omicron-adapted bivalent vaccines in a manner similar to live-virus NTs. Indeed, by using the live-virus NTs as a reference, we found a significant correlation between the variant-specific NT titers and levels of ACE2-RBD binding inhibition ($p < 0.0001$, $r \leq 0.78$ respectively). Furthermore, the sVNTs identified higher inhibition values against BA.5 and BA.1 in individuals vaccinated with Omicron-adapted vaccines than in those with monovalent wild-type vaccines. Our data thus demonstrate the ability of sVNTs to detect variant-specific nAbs following a booster with bivalent vaccines.

Keywords: SARS-CoV-2; Omicron; bivalent vaccination; neutralization; assay; surrogate

1. Introduction

With the emergence of immune-evasive variants of the severe acute respiratory syndrome coronavirus type 2 (SARS-CoV-2), such as Delta and Omicron, the assessment of vaccine-induced humoral immunity has become increasingly complex. In addition, due to multiple previous infections with earlier variants and the sequential application of mono- and bivalent vaccines, the neutralizing activity of serum antibodies against those variants may vary considerably among individuals [1–7]. Consequently, the profiles of variant-specific antibodies cannot be adequately assessed by commercial assays that only measure antibody binding to the wild-type (WT) spike protein [8].

Live-virus neutralization tests (NTs) are considered the gold standard for analyzing neutralizing antibodies (nAbs) against multiple variants, including the Omicron subvariants (e.g., BA.1, BA.2, and BA.5) [1–3,5,6,9]. However, these tests are work intensive, slow, and require laboratories with high biosafety levels. Therefore, we and others evaluated variant-adapted surrogate-virus neutralization tests (sVNTs) that quantify the antibody-mediated binding inhibition of the viral receptor-binding domain (RBD) to its cellular receptor [angiotensin-converting enzyme 2 (ACE2)] [8,10,11].

In that regard, microarrays are particularly suitable as sVNT because the combination of multiple RBD proteins of different SARS-CoV-2 variants can be plotted as the target antigens into a single well, providing simultaneous measurement of antibody-mediated inhibition of ACE2 binding to the RBDs, which acts as a substitute for multiple variant-specific virus neutralization assays [8,10,11].

However, nAb titers against different SARS-CoV-2 variants span a wide concentration range, which are assessed by using dilution series of the serum samples in live-virus NTs. In contrast, in sVNTs, only a single dilution of the sample is used to measure inhibition. The concentration of antibodies against the RBD of one specific variant in one sample may already reach saturation and cause complete inhibition of RBD-ACE2 binding, whereas antibodies against the RBD of another variant may still be in an optimal quantification range.

Thus, multivariant sVNTs require thorough evaluation with multiple serial serum dilutions and correlation between the variant-specific measurements with the respective titers of live-virus NTs in order to identify correct cut-off concentrations within the sVNT's linear test range.

To this end, we first evaluated a novel multivariant sVNT microarray using multiple variant-specific live-virus NTs as the reference. Then, we analyzed whether the increase in the breadth of variant-specific nAbs following Omicron-adapted vs. monovalent wild-type booster was similarly quantifiable by the sVNT as compared with live-virus NTs [6,12,13].

2. Materials and Methods

2.1. Sample Cohort

The study included 65 serum samples from our serum biobank of both SARS-CoV-2 vaccinated and infected individuals, as well as 30 additional pre-pandemic control samples from healthy individuals. The samples were drawn 15–40 days after the last vaccination or infection.

The first cohort included samples from individuals hospitalized for a SARS-CoV-2 wild-type infection in the early pandemic before vaccines were available (n = 18, median age: 48 years (y), range: 22–77 y). The second and third cohorts included sera drawn approximately one month after the individuals' respective second (n = 11, median age: 40 y, range 20–81) or third (n = 14, median age: 46 y, range: 27–64 y) mRNA monovalent wild-type vaccination. The fourth and fifth cohorts included individuals who received three monovalent wild-type vaccinations plus one bivalent Omicron-adapted BA.1/WT booster vaccination (n = 9, median age: 52 years, range: 38–62 y) or BA.5/WT (n = 13, median age: 50 y, range: 32–57 y). Only one sample was included from each individual. Detailed cohort characteristics are provided in Tables S1 and S2.

All vaccinated individuals reported that they had not been infected with SARS-CoV-2, and anti-nucleocapsid antibodies were not detected in those cohorts (Anti-NC-IgG-ELISA, Euroimmun, Lübeck, Germany). None of the individuals reported any immunosuppressive therapy or condition. Data from the live-virus NTs were previously reported and were included only as reference for the sVNT evaluated in this study [2,6].

2.2. Live-Virus NTs

As previously described, each serum was tested for nAbs against the WT (with the D614G mutation), Delta, and Omicron BA.1, BA.2, and BA.5 SARS-CoV-2 subvariants [2,6].

In brief, SARS-CoV-2 variants were isolated from respiratory swabs taken from infected individuals using VeroE6 or VeroE6 TMPRSS2 cells (kindly provided by Anna Ohradanova-Repic). Virus sequences were determined using next-generation sequencing (Illumina, Minato, Tokyo) and uploaded to the GISAID database (WT, B.1.1 with the D614G mutation EPI_ISL_438123; Delta, B.1.617.2-like, sub-lineage AY.122:EPI_ISL_4172121; Omicron, B.1.1.529 + BA.*, sub-lineage BA.1.17:EPI_ISL_9110894; Omicron, B.1.1.529 + BA.*, sublineage BA.2:EPI_ISL_11110193; and Omicron, B.1.1.529 + BA.*, sub-lineage BA.5.3:EPI_ISL_15982848). The lineages were determined using Pango 4.1.3, Pango-data v.1.17.

Serum samples were serially diluted (two-fold) and incubated with 50–100 TCID50 SARS-CoV-2 for one hour at 37 °C. The dilution series ranged from 1:10 to 1:10,240. This mixture was then applied to VeroE6 cells and incubated for 3–5 days at 37 °C. Then, the cytopathic effect (CPE) was microscopically assessed, and the final titers were calculated as the inverse of the last titration at which CPE was prevented by serum-neutralizing activity. In all samples, the NTs were performed in duplicates.

2.3. Multivariant Surrogate-Virus Neutralization Test

The multivariant surrogate sVNT was performed using a similar protocol as in a previous study [8]. In brief, the basic framework of this assay is a commercial SARS-CoV-2 VoC ViraChip® IgG microarray developed by Viramed (Planegg, Germany). However, in contrast to the standard version of the commercial microarray, the manufacturer plotted the RBD proteins of the SARS-CoV-2 wild-type (WT), the Delta variant, and the Omicron subvariants BA.1, BA.2, and BA.5 in triplets as the target antigens on the solid phase of each well, spatially separated into microspots.

After incubation of the wells with the diluted serum samples, recombinant ACE2 bound to alkaline phosphatase (ACE2-AP; also obtained from Viramed, Planegg, Germany) was added to each well, which could only bind to the RBD proteins in inverse correlation to the levels of neutralizing antibodies against the specific SARS-CoV-2 variants and Omicron subvariants present in the samples (until complete inhibition was achieved at the specific dilution).

Finally, after a washing step, the bound ACE2-AP (if present) was made visible via a colorimetric reaction of a chromogen substrate and assessed using the Viramed plate reader. For each dilution step performed with each sample, the variant-specific ACE2-RBD binding inhibition was calculated as a percentage of the reduction in the inhibited color reaction relative to the maximum uninhibited color reaction obtained from a negative buffer control sample (incubation without serum but with ACE2-AP) after subtraction of the background (i.e., 100% represents complete inhibition of the binding).

Since neutralizing antibody titers against multiple variants in different serum samples covered a wide range (1:10 up to 1:2560), no single serum dilution was suitable to obtain valid ACE2-RBD inhibition values for all variants simultaneously. Therefore, each sample was tested in five serial two-fold dilutions, starting from 1:20 up to 1:320. In some instances, additional dilutions were required (up to 1:2560). Next, all values representing a total inhibition (=100%) were discarded. Additionally, all values below 20% were discarded for all dilutions except for the 1:20 one, since the preliminary data showed that this was below the assay's linear range (and the error would have been amplified due to the dilutions). Then, we accounted for the additional dilution steps in the quantification and multiplied the corresponding values with the dilution factor relative to the starting dilution of 1:20. Finally, the result was the median of all valid dilution-corrected inhibition values for each variant. Supplementary Figure S1 provides a schematic overview of the described calculations.

2.4. Statistical Analysis

Statistical analyses were performed and figures were created with R version 4.2.0. Correlations between the NT titer and the sVNT results for all tested virus variants were calculated and reported as Spearman's r (WT, Delta, BA.1, BA.2, and BA.5). Additionally, a Bland–Altman analysis was performed to assess the level of agreement between the two methods at different antibody concentrations [14].

Furthermore, cross-neutralizing activity was assessed by calculating the ratio of the WT neutralizing activity (log2 of the NT titer or sVNT result) to the variant-specific (Delta, BA.1, BA.2, and BA.5) NT titer or sVNT result (also log2 transformed) for each variant and each cohort. Therefore, NT ratios or sVNT ratios close to one represent strong cross-neutralization, whereas a value between zero and one represents the dominance of the WT activity over the variant-neutralizing activity. Finally, NT-ratios for each variant were compared between the cohorts using pairwise Wilcoxon ranked tests (multiplicity-adjusted for the cohorts using the Bonferroni–Holm method). Alpha was set to 0.05. Possible sVNT cut-off values that represent a positive NT titer (≥10) were calculated using ROC analysis and Youden's index for each variant.

3. Results

3.1. Correlation of the Multivariant sVNT with Variant-Specific Live-Virus NTs

In this study, we analyzed whether differences in the variant-specific neutralizing activity of antibodies induced by different vaccination regimes (mono- and bivalent SARS-CoV-2 mRNA vaccinations) could be assessed with a dilution-corrected multivariant sVNT similar to live-virus NTs.

Therefore, we first analyzed the correlation of variant-specific sVNT values to the respective NT titers assessed by live-virus NTs (Figure 1A). We observed a robust correlation for all SARS-CoV-2 variants, with a higher correlation for WT-, Delta-, and BA.2-specific neutralization than for Omicron BA.1 and BA.5 (Spearman: WT r = 0.87, Delta r = 0.90, BA.1 r = 0.79, BA.2 r = 0.87, and BA.5 r = 0.78). Then, we calculated cut-off values for the sVNT that correspond to positive NT titers (≥10) against the respective variants (Figure 1A, Supplementary Table S3).

Finally, we also conducted a Bland–Altman analysis [14] (Supplementary Figure S2). The results of this analysis suggested a certain degree of non-linearity in the correlation between the two methods, as they indicated a trend that the sVNT results underestimated higher NT titer results and overestimated lower NT titers, although most of the sera were still within the bounds of the 95% agreement intervals.

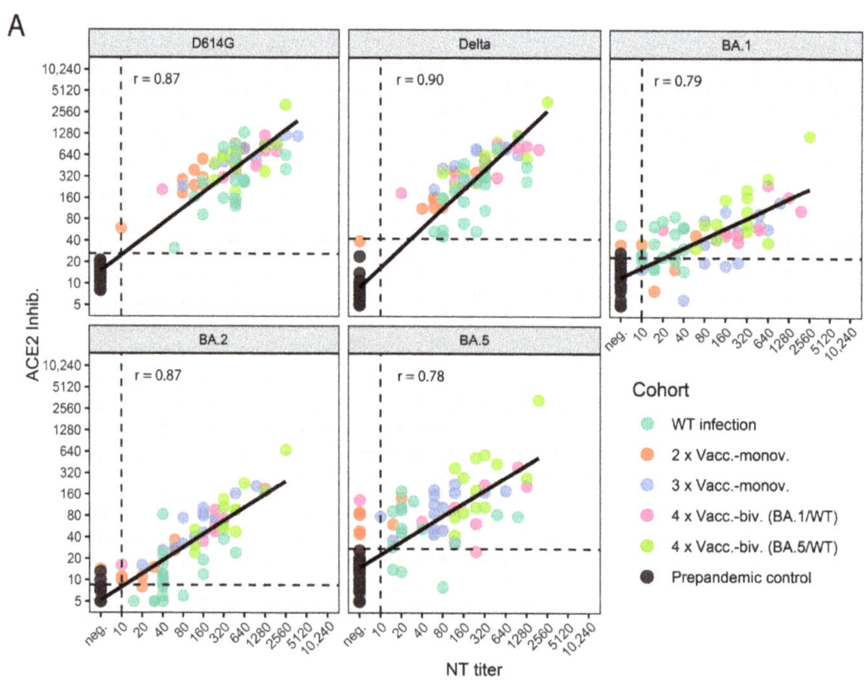

Figure 1. *Cont.*

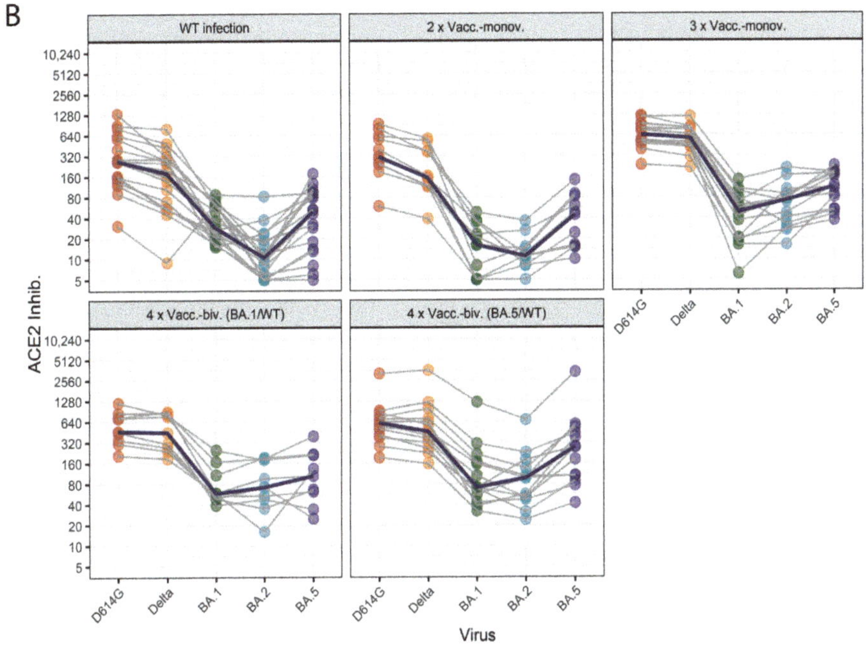

Figure 1. Correlation of sVNT values to NT titers against D614G (WT), Delta, and Omicron BA.1, BA.2, and BA.5. (**A**) Each dot represents a single serum (n = 95). The *x*-axis shows live-virus NT titers; the *y*-axis shows the neutralizing activity assessed by the sVNT ACE2-Inhib. (ACE2-RBD inhibition). The cohorts are color coded as indicated. Spearman's r is indicated in the top-left corner of each panel. The dotted lines indicate the cutoff values of the respective tests. The black line indicates the linear regression slope. (**B**) Neutralization profiles of vaccinated and infected cohorts. Each serum is represented by five dots (one for each variant), connected by a grey line. The bold blue lines indicate the median titer for each variant. WT infection: sera drawn from unvaccinated patients hospitalized for a wild-type SARS-CoV-2 infection (n = 18). 2 × Vacc.-monov./3 × Vacc.-monov. Sera drawn approximately one month after two (n = 11)/three vaccinations (n = 14) with a monovalent wild-type vaccine. 4 × Vacc.-biv. (BA.1/WT)/4 × Vacc.-biv. (BA.5/WT): Sera were drawn after a booster with a bivalent Omicron-adapted vaccine containing mRNA encoding for the Spike of either BA.1 and WT (n = 9) or BA.5 and WT (n = 13).

3.2. *Profiles of Neutralizing Activity after SARS-CoV-2 Wild-Type Infection and Vaccinations*

Next, we analyzed neutralization profiles for the cohorts of WT-infected and vaccinated individuals (Figure 1B). The cohorts of WT-infected individuals and two-times vaccinated subjects showed moderate WT- and Delta-specific and weak Omicron-specific neutralizing activity. In contrast, higher levels of nAbs and a less pronounced difference between WT- and Omicron-specific neutralizing activities were observed for three-times vaccinated individuals and individuals who received a bivalent (WT/BA.1 or WT/BA.5) booster (fourth dose) vaccination. Indeed, median nAb levels against Omicron BA.2 and BA.5 were significantly higher in individuals who received a bivalent booster dose (WT/BA.1 or WT/BA5) as compared to subjects after WT-infection and individuals vaccinated only twice ($p < 0.001$ respectively).

Notably, a BA.5-specific RBD-ACE2-binding inhibition, as assessed by the sVNT, was higher in all cohorts than for BA.2 and BA.1, which was inverse to the relation measured by the NT.

3.3. Effect of Mono- and Bivalent Booster Vaccinations on the Neutralizing Activity

Finally, we analyzed the effect of monovalent versus bivalent booster vaccinations on the breadth of the neutralizing activity. To this end, we calculated the ratios of Delta, BA.1, BA.2, and BA.5 to the WT surrogate neutralization results, respectively. As shown in Figure 2, variant cross-neutralization (as expressed by ratios closer to one) was increased after the third monovalent vaccination compared to only two-times vaccinated or WT-infected individuals. Indeed, the three-times vaccinated individuals had significantly higher Omicron BA.2- and BA.5-neutralizing activities than those after two vaccinations, with median ratios closer to one (for *p*-values, see Supplementary Table S4).

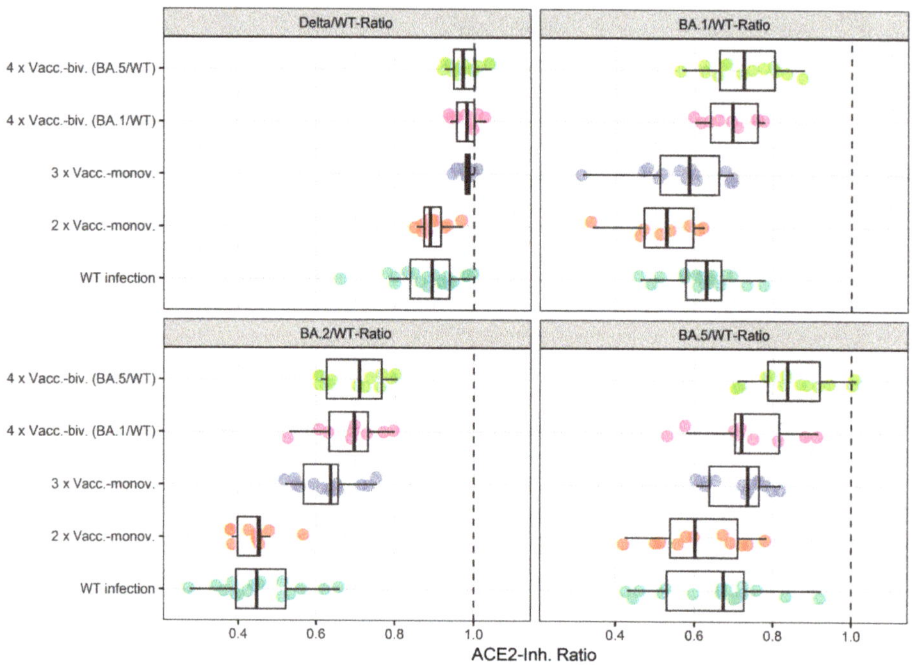

Figure 2. Neutralizing activity of the vaccinated and infected cohorts. Cross-neutralizing activity was assessed by calculating the ratio of sVNT values (ACE2-Inhib.) of Delta, BA.1, BA.2, and BA.5 divided by the ACE2-Inhibition against WT, respectively. A single dot represents one serum. The median is shown as a solid line, and boxes and whiskers represent the quartiles and range, respectively. ACE2-Inhib. (ACE2-RBD Inhibition). WT infection: sera drawn from unvaccinated patients hospitalized for a wild-type infection (n = 18). 2 × Vacc.-monov./3 × Vacc.-monov. Sera drawn approximately one month after two (n = 11)/three vaccinations (n = 14) with a monovalent wild-type vaccine. 4 × Vacc.-biv. (BA.1/WT)/4 × Vacc.-biv. (BA.5/WT): Sera were drawn after a booster with a bivalent Omicron-adapted vaccine containing mRNA encoding for the Spike of either BA.1 and WT (n = 9) or BA.5 and WT (n = 13). The cohorts are color coded as in Figure 1A.

A further increase in the neutralization breadth was observed in individuals with a bivalent vaccination compared to those with monovalent vaccinations (Figure 2). Notably, the bivalent BA.5/WT-vaccinated individuals had significantly higher Omicron BA.1 and BA.5 cross-neutralizing activity than all other cohorts except for the bivalent WT/BA.1-vaccinated individuals who exhibited higher Delta and BA.2 cross-neutralization than the two-times vaccinated and wild-type-infected cohorts. Similarly, the BA.1/WT bivalent-vaccinated cohort displayed more Delta and BA.2 cross-neutralization than the WT infected and two-times vaccinated individuals, as well as more BA.1 cross-neutralization than the two- or three-times monovalent vaccinated cohorts.

4. Discussion

In this study, we evaluated whether a novel multivariant sVNT that is based on a commercial microarray and quantifies the antibody-mediated inhibition of binding between ACE2 and variant-specific RBD proteins can identify differences in the cross-neutralization after monovalent versus bivalent booster vaccinations similarly to variant-specific live-virus NTs.

Indeed, in earlier studies that used live-virus NTs, individuals vaccinated three times with monovalent mRNA WT vaccines showed a significantly broader neutralizing activity than WT-infected or twice-vaccinated subjects [15,16]. However, no further improvement in the cross-neutralizing activity was found after a fourth monovalent wild-type vaccine dose in comparable cohorts [6,17–20]. Hence, bivalent vaccines containing mRNA encoding both the SARS-CoV-2 WT, as well as Omicron BA.1 or BA.5, have been developed to increase the (cross)-neutralizing antibody activity against circulating and possibly future variants [21].

Thus, measuring cross-neutralizing antibody profiles after bivalent booster vaccinations may be essential, e.g., for vaccine efficacy studies and routine diagnostics. However, this is limited by the laborious setup of performing multiple different variant-specific live-virus NTs with the same samples (e.g., [2–6]). Using this labor-intensive approach, our group previously reported that bivalent vaccination moderately increased neutralizing activity against the respective Omicron variant applied with the vaccine [6].

In the present study, we used the multivariant sVNT and similarly demonstrated that the bivalent BA.1/WT-boosted cohort displayed increased cross-neutralization of BA.1 compared to the individuals three-times vaccinated with monovalent vaccines. Also, similar to the live-virus NTs, individuals vaccinated with BA.5/WT-adapted vaccines showed an increased cross-neutralization of both BA.1 and BA.5. Yet, in both bivalent vaccinated cohorts, wild-type titers were generally higher than Omicron titers, which has been attributed to immune imprinting to WT [22].

Wild-type-infected and twice-vaccinated individuals showed similar cross-neutralization (Figure 2), as also seen in earlier studies [2,3,23]. Unexpectedly, the sVNT detected stronger cross-neutralization against BA.1 in the WT-infected cohort than in individuals vaccinated twice, but this may be related to the large number of individuals with low nAb levels against BA.1 in the latter cohort. Test-inherent differences, in the sensitivity of detecting nAbs against different Omicron subvariants by the sVNT, might also contribute to this difference that we observed. Furthermore, the sVNT generally measured a higher neutralizing activity against BA.5 than against BA.2, which was inverse to NT titer results, as reported earlier [6]. However, the relation of neutralizing activity against different variants may also vary between different neutralization assays (e.g., depending on the kinetics of virus growth), as greater BA.5 than BA.2 NT titers in vaccinated individuals have also been reported previously [5].

If antibody neutralization against different Omicron subvariants is not assessed with a similar sensitivity by the sVNT microarray (e.g., due to differences in the concentration or conformation of the RBDs), this might pose a limitation, complicating a comparison of quantitative test results by the sVNT with the respective NT titers. While the issue of assay oversaturation at high antibody concentrations can be overcome by performing dilution series, such possible test-inherent differences in sensitivity would require fine tuning the microarray platform, such as applying different antigen concentrations. However, in this study, we evaluated a commercially available immunoassay that served as a basis for the sVNT, wherefore such adaptations were beyond the scope of this study. Another limitation of this study, particularly with regard to comparing different vaccination regimens, is the modest sample size for the different cohorts. However, this study mainly served as a proof-of-concept of which results similar to those obtained by live-virus NTs [5,6,12,13] are also reproducible with a lower-effort sVNT.

Overall, the sVNTs gave very similar results to the live-virus NTs, which enables an analysis of cross-neutralizing antibodies against different SARS-CoV-2 variants in seroprevalence studies involving large parts of the population. Indeed, due to individual vaccination and infection histories, the profiles of nAbs against different variants and Omicron subvariants in the population will be highly heterogeneous, as has been demonstrated by Zaballa and colleagues [10].

For this reason, future studies investigating the benefits of adapting COVID-19 vaccines to newly circulating strains (such as XBB) might measure antibody profiles in large study cohorts using sVNTs that allow for a higher and faster turnover than live-virus NTs.

Thus, our data evaluating a sVNT microarray, including multiple dilution steps to identify optimal cut-offs with variant-specific live-virus NTs as the reference, may be relevant for applying such immunoassays in vaccine efficacy studies.

5. Conclusions

In this study, we demonstrate that not only laborious variant-specific live-virus NTs but also a more feasible multi-variant sVNT can identify distinct profiles of nAbs against different SARS-CoV-2 variants and Omicron sub-variants. Thus, the multivariant sVNT was able to assess the effect of Omicron-adapted SARS-CoV-2 mRNA vaccines on the breadth of nAbs against different variants.

Supplementary Materials: The following supporting information can be downloaded at: https://www.mdpi.com/article/10.3390/vaccines12010094/s1: Figure S1: Schematic overview of the sVNT result calculation; Figure S2: Bland–Altman analysis; Table S1: Cohort characteristics; Table S2: Vaccination history; Table S3: sVNT cutoff values obtained by ROC and Youden's Index; Table S4: Cohort-wise comparisons of variant/WT ratios (Wilcoxon test).

Author Contributions: All authors have made substantial contributions to the study, have approved the submitted version of the manuscript, and have agreed to be personally accountable for the author's own contributions and for ensuring that questions related to the accuracy or integrity of any part of the work. Conceptualization, D.N.S. and L.W.; Methodology, D.N.S., K.P. and K.S.; Software, D.N.S.; Validation, D.N.S., K.P. and K.S.; Formal Analysis, D.N.S. and L.W.; Investigation, D.N.S., E.H., K.P., K.S. and L.W.; Resources, E.H. and E.P.-S.; Data Curation, D.N.S. and K.P.; Writing—Original Draft Preparation, D.N.S. and L.W.; Writing—Review and Editing, D.N.S., J.H.A., K.S. and L.W.; Visualization, D.N.S.; Supervision, L.W.; Project Administration, L.W.; Funding Acquisition: L.W. All authors have read and agreed to the published version of the manuscript.

Funding: This research received no external funding.

Institutional Review Board Statement: The study was conducted according to the guidelines of the Declaration of Helsinki and following national legislation and institutional requirements. The study was approved by the local ethics committee of the Medical University of Vienna, Austria (EK 1035/2016, EK 1513/2016, EK 1926/2020, EK 1291/2021). All samples were stored in the biobank according to established protocols approved by the local ethics committee.

Informed Consent Statement: All participants consented to antibody testing at the Center for Virology. As the study used only anonymized residual material from routine diagnostics, the ethics committee waived the necessity for written consent.

Data Availability Statement: Data from this study has not been publicly archived.

Acknowledgments: We thank Jutta Hutecek, Christina Tratberger, and Elke Peil for their excellent technical assistance.

Conflicts of Interest: The authors declare no conflicts of interest.

References

1. Hachmann, N.P.; Miller, J.; Collier, A.Y.; Ventura, J.D.; Yu, J.; Rowe, M.; Bondzie, E.A.; Powers, O.; Surve, N.; Hall, K.; et al. Neutralization Escape by SARS-CoV-2 Omicron Subvariants BA.2.12.1, BA.4, and BA.5. *N. Engl. J. Med.* **2022**, *387*, 86–88. [CrossRef]
2. Medits, I.; Springer, D.N.; Graninger, M.; Camp, J.V.; Höltl, E.; Aberle, S.W.; Traugott, M.T.; Hoepler, W.; Deutsch, J.; Lammel, O.; et al. Different Neutralization Profiles After Primary SARS-CoV-2 Omicron BA.1 and BA.2 Infections. *Front. Immunol.* **2022**, *13*, 946318. [CrossRef] [PubMed]
3. Rössler, A.; Knabl, L.; von Laer, D.; Kimpel, J. Neutralization Profile after Recovery from SARS-CoV-2 Omicron Infection. *N. Engl. J. Med.* **2022**, *386*, 1764–1766. [CrossRef] [PubMed]
4. Rössler, A.; Netzl, A.; Knabl, L.; Schäfer, H.; Wilks, S.H.; Bante, D.; Falkensammer, B.; Borena, W.; von Laer, D.; Smith, D.J.; et al. BA.2 and BA.5 omicron differ immunologically from both BA.1 omicron and pre-omicron variants. *Nat. Commun.* **2022**, *13*, 7701. [CrossRef] [PubMed]
5. Rössler, A.; Netzl, A.; Knabl, L.; Bante, D.; Wilks, S.H.; Borena, W.; von Laer, D.; Smith, D.J.; Kimpel, J. Characterizing SARS-CoV-2 neutralization profiles after bivalent boosting using antigenic cartography. *Nat. Commun.* **2023**, *14*, 5224. [CrossRef] [PubMed]
6. Springer, D.N.; Bauer, M.; Medits, I.; Camp, J.V.; Aberle, S.W.; Burtscher, C.; Höltl, E.; Weseslindtner, L.; Stiasny, K.; Aberle, J.H. Bivalent COVID-19 mRNA booster vaccination (BA.1 or BA.4/BA.5) increases neutralization of matched Omicron variants. *npj Vaccines* **2023**, *8*, 110. [CrossRef]
7. Dejnirattisai, W.; Huo, J.; Zhou, D.; Zahradník, J.; Supasa, P.; Liu, C.; Duyvesteyn, H.M.E.; Ginn, H.M.; Mentzer, A.J.; Tuekprakhon, A.; et al. SARS-CoV-2 Omicron-B.1.1.529 leads to widespread escape from neutralizing antibody responses. *Cell* **2022**, *185*, 467–484.e415. [CrossRef] [PubMed]
8. Springer, D.N.; Traugott, M.; Reuberger, E.; Kothbauer, K.B.; Borsodi, C.; Nägeli, M.; Oelschlägel, T.; Kelani, H.; Lammel, O.; Deutsch, J.; et al. A Multivariant Surrogate Neutralization Assay Identifies Variant-Specific Neutralizing Antibody Profiles in Primary SARS-CoV-2 Omicron Infection. *Diagnostics* **2023**, *13*, 2278. [CrossRef]
9. Cao, Y.; Yisimayi, A.; Jian, F.; Song, W.; Xiao, T.; Wang, L.; Du, S.; Wang, J.; Li, Q.; Chen, X.; et al. BA.2.12.1, BA.4 and BA.5 escape antibodies elicited by Omicron infection. *Nature* **2022**, *608*, 593–602. [CrossRef]
10. Zaballa, M.E.; Perez-Saez, J.; de Mestral, C.; Pullen, N.; Lamour, J.; Turelli, P.; Raclot, C.; Baysson, H.; Pennacchio, F.; Villers, J.; et al. Seroprevalence of anti-SARS-CoV-2 antibodies and cross-variant neutralization capacity after the Omicron BA.2 wave in Geneva, Switzerland: A population-based study. *Lancet Reg. Health Eur.* **2023**, *24*, 100547. [CrossRef]
11. Santos da Silva, E.; Servais, J.Y.; Kohnen, M.; Arendt, V.; Staub, T.; CON-VINCE Consortium; CoVaLux Consortium; Krüger, R.; Fagherazzi, G.; Wilmes, P.; et al. Validation of a SARS-CoV-2 Surrogate Neutralization Test Detecting Neutralizing Antibodies against the Major Variants of Concern. *Int. J. Mol. Sci.* **2023**, *24*, 14965. [CrossRef]
12. Collier, A.Y.; Miller, J.; Hachmann, N.P.; McMahan, K.; Liu, J.; Bondzie, E.A.; Gallup, L.; Rowe, M.; Schonberg, E.; Thai, S.; et al. Immunogenicity of BA.5 Bivalent mRNA Vaccine Boosters. *N. Engl. J. Med.* **2023**, *388*, 565–567. [CrossRef] [PubMed]
13. Chalkias, S.; Harper, C.; Vrbicky, K.; Walsh, S.R.; Essink, B.; Brosz, A.; McGhee, N.; Tomassini, J.E.; Chen, X.; Chang, Y.; et al. A Bivalent Omicron-Containing Booster Vaccine against COVID-19. *N. Engl. J. Med.* **2022**, *387*, 1279–1291. [CrossRef] [PubMed]
14. Altman, D.G.; Bland, J.M. Measurement in Medicine: The Analysis of Method Comparison Studies. *J. R. Stat. Soc. Ser. D Stat.* **1983**, *32*, 307–317. [CrossRef]
15. Gruell, H.; Vanshylla, K.; Tober-Lau, P.; Hillus, D.; Schommers, P.; Lehmann, C.; Kurth, F.; Sander, L.E.; Klein, F. mRNA booster immunization elicits potent neutralizing serum activity against the SARS-CoV-2 Omicron variant. *Nat. Med.* **2022**, *28*, 477–480. [CrossRef] [PubMed]
16. Quandt, J.; Muik, A.; Salisch, N.; Lui, B.G.; Lutz, S.; Krüger, K.; Wallisch, A.K.; Adams-Quack, P.; Bacher, M.; Finlayson, A.; et al. Omicron BA.1 breakthrough infection drives cross-variant neutralization and memory B cell formation against conserved epitopes. *Sci. Immunol.* **2022**, *7*, eabq2427. [CrossRef]
17. Link-Gelles, R.; Levy, M.E.; Gaglani, M.; Irving, S.A.; Stockwell, M.; Dascomb, K.; DeSilva, M.B.; Reese, S.E.; Liao, I.C.; Ong, T.C.; et al. Effectiveness of 2, 3, and 4 COVID-19 mRNA Vaccine Doses Among Immunocompetent Adults During Periods when SARS-CoV-2 Omicron BA.1 and BA.2/BA.2.12.1 Sublineages Predominated—VISION Network, 10 States, December 2021–June 2022. *Morb. Mortal. Wkly. Rep.* **2022**, *71*, 931–939. [CrossRef]
18. Regev-Yochay, G.; Gonen, T.; Gilboa, M.; Mandelboim, M.; Indenbaum, V.; Amit, S.; Meltzer, L.; Asraf, K.; Cohen, C.; Fluss, R.; et al. Efficacy of a Fourth Dose of COVID-19 mRNA Vaccine against Omicron. *N. Engl. J. Med.* **2022**, *386*, 1377–1380. [CrossRef]
19. Chong, Y.; Goto, T.; Watanabe, H.; Tani, N.; Yonekawa, A.; Ikematsu, H.; Shimono, N.; Tanaka, Y.; Akashi, K. Achievement of sufficient antibody response after a fourth dose of wild-type SARS-CoV-2 mRNA vaccine in nursing home residents. *Immun. Inflamm. Dis.* **2023**, *11*, e962. [CrossRef]
20. Xie, X.; Zou, J.; Kurhade, C.; Liu, M.; Ren, P.; Pei-Yong, S. Neutralization of SARS-CoV-2 Omicron sublineages by 4 doses of the original mRNA vaccine. *Cell Rep.* **2022**, *41*, 111729. [CrossRef]
21. Krammer, F.; Ellebedy, A.H. Variant-adapted COVID-19 booster vaccines. *Science* **2023**, *382*, 157–159. [CrossRef] [PubMed]

22. Carreño, J.M.; Singh, G.; Simon, V.; Krammer, F. Bivalent COVID-19 booster vaccines and the absence of BA.5-specific antibodies. *Lancet. Microbe* **2023**, *8*, E569. [CrossRef]
23. Wang, W.; Lusvarghi, S.; Subramanian, R.; Epsi, N.J.; Wang, R.; Goguet, E.; Fries, A.C.; Echegaray, F.; Vassell, R.; Coggins, S.A.; et al. Antigenic cartography of well-characterized human sera shows SARS-CoV-2 neutralization differences based on infection and vaccination history. *Cell Host Microbe* **2022**, *30*, 1745–1758.e7. [CrossRef] [PubMed]

vaccines MDPI

Article

Clinical Characteristics of Mild Patients with Breakthrough Infection of Omicron Variant in China after Relaxing the Dynamic Zero COVID-19 Policy

Yingyu He [1,2,†], Fang Zhang [3,†], Yan Liu [2], Zhou Xiong [2], Shangen Zheng [2], Wanbing Liu [1,2] and Lei Liu [1,2,*]

1 Hubei Key Laboratory of Central Nervous System Tumor and Intervention, Wuhan 430070, China; wanbing1203x@163.com (W.L.)
2 Department of Transfusion Medicine, General Hospital of Central Theater Command of the PLA, Wuhan 430070, China
3 Department of Anesthesiology, General Hospital of Central Theater Command of the PLA, Wuhan 430070, China
* Correspondence: liulei890207@163.com
† These authors contributed equally to this work.

Abstract: For SARS-CoV-2 mutants, the effectiveness of the COVID-19 vaccines is still controversial. In this study, we aimed to investigate the clinical characteristics of Omicron-infected patients who completed primary immunization and booster immunization, respectively, during the rapid propagation of the Omicron variant in China. A total of 932 patients with confirmed SARS-CoV-2 infection from 18 December 2022 to 1 January 2023 were included in this survey by filling out questionnaires online. The enrolled patients were divided into the primary immunization group and the booster immunization group according to their vaccination status. During the whole course of disease, the most frequent symptoms were fever (90.6%), cough (84.3%), weakness (77.4%), headache and dizziness (76.1%), and myalgia (73.9%). Nearly 90% of the patients had symptoms lasting for less than 10 days, and 39.8% of the patients ended the course of the disease in 4–6 days. A total of 58.8% of these patients had a fever with a maximum body temperature of over 38.5 °C. Moreover, 61.4% of the patients had a fever that lasted less than 2 days. There were no obvious differences in initial symptoms, cardinal symptoms, symptom duration time, maximum body temperature, and fever duration time between the two groups of patients. In addition, no significant difference was found in the positive or negative conversion time of SARS-CoV-2 antigen/nucleic acid between the two groups of patients. For mild patients with Omicron breakthrough infection, enhanced immunization has no significant impact on the clinical performance and duration of viral infection compared with primary immunization. The reasons behind the different clinical manifestations of patients with mild symptoms after the breakthrough infection of the Omicron strain are still worth further research. Heterologous vaccination may be a better strategy for enhanced immunization, which can help improve the immune protection ability of the population. Further research should be carried out on vaccines against mutant strains and spectral anti-COVID-19 vaccines.

Keywords: COVID-19; omicron; breakthrough infection; booster vaccination; primary vaccination

1. Introduction

Coronavirus disease 2019 (COVID-19) is caused by severe acute respiratory syndrome coronavirus 2 (SARS-CoV-2). SARS-CoV-2 has evolved, and new variants such as Alpha, Delta, and Omicron have emerged since the onset of COVID-19. The Omicron variant is currently widespread worldwide and has a high degree of mutation in infectivity and immune escape [1]. As of the end of April 2023, more than 680 million people worldwide have been infected with COVID-19, resulting in more than 6.8 million deaths. Vaccination is an important means of establishing an immune barrier. However, breakthrough infections in vaccinated populations, as well as repeat infections in previously infected populations, often occur due to the strong infectivity of Omicron and immune regression [2,3]. The protective effect of the COVID-19 vaccine against Omicron variant infection is limited and rapidly decreases over time [4–8], while the protective effect of the COVID-19 vaccine against severe illness or death after Omicron variant infection is significant [9–12]. Timely completion of enhanced immunization is an effective way to improve the immune protection ability to prevent the disease from developing into severe illness after infection. Nevertheless, it is still unknown whether the completion of enhanced immunization will affect the clinical performances of mild patients with Omicron breakthrough infection.

At the beginning of 2021, the Chinese government began to scientifically deploy and orderly promote the nationwide COVID-19 vaccination work. The main vaccines widely administered to the domestic population are three inactivated vaccines (made in Beijing or Wuhan Institute of Biological Products Co., Ltd., Beijing or Wuhan, China or Sinovac Life Sciences Co., Ltd., Beijing, China; respectively), adenovirus vector vaccines (made in CanSino Biologics Co., Ltd., Tianjin, China), and recombinant subunit vaccines (made in Anhui Zhifei Longcom Biopharmaceutical Co., Ltd., Hefei, China). Among them, the Chinese population receiving the three inactivated vaccines is the highest and most widespread. As of 2 March 2023, Chinese people have received a total of 3.49 billion vaccine doses, accounting for 90.58% of the total population. On 6 December 2022, China announced the cancellation of the dynamic zero COVID-19 policy after taking into account the epidemic characteristics of Omicron variant strains in China, the active immune level of the population, and the needs of social and economic development. Then, the Omicron variant strains quickly spread among the population. Up to now, it is estimated that nearly 80% of the population in China has been infected by Omicron, and the vast majority of infected people are cases of breakthrough infection that occur after vaccination and always show mild symptoms. However, we also found that the significant differences in the duration and manifestations of the disease, as well as the severity of the disease, existed even among these patients with mild infections. These different manifestations such as duration of virus infection may further have effects on the transmission of the Omicron variant. The duration and severity of illness can also have a significant impact on an individual's work and life. Therefore, we want to clarify whether completing the booster vaccination will have a significant impact on the clinical manifestations of these mild patients. Additionally, immunological memory extinction is a common phenomenon after natural immunity or artificial immunity, so we further analyzed the impact of time intervals for strengthening immunization.

In this study, we made a comparative analysis about the clinical characteristics between the mild patients with Omicron infection who completed primary immunization and booster immunization, respectively, during the rapid spread of Omicron in China and after relaxing the dynamic zero COVID-19 policy. We hope that our research results can provide some basis for the prevention and control policies of COVID-19 and COVID-19 vaccination policies in the next stage.

2. Materials and Methods

2.1. Study Design and Participants

This investigation was performed between 18 December 2022 and 1 January 2023 (during the rapid spread of Omicron in China). All participants who were infected with Omicron and the course of disease ended were recruited online and completed an online self-administered questionnaire through Wenjuanxing. The general information and clinical data of 1551 patients were collected in total. Of these, 619 cases were excluded (403 patients had no results for SARS-CoV-2 nucleic acid or antigen test during the course of disease, 71 patients had incomplete information, 69 patients failed to vaccinate or completed primary immunization, the inoculated vaccines of 53 patients were not inactivated vaccines, 16 patients were hospitalized, and the permanent residence of 7 patients was not in China). The remaining 932 patients were enrolled in this study (Figure 1). These patients all showed mild symptoms after being infected with Omicron and at least completed the primary immunization of COVID-19 inactivated vaccines (made in Beijing or Wuhan Institute of Biological Products Co., Ltd., Beijing or Wuhan, China or Sinovac Life Sciences Co., Ltd., Beijing, China). Completing primary immunization means receiving two doses of inactivated vaccines. Completing booster immunization means receiving three doses of inactivated vaccines. That is to say, these enrolled research subjects all used homologous enhanced immunization. The vaccination dose of each dose of inactivated vaccine is in accordance with the standards formulated by the health departments in China. The time interval between booster immunization and primary immunization is generally more than six months. According to whether they had finished booster immunization or not, the enrolled patients were divided into two groups (the primary immunization group and the booster immunization group). Moreover, the patients in the booster immunization group were further divided into two groups (the less than 1 year group and the more than 1 year group) according to whether the time interval between booster vaccination and breakthrough infection was more than one year.

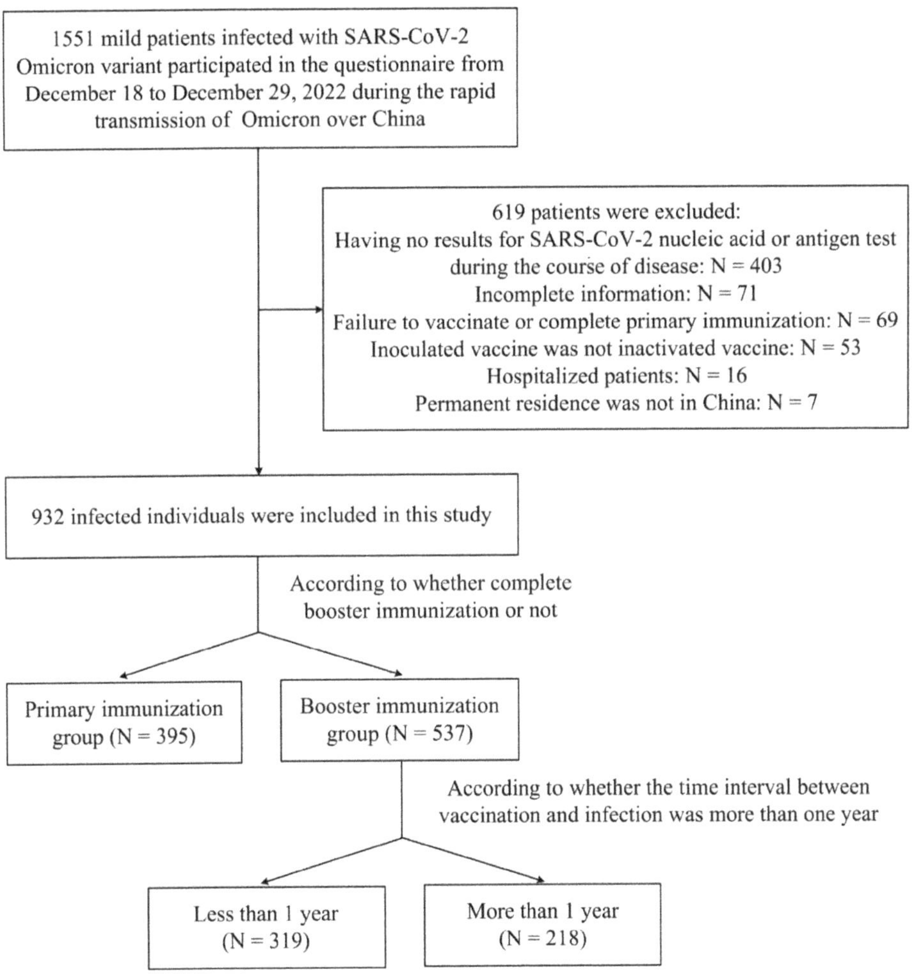

Figure 1. Study population.

2.2. Data Collection

A self-designed structured questionnaire was used to obtain information about the enrolled subjects. The general information including age, sex, occupation, coexisting disorders, past history of COVID-19, medical treatment, used drugs, and the most likely infection channels were obtained. The clinical manifestations including typical clinical symptoms, initial symptoms, symptom duration time, maximum body temperature, fever duration time, days of antigen/nucleic acid positive after symptom onset, and days of antigen/nucleic acid negative after symptom onset were carefully recorded. Some of the patients only underwent SARS-CoV-2 nucleic acid or antigen testing during the course of the disease, while others underwent both tests. For the patients who underwent both SARS-CoV-2 nucleic acid and antigen testing, the days of antigen/nucleic acid positive after symptom onset were calculated as the time when antigen or nucleic acid first appeared positive. The days of antigen/nucleic acid negative after symptom onset were calculated as the time when antigen or nucleic acid last appeared negative.

2.3. Statistical Analyses

Classification variables are expressed as numbers/percentages. Variables were compared using the chi-square test, while Fisher's exact test was used when data were limited. SPSS 26.0 software was used for statistical analysis. A two-sided test of α less than 0.05 was considered statistically significant.

3. Results

3.1. Demographics and Characteristics of Patients Infected with SARS-CoV-2 Omicron Variant

A total of 932 subjects were enrolled in the cohort. Among them, 537 cases completed booster immunization, while 395 cases just completed primary immunization. The general characteristics of these patients were shown in Table 1. There were no significant differences both in age and sex between the two groups. The proportion who completed enhanced immunization was higher than that who completed basic immunization among teachers, which was the opposite to the case among company staff ($p < 0.05$). Additionally, no significant differences were found in combined basic diseases and past history of COVID-19 between the two groups of patients. In total, 80.8% of the patients treated diseases through self-purchased medicines. Antipyretic and analgesic drugs were the most frequently used drugs, accounting for 84.2%. In addition, 31.7% of the patients chose traditional Chinese medicine treatment. Regarding the most likely infection mode, 29.7%, 27.6%, and 22.2% of the patients were presumed to be infected by co-workers, family members, and strangers in public places, respectively.

Table 1. Demographics and characteristics of patients infected with SARS-CoV-2 Omicron variant.

Characteristics	All Patients (N = 932)	Primary Immunization (N = 395)	Booster Immunization (N = 537)	*p* Value
Age groups (years)				
<14	5 (0.5)	5 (1.2)	0 (0.0)	
14–29	392 (42.1)	165 (41.8)	227 (42.3)	
30–45	442 (47.4)	188 (47.6)	254 (47.3)	0.088
46–59	84 (9.0)	32 (8.1)	52 (9.7)	
>59	9 (1.0)	5 (1.3)	4 (0.7)	
Sex				
Male	362 (38.8)	141 (35.7)	221 (41.2)	0.091
Female	570 (61.2)	254 (64.3)	316 (58.8)	
Occupation				
Teacher **	49 (5.3)	11 (2.8)	38 (7.1)	
Student	110 (11.8)	46 (11.6)	64 (11.9)	0.019 *
Self-employed	44 (4.7)	22 (5.6)	22 (4.0)	

Table 1. *Cont.*

Characteristics	All Patients (N = 932)	Primary Immunization (N = 395)	Booster Immunization (N = 537)	p Value
Company Staff **	216 (23.2)	106 (26.8)	110 (20.5)	
Public servants	47 (5.0)	17 (4.3)	30 (5.6)	
Healthcare workers	201 (21.6)	89 (22.5)	112 (20.9)	
Others	265 (28.4)	104 (26.4)	161 (30.0)	
Coexisting disorders				
Hypertension	39 (4.2)	21 (5.3)	18 (3.4)	0.139
Diabetes	13 (1.4)	7 (1.8)	6 (1.1)	0.400
Respiratory diseases	15 (1.6)	8 (2.8)	7 (1.3)	0.387
Cardiovascular and cerebrovascular diseases	9 (1.0)	5 (1.3)	4 (0.7)	0.422
Tumor	5 (0.5)	4 (1.0)	1 (0.2)	0.088
Past history of COVID-19				
Yes	21 (2.3)	12 (3.0)	9 (1.7)	0.166
No	911 (97.7)	383 (97.0)	528 (98.3)	
Medical treatment				
Untreated	141 (15.1)	57 (14.4)	84 (15.6)	
Self-purchased medicine	753 (80.8)	322 (81.5)	431 (80.3)	0.871
Fever outpatient treatment	38 (4.1)	16 (4.1)	22 (4.1)	
Drugs use				
Antipyretics and analgesics	666 (84.2)	282 (83.4)	384 (84.8)	0.610
Antiviral drugs	202 (25.5)	80 (23.7)	122 (26.9)	0.298
Antibiotics	127 (16.1)	55 (16.3)	72 (15.9)	0.886
Chinese medicine	251 (31.7)	81 (24.0)	170 (37.5)	<0.001*
Fluid administration	15 (1.9)	5 (1.5)	10 (2.2)	0.458
Most likely infection mode				
Family	257 (27.6)	107 (27.1)	150 (27.9)	
Colleagues **	277 (29.7)	102 (25.8)	175 (32.6)	
Confronting and treating patients	174 (18.7)	70 (17.7)	104 (19.4)	<0.001 *
Public places **	207 (22.2)	109 (27.6)	98 (18.2)	
Takeaway delivery	17 (1.8)	7 (1.8)	10 (1.9)	

* For overall comparison, $p < 0.05$. ** For individual item comparison, $p < 0.05$.

3.2. Clinical Manifestations of Patients Infected with SARS-CoV-2 Omicron Variant

The clinical performances of the two groups of patients were shown in Table 2. The most common initial symptoms were fever (36.2%), pharyngalgia (23.2%), and myalgia (11.9%). During the whole course of disease, the most frequent symptoms were fever (90.6%), cough (84.3%), weakness (77.4%), headache and dizziness (76.1%), and myalgia (73.9%). There were no obvious differences in cardinal symptoms except for pharyngalgia, ageusia, and nausea between the two groups of patients. Nearly 90% of the patients had symptoms lasting for less than 10 days, and 39.8% of the patients ended the course of the disease in 4–6 days. A total of 58.8% of these patients had a fever with a maximum body temperature of over 38.5 °C. Moreover, 61.4% of the patients had a fever that lasted less than 2 days. However, there were no significant differences in symptom duration time, maximum body temperature, and fever duration time between the two groups of patients. The antigen/nucleic acid results of most patients (74.6%) turned positive within 2 days of the onset of the disease. Additionally, the antigen/nucleic acid results of 87.3% of the patients turned negative within 10 days of the onset of the disease.

Table 2. Clinical manifestations of patients infected with SARS-CoV-2 Omicron variant.

Characteristics	All Patients (N = 932)	Primary Immunization (N = 395)	Booster Immunization (N = 537)	p Value
Initial symptom				
Fever	337 (36.2)	129 (32.7)	208 (38.7)	0.160
Pharyngalgia	216 (23.2)	85 (21.4)	131 (24.4)	0.304
Myalgia	111 (11.9)	53 (13.4)	58 (10.8)	0.223
Headache and dizziness	92 (9.8)	47 (11.9)	45 (8.4)	0.075
Cough and expectoration	65 (7.0)	37 (9.4)	28 (5.2)	0.014 *
Cardinal symptoms				
Fever	844 (90.6)	355 (89.9)	489 (91.1)	0.540
Cough	786 (84.3)	336 (85.1)	450 (83.8)	0.600
Weakness	721 (77.4)	297 (75.2)	424 (79.0)	0.174
Dizziness and headache	709 (76.1)	300 (75..9)	409 (76.2)	0.940
Myalgia	689 (73.9)	291 (73.7)	398 (74.1)	0.897
Pharyngalgia	647 (69.4)	256 (64.8)	391 (72.8)	0.009 *
Expectoration	626 (67.2)	264 (66.8)	362 (67.4)	0.853
Poor appetite	494 (53.0)	215 (54.4)	279 (52.0)	0.454
Hoarseness	412 (44.2)	172 (43.5)	240 (44.7)	0.727
Ageusia	328 (35.2)	162 (41.0)	166 (30.9)	0.001 *
Anosmia	276 (29.6)	123 (31.1)	153 (28.5)	0.382
Nausea	252 (27.0)	92 (23.3)	160 (29.8)	0.027 *
Diarrhea	205 (22.0)	82 (20.8)	123 (22.9)	0.435
Vomiting	172 (18.5)	72 (18.2)	100 (18.6)	0.878
Dyspnea	116 (12.4)	57 (14.4)	59 (11.0)	0.116
Symptom duration (d)				
1–3	216 (23.3)	100 (25.3)	116 (21.6)	
4–6	372 (39.8)	141 (35.7)	231 (43.0)	
7–10	245 (26.2)	105 (26.6)	140 (26.1)	0.150
11–14	49 (5.3)	23 (5.8)	26 (4.8)	
≥15	50 (5.4)	26 (6.6)	24 (4.5)	
Maximum body temperature (°C)				
<37.0	49 (5.3)	18 (4.6)	31 (5.8)	
37.0–38.5	335 (35.9)	150 (38.0)	185 (34.5)	0.202
38.6–39.9	520 (55.8)	211 (53.4)	309 (57.5)	
≥40.0	28 (3.0)	16 (4.0)	12 (2.2)	
Fever duration (d)				
0–2	542 (61.4)	242 (64.2)	300 (59.3)	
3–4	310 (35.1)	122 (32.4)	188 (37.2)	0.500
5–6	28 (3.2)	12 (3.2)	16 (3.2)	
≥7	3 (0.3)	1 (0.2)	2 (0.3)	
Days of antigen/nucleic acid positive after symptom onset (d)				
0	123 (13.2)	40 (10.1)	83 (15.5)	
1	287 (30.8)	124 (31.4)	163 (30.4)	
2	285 (30.6)	128 (32.4)	157 (29.2)	0.191
3	150 (16.1)	68 (17.2)	82 (15.3)	
4	38 (4.1)	13 (3.3)	25 (4.7)	
≥5	49 (5.3)	22 (5.6)	27 (5.0)	
Days of antigen/nucleic acid negative after symptom onset (d)				
1–3	65 (7.0)	26 (6.6)	39 (7.3)	
4–6	211 (22.6)	87 (22.0)	124 (23.1)	
7–10	538 (57.7)	230 (58.2)	308 (57.4)	0.974
11–14	78 (8.4)	35 (8.9)	43 (8.0)	
≥15	40 (4.3)	17 (4.3)	23 (4.3)	

* For overall comparison, $p < 0.05$.

3.3. Clinical Manifestations of Patients with Different Booster Immunization Time Interval

Among the patients who completed booster immunization, the time interval between vaccination and breakthrough infection of 319 cases was less than one year, and that of the other 218 cases was more than one year. We observed no significant differences in both initial symptoms and cardinal symptoms during the whole illness between the two groups of patients (Table 3). Similarly, no obvious differences were found in symptom duration time, maximum body temperature, and fever duration time between the two groups of patients. In addition, the time for antigen/nucleic acid to turn positive or negative also did not show significant differences.

Table 3. Clinical manifestations of patients with different booster immunization time interval.

Characteristics	All Patients (N = 537)	≤1 Year (N = 319)	>1 Year (N = 218)	*p* Value
Initial symptom				
Fever	209 (38.9)	121 (37.9)	88 (40.4)	0.570
Pharyngalgia	131 (24.4)	75 (23.5)	56 (25.7)	0.564
Myalgia	58 (10.8)	40 (12.5)	18 (8.3)	0.116
Headache and dizziness	45 (8.4)	25 (7.8)	20 (9.2)	0.583
Cough and expectoration	28 (5.2)	15 (4.8)	13 (6.0)	0.519
Cardinal symptoms				
Fever	489 (91.1)	290 (90.9)	199 (91.3)	0.881
Cough	450 (83.8)	272 (85.3)	178 (81.7)	0.264
Weakness	424 (79.0)	256 (80.3)	168 (77.1)	0.374
Dizziness and headache	409 (76.2)	244 (76.5)	165 (75.7)	0.831
Myalgia	398 (74.1)	231 (72.4)	167 (76.6)	0.276
Pharyngalgia	391 (72.8)	235 (73.7)	156 (71.6)	0.590
Expectoration	362 (67.4)	221 (69.3)	141 (64.7)	0.264
Poor appetite	279 (52.0)	169 (53.0)	110 (50.5)	0.566
Hoarseness	240 (44.7)	144 (45.1)	96 (44.0)	0.800
Ageusia	166 (30.9)	105 (32.9)	61 (28.0)	0.224
Anosmia	153 (28.5)	101 (31.7)	52 (23.9)	0.050
Nausea	160 (29.8)	101 (31.7)	59 (27.1)	0.253
Diarrhea	123 (22.9)	75 (23.5)	48 (22.0)	0.686
Vomiting	100 (18.6)	62 (19.4)	38 (17.4)	0.558
Dyspnea	59 (11.0)	38 (11.9)	21 (9.6)	0.407
Symptom duration (d)				
1–3	116 (21.6)	72 (22.6)	44 (20.2)	
4–6	231 (43.0)	140 (43.9)	91 (41.7)	
7–10	140 (26.1)	79 (24.8)	61 (28.0)	0.728
11–14	26 (4.8)	13 (4.0)	13 (6.0)	
≥15	24 (4.5)	15 (4.7)	9 (4.1)	
Maximum body temperature (°C)				
<37.0	31 (5.8)	20 (6.3)	11 (5.0)	
37.0–38.5	185 (34.5)	113 (35.4)	72 (33.0)	0.762
38.6–39.9	309 (57.5)	180 (56.4)	129 (59.2)	
≥40.0	12 (2.2)	6 (1.9)	6 (2.8)	
Fever duration (d)				
0–2	300 (59.2)	173 (57.9)	127 (61.4)	
3–4	188 (37.2)	114 (38.1)	74 (35.7)	0.771
5–6	16 (3.2)	11 (3.7)	5 (2.4)	
≥7	2 (0.4)	1 (0.3)	1 (0.5)	
Days of antigen/nucleic acid positive after symptom onset (d)				
0	123 (13.2)	40 (10.1)	83 (15.5)	
1	287 (30.8)	124 (31.4)	163 (30.4)	0.410
2	285 (30.6)	128 (32.4)	157 (29.2)	
3	150 (16.1)	68 (17.2)	82 (15.3)	

Table 3. *Cont.*

Characteristics	All Patients (N = 537)	≤1 Year (N = 319)	>1 Year (N = 218)	*p* Value
4	38 (4.1)	13 (3.3)	25 (4.7)	
≥5	49 (5.3)	22 (5.6)	27 (5.0)	
Days of antigen/nucleic acid negative after symptom onset (d)				
1–3	44 (8.2)	28 (8.8)	16 (7.3)	
4–6	119 (22.2)	75 (23.5)	44 (20.2)	
7–10	305 (56.8)	183 (57.4)	122 (56.0)	0.162
11–14	41 (7.6)	17 (5.3)	24 (11.0)	
≥15	28 (5.2)	16 (5.0)	12 (5.5)	

4. Discussion

The wide spread of the epidemic is found to be closely related to the continuous evolution and variation of SARS-CoV-2 since the beginning of COVID-19 [13,14], which brings great challenges to epidemic prevention and control globally. A great deal of evidence indicates that COVID-19 vaccination against the original strain can reduce the number of infected cases and the severity of the disease [15–18]. However, for mutants, the effectiveness of the vaccination is still controversial [19–21]. The mutants, especially the Omicron variant, dramatically increase the transmissibility and the ability to escape from natural and vaccine-induced immunity. This raises concerns about the ability of the original COVID-19 vaccination to prevent Omicron infection. The latest studies show that vaccination, especially inoculating booster doses, and previous infections still have a certain role in controlling new infections during the pandemic of the Omicron variant [22,23]. In the context of the global pandemic of the Omicron strain, many people who have been vaccinated or previously infected with COVID-19 still have breakthrough infections. These people with breakthrough infection mainly show mild symptoms such as fever, cough, fatigue, sore throat, etc. This is mainly due to the reduction of the virulence of the Omicron strain and the existence of a certain degree of immune response and immunological memory in the population. However, as mild patients, there are still significant differences in the symptoms and duration of these infected individuals. Some patients even have long COVID symptoms. It is still unknown whether these differentiated performances are related to the patients' immune response status and level.

In this study, we plan to analyze the impact of enhanced immunization on the clinical performance and disease transmission ability in the mild patients with breakthrough infection of the Omicron variant. On 6 December 2022, the Omicron variant began to spread quickly among community populations in China after relaxing the dynamic zero COVID-19 policy. We found that the vast majority of patients could cure their diseases by purchasing drugs themselves and that the course of the disease ended quickly. The vast majority of patients had a disease course of about 7 days, usually no more than 10 days. This result further confirms that the Omicron variant is extremely infectious and weakly pathogenic and that original vaccination is likely to still have an effect on severe prevention. More than 90% of these patients experienced fever symptoms, but the duration of the fever was generally not long (mostly not exceeding 3 days). Nevertheless, over half of the patients had a maximum body temperature above 38.5 °C. Effectively controlling body temperature is a concern to prevent sustained high fever from causing other complications. Most patients experienced a positive conversion of SARS-CoV-2 antigen or nucleic acid within two days of symptom onset, and a negative conversion occurs around 7 days after symptom onset. Overall, mild patients often have a duration of virus infection of no more than one week. However, we also found that there were no significant impacts on symptom manifestations and duration of virus infection (according to the time of antigen/nucleic acid positive/negative after symptom onset) in people who had completed enhanced immunization compared to those who had only completed primary immunization. We consider that the reasons may be that the long time between strengthening immunization and breakthrough infection leads to obvious immune regression. This is due to a rapid decrease in vaccine-induced peak IgA, which is a mucosal antibody with a stronger neutralizing activity against spike protein than IgG [24,25].

Several studies have shown that the protective effect of booster immunization on severe COVID-19 can be maintained steadily for a year, while the protection against symptomatic infection decreased significantly over time after the last vaccination [26–28]. Therefore, we further grouped according to the time interval for enhanced immunization. Nevertheless, we also found no significant differences in the symptoms' manifestations and the duration of virus infection between patients receiving enhanced vaccination within one year and more than one year. We analyze the reasons behind this, which may be that the rapid decline in immune response induced by enhanced immunization is insufficient to effectively prevent Omicron variant infection and remove the virus. Vaccine-induced protection against SARS-CoV-2 infection is temporary and is substantially affected by evolution and variation of the virus. Although the primary purpose of vaccination is to protect individuals from severe COVID-19 and its consequences, the extent to which vaccines can reduce the continued spread of COVID-19 is key to containing the epidemic. It depends on the ability of the vaccine to prevent infection and the extent to which vaccination reduces the infectiousness of breakthrough infections. Considering the diminished effectiveness of the vaccine against infection of SARS-CoV-2 new mutant strains, we should establish a scientific enhanced immunization cycle to provide more extensive protection. The selection of enhanced immunization strategies is also an important issue. Heterologous enhanced immunization usually achieves a stronger level of immune response compared to homologous enhanced immunization. Based on the fact that the vaccines administered to Chinese citizens in the early stage are mainly inactivated vaccines, other vaccines such as adenovirus vector vaccines and recombinant subunit vaccines can be chosen for future booster immunization. In addition, it is also necessary to conduct targeted vaccine research on new mutant strains or more spectral vaccine research. Despite the significant benefits in terms of reduced risk of hospitalization and death, different adverse events such as pain, soreness, itching at the vaccination site, and fever, headache, and muscle soreness may present after vaccination. A systematic literature review and meta-analysis demonstrates that vaccines are associated to a two-fold risk of developing a headache within 7 days of injection, and the lack of difference between vaccine types enables us to hypothesize that headache is secondary to systemic immunological reaction rather than to a vaccine-type specific reaction [29]. However, the mechanism of headache after vaccination is not clear but may also be related to dangerous complications.

5. Conclusions

For people who exhibit mild symptoms after breakthrough infection with Omicron, the homologous enhanced immunization using inactivated vaccines seems to have no significant impact on the symptom manifestations, duration, severity, and duration of virus infection compared to primary immunization. Therefore, using allogeneic immunization for enhanced vaccination may be a better option. We should strengthen research on vaccines against mutant strains and spectral anti-COVID-19 vaccines.

Author Contributions: Conceptualization, L.L. and W.L.; methodology, Y.H., F.Z. and Y.L.; data analysis, Z.X. and S.Z.; investigation, Y.H. and F.Z.; writing—original draft, L.L.; writing—review and editing, L.L. and W.L.; funding acquisition, L.L. All authors have read and agreed to the published version of the manuscript.

Funding: This work was supported by the Natural Science Foundation of Hubei Province (2021CFB501) and the Open Foundation of Hubei Province Key Laboratory (ZZYKF202201).

Institutional Review Board Statement: We conducted this study according to the Declaration of Helsinki. This study was deemed exempt by the Ethics Review Committee of General Hospital of Central Theater Command for the following reasons: (1) the survey was anonymous, and no personal identifiers were collected in our questionnaire; (2) participation was voluntary, and participants could exit the survey at any time before submission; (3) the participants were aware of the aim of the study and how the data would be used.

Informed Consent Statement: Informed consent was obtained from all subjects involved in the study.

Data Availability Statement: The data that support the findings of this study are available from the corresponding author upon reasonable request.

Conflicts of Interest: The authors declare that they have no competing interest.

References

1. Tan, S.T.; Kwan, A.T.; Rodriguez-Barraquer, I.; Singer, B.J.; Park, H.J.; Lewnard, J.A.; Sears, D.; Lo, N.C. Infectiousness of SARS-CoV-2 Breakthrough Infections and Reinfections During the Omicron Wave. *Nat. Med.* **2023**, *29*, 358–365. [CrossRef] [PubMed]
2. Mugisha, J.; Mpairwe, B.; Newton, R.; Cotten, M.; Phan, M.V.T. SARS-CoV-2 Omicron Ba.5 Infections in Vaccinated Persons, Rural Uganda. *Emerg. Infect. Dis.* **2023**, *29*, 224–226. [CrossRef] [PubMed]
3. Harris, J.R.; Owusu, D.; O'Laughlin, K.; Cohen, A.L.; Ben Hamida, A.; Patel, J.C.; Freeman, M.M.; Nsibambi, T.; Nieves, R.; Marston, B.J.; et al. SARS-CoV-2 Breakthrough Infections among Us Embassy Staff Members, Uganda, May–June 2021. *Emerg. Infect. Dis.* **2022**, *28*, 1279–1280. [CrossRef]
4. Tseng, H.F.; Ackerson, B.K.; Luo, Y.; Sy, L.S.; Talarico, C.A.; Tian, Y.; Bruxvoort, K.J.; Tubert, J.E.; Florea, A.; Ku, J.H.; et al. Effectiveness of Mrna-1273 against SARS-CoV-2 Omicron and Delta Variants. *Nat. Med.* **2022**, *28*, 1063–1071. [CrossRef]
5. Jara, A.; Undurraga, E.A.; Zubizarreta, J.R.; Gonzalez, C.; Acevedo, J.; Pizarro, A.; Vergara, V.; Soto-Marchant, M.; Gilabert, R.; Flores, J.C.; et al. Effectiveness of Coronavac in Children 3–5 Years of Age During the SARS-CoV-2 Omicron Outbreak in Chile. *Nat. Med.* **2022**, *28*, 1377–1380. [CrossRef]
6. Andrews, N.; Stowe, J.; Kirsebom, F.; Toffa, S.; Rickeard, T.; Gallagher, E.; Gower, C.; Kall, M.; Groves, N.; O'Connell, A.M.; et al. COVID-19 Vaccine Effectiveness against the Omicron (B.1.1.529) Variant. *N. Engl. J. Med.* **2022**, *386*, 1532–1546. [CrossRef]
7. Collie, S.; Champion, J.; Moultrie, H.; Bekker, L.G.; Gray, G. Effectiveness of Bnt162b2 Vaccine against Omicron Variant in South Africa. *N. Engl. J. Med.* **2022**, *386*, 494–496. [CrossRef]
8. Kirsebom, F.C.M.; Andrews, N.; Sachdeva, R.; Stowe, J.; Ramsay, M.; Bernal, J.L. Effectiveness of Chadox1-S COVID-19 Booster Vaccination against the Omicron and Delta Variants in England. *Nat. Commun.* **2022**, *13*, 7688. [CrossRef] [PubMed]
9. Price, A.M.; Olson, S.M.; Newhams, M.M.; Halasa, N.B.; Boom, J.A.; Sahni, L.C.; Pannaraj, P.S.; Irby, K.; Bline, K.E.; Maddux, A.B.; et al. Bnt162b2 Protection against the Omicron Variant in Children and Adolescents. *N. Engl. J. Med.* **2022**, *386*, 1899–1909. [CrossRef] [PubMed]
10. Martellucci, C.A.; Flacco, M.E.; Soldato, G.; Di Martino, G.; Carota, R.; Caponetti, A.; Manzoli, L. Effectiveness of COVID-19 Vaccines in the General Population of an Italian Region before and During the Omicron Wave. *Vaccines* **2022**, *10*, 662. [CrossRef]
11. McMenamin, M.E.; Nealon, J.; Lin, Y.; Wong, J.Y.; Cheung, J.K.; Lau, E.H.Y.; Wu, P.; Leung, G.M.; Cowling, B.J. Vaccine Effectiveness of One, Two, and Three Doses of Bnt162b2 and Coronavac against COVID-19 in Hong Kong: A Population-Based Observational Study. *Lancet Infect. Dis.* **2022**, *22*, 1435–1443. [CrossRef] [PubMed]
12. Magen, O.; Waxman, J.G.; Makov-Assif, M.; Vered, R.; Dicker, D.; Hernán, M.A.; Lipsitch, M.; Reis, B.Y.; Balicer, R.D.; Dagan, N. Fourth Dose of Bnt162b2 Mrna COVID-19 Vaccine in a Nationwide Setting. *N. Engl. J. Med.* **2022**, *386*, 1603–1614. [CrossRef]
13. Zhang, Y.; He, L.; Chen, H.; Lu, S.; Xiong, Y.; Liu, J.; Zheng, Y.; Wang, S.; Liu, L. Manifestations of Blood Coagulation and Its Relation to Clinical Outcomes in Severe COVID-19 Patients: Retrospective Analysis. *Int. J. Lab. Hematol.* **2020**, *42*, 766–772. [CrossRef] [PubMed]
14. Zaidi, A.; Harris, R.; Hall, J.; Woodhall, S.; Andrews, N.; Dunbar, K.; Lopez-Bernal, J.; Dabrera, G. Effects of Second Dose of SARS-CoV-2 Vaccination on Household Transmission, England. *Emerg. Infect. Dis.* **2023**, *29*, 127–132. [CrossRef] [PubMed]
15. Vitiello, A.; Ferrara, F.; Troiano, V.; La Porta, R. COVID-19 Vaccines and Decreased Transmission of SARS-CoV-2. *Inflammopharmacology* **2021**, *29*, 1357–1360. [CrossRef] [PubMed]
16. Santos da Silva, E.; Servais, J.Y.; Kohnen, M.; Arendt, V.; Gilson, G.; Staub, T.; Seguin-Devaux, C.; Perez-Bercoff, D. Vaccine- and Bti-Elicited Pre-Omicron Immunity More Effectively Neutralizes Omicron Sublineages Ba.1, Ba.2, Ba.4 and Ba.5 Than Pre-Omicron Infection Alone. *Curr. Issues Mol. Biol.* **2023**, *45*, 1741–1761. [CrossRef]
17. Higdon, M.M.; Wahl, B.; Jones, C.B.; Rosen, J.G.; Truelove, S.A.; Baidya, A.; Nande, A.A.; ShamaeiZadeh, P.A.; Walter, K.K.; Feikin, D.R.; et al. A Systematic Review of Coronavirus Disease 2019 Vaccine Efficacy and Effectiveness against Severe Acute Respiratory Syndrome Coronavirus 2 Infection and Disease. *Open Forum Infect. Dis.* **2022**, *9*, ofac138. [CrossRef]
18. Ghazy, R.M.; Ashmawy, R.; Hamdy, N.A.; Elhadi, Y.A.M.; Reyad, O.A.; Elmalawany, D.; Almaghraby, A.; Shaaban, R.; Taha, S.H.N. Efficacy and Effectiveness of SARS-CoV-2 Vaccines: A Systematic Review and Meta-Analysis. *Vaccines* **2022**, *10*, 350. [CrossRef]
19. Kislaya, I.; Rodrigues, E.F.; Borges, V.; Gomes, J.P.; Sousa, C.; Almeida, J.P.; Peralta-Santos, A.; Nunes, B. Comparative Effectiveness of Coronavirus Vaccine in Preventing Breakthrough Infections among Vaccinated Persons Infected with Delta and Alpha Variants. *Emerg. Infect. Dis.* **2022**, *28*, 331–337. [CrossRef]

20. Nielsen, K.F.; Moustsen-Helms, I.R.; Schelde, A.B.; Gram, M.A.; Emborg, H.D.; Nielsen, J.; Hansen, C.H.; Andersen, M.A.; Meaidi, M.; Wohlfahrt, J.; et al. Vaccine Effectiveness against SARS-CoV-2 Reinfection During Periods of Alpha, Delta, or Omicron Dominance: A Danish Nationwide Study. *PLoS Med.* **2022**, *19*, e1004037. [CrossRef]
21. Kissler, S.M.; Fauver, J.R.; Mack, C.; Tai, C.G.; Breban, M.I.; Watkins, A.E.; Samant, R.M.; Anderson, D.J.; Metti, J.; Khullar, G.; et al. Viral Dynamics of SARS-CoV-2 Variants in Vaccinated and Unvaccinated Persons. *N. Engl. J. Med.* **2021**, *385*, 2489–2491. [CrossRef] [PubMed]
22. Accorsi, E.K.; Britton, A.; Fleming-Dutra, K.E.; Smith, Z.R.; Shang, N.; Derado, G.; Miller, J.; Schrag, S.J.; Verani, J.R. Association between 3 Doses of Mrna COVID-19 Vaccine and Symptomatic Infection Caused by the SARS-CoV-2 Omicron and Delta Variants. *Jama* **2022**, *327*, 639–651. [CrossRef] [PubMed]
23. Fleming-Dutra, K.E.; Britton, A.; Shang, N.; Derado, G.; Link-Gelles, R.; Accorsi, E.K.; Smith, Z.R.; Miller, J.; Verani, J.R.; Schrag, S.J. Association of Prior Bnt162b2 COVID-19 Vaccination with Symptomatic SARS-CoV-2 Infection in Children and Adolescents During Omicron Predominance. *Jama* **2022**, *327*, 2210–2219. [CrossRef] [PubMed]
24. Sterlin, D.; Mathian, A.; Miyara, M.; Mohr, A.; Anna, F.; Claër, L.; Quentric, P.; Fadlallah, J.; Devilliers, H.; Ghillani, P.; et al. Iga Dominates the Early Neutralizing Antibody Response to SARS-CoV-2. *Sci. Transl. Med.* **2021**, *13*, eabd2223. [CrossRef] [PubMed]
25. Wisnewski, A.V.; Luna, J.C.; Redlich, C.A. Human Igg and Iga Responses to COVID-19 Mrna Vaccines. *PLoS ONE* **2021**, *16*, e0249499. [CrossRef] [PubMed]
26. Ng, O.T.; Marimuthu, K.; Lim, N.; Lim, Z.Q.; Thevasagayam, N.M.; Koh, V.; Chiew, C.J.; Ma, S.; Koh, M.; Low, P.Y.; et al. Analysis of COVID-19 Incidence and Severity among Adults Vaccinated with 2-Dose Mrna COVID-19 or Inactivated SARS-CoV-2 Vaccines with and without Boosters in Singapore. *JAMA Netw. Open* **2022**, *5*, e2228900. [CrossRef]
27. Šmíd, M.; Berec, L.; Přibylová, L.; Májek, O.; Pavlík, T.; Jarkovský, J.; Weiner, J.; Barusová, T.; Trnka, J. Protection by Vaccines and Previous Infection against the Omicron Variant of Severe Acute Respiratory Syndrome Coronavirus 2. *J. Infect. Dis.* **2022**, *226*, 1385–1390. [CrossRef]
28. Külper-Schiek, W.; Piechotta, V.; Pilic, A.; Batke, M.; Dreveton, L.S.; Geurts, B.; Koch, J.; Köppe, S.; Treskova, M.; Vygen-Bonnet, S.; et al. Facing the Omicron Variant-How Well Do Vaccines Protect against Mild and Severe COVID-19? Third Interim Analysis of a Living Systematic Review. *Front. Immunol.* **2022**, *13*, 940562. [CrossRef]
29. Castaldo, M.; Waliszewska-Prosół, M.; Koutsokera, M.; Robotti, M.; Straburzyński, M.; Apostolakopoulou, L.; Capizzi, M.; Çibuku, O.; Ambat, F.D.F.; Frattale, I.; et al. Headache onset after vaccination against SARS-CoV-2: A systematic literature review and meta-analysis. *J. Headache Pain* **2022**, *23*, 41. [CrossRef]

vaccines

MDPI

Article

The Quantification of Spike Proteins in the Inactivated SARS-CoV-2 Vaccines of the Prototype, Delta, and Omicron Variants by LC–MS

Kangwei Xu [1,†], Huang Sun [2,†], Kaiqin Wang [1], Yaru Quan [1], Zhizhong Qiao [1], Yaling Hu [2,*] and Changgui Li [1,*]

1 NHC Key Laboratory of Research on Quality and Standardization of Biotech Products, NMPA Key Laboratory for Quality Research and Evaluation of Biological Products, National Institutes for Food and Drug Control, No. 2, Tiantan Xili, Dongcheng District, Beijing 100050, China; xkwnifdc@163.com (K.X.)

2 Sinovac Life Sciences Co., Ltd., No. 21, Tianfu St., Daxing Biomedicine Industrial Base of Zhongguancun Science Park, Daxing District, Beijing 100050, China

* Correspondence: huyl@sinovac.com (Y.H.); changguili@aliyun.com (C.L.)

† These authors contributed equally to this work.

Abstract: Developing variant vaccines or multivalent vaccines is a feasible way to address the epidemic as the SARS-CoV-2 variants of concern (VOCs) posed an increased risk to global public health. The spike protein of the SARS-CoV-2 virus was usually used as the main antigen in many types of vaccines to produce neutralizing antibodies against the virus. However, the spike (S) proteins of different variants were only differentiated by a few amino acids, making it difficult to obtain specific antibodies that can distinguish different VOCs, thereby challenging the accurate distinction and quantification of the variants using immunological methods such as ELISA. Here, we established a method based on LC–MS to quantify the S proteins in inactivated monovalent vaccines or trivalent vaccines (prototype, Delta, and Omicron strains). By analyzing the S protein sequences of the prototype, Delta, and Omicron strains, we identified peptides that were different and specific among the three strains and synthesized them as references. The synthetic peptides were isotopically labeled as internal targets. Quantitative analysis was performed by calculating the ratio between the reference and internal target. The verification results have shown that the method we established had good specificity, accuracy, and precision. This method can not only accurately quantify the inactivated monovalent vaccine but also could be applied to each strain in inactivated trivalent SARS-CoV-2 vaccines. Hence, the LC–MS method established in this study can be applied to the quality control of monovalent and multivalent SARS-CoV-2 variation vaccines. By enabling more accurate quantification, it will help to improve the protection of the vaccine to some extent.

Keywords: SARS-CoV-2; spike; standardization; LC–MS; vaccine

1. Introduction

Since the emergence of the COVID-19 pandemic, its pathogen, a novel coronavirus (SARS-CoV-2), has continued to undergo genetic mutations [1–5]. SARS-CoV-2 is an enveloped, positive-sense, and single-stranded RNA viruse belonging to the Beta coronavirus genus of coronaviruses [6]. SARS-CoV-2 encodes 16 predicted non-structural proteins (NSP1–NSP16) and 4 structural proteins: spike (S), envelope (E), membrane (M), and nucleocapsid (N) proteins [7,8]. The S protein is a glycoprotein located on the surface of the coronavirus which can be cleaved by host cell serine protease TMPRSS2 at the S1/S2 cleavage site. The receptor binding domain (RBD) on the S1 subunit is able to bind to the ACE2 receptor on the surface of the host cells, allowing the virus to enter and infect the host [9]. Montenegro, R. C. et al. investigated the differences between coronaviruses, including variations in the S protein that make SARS-Cov-2 more virulent than other coronaviruses [10]. As the S protein is an essential neutralizing epitope for both inactivated and recombinant COVID-19 vaccines, the primary active ingredient is the S protein which is quantified using immunological techniques such as ELISA.

Vaccination is an effective means of preventing and controlling COVID-19. However, the continuous mutation of the virus has led to a reduction in the protective efficacy of the vaccines. The Delta variant, which was first detected in India in March 2021, quickly became the dominant global epidemic strain until it was displaced by the Omicron variant at the end of 2021. The Delta variant is known to be more transmissible and pathogenic than the prototype strain [11–14]. The Omicron strain has amassed more mutations than any other variant which makes it difficult for the body to eliminate the virus [15]. The protective efficacy of many existing vaccines against Omicron variants has been significantly reduced. Alex Sigal et al. found that the serum neutralizing antibody titer targeting Omicron of people previously vaccinated with the Pfizer BNT162b2 mRNA vaccine showed a significant decrease, with an average of a 41-fold decrease by being vaccinated alone [16]. Sandra Ciesek et al. investigated the effect of dosing regimens of mRNA vaccines on the neutralizing capacity against Omicron variants, including BNT162b2 (Pfizer, 2 dose), Spikevax (Moderna, 2 dose), and Vaxzevria (AstraZeneca, 1 dose) + BNT162b2 (Pfizer, 1 dose), and found that the serum antibodies from subjects vaccinated with three regimens cannot neutralize Omicron strain after 6 months. After an additional booster dose of BNT162b2, the serum neutralization rate only increased to 25% [17]. The protective efficacy of the inactivated vaccine was also substantially reduced. A total of 28 days after receiving the third dose of the inactivated vaccine (Sinopharm BIBP), 228 volunteers (78.08%) showed detectable serum neutralizing activity against the Omicron variant, with a 20.1-fold decrease in the geometric mean titer of neutralizing antibody compared to the prototype strain [18].

It becomes very necessary to have more effective vaccines to protect humans from Omicron or other variants. Developing broad-spectrum vaccines is a potential solution to the problem of reduced protective efficacy against VOCs [19–23]. Hwang et al. proposed the design of developing a universal vaccine (pan SARS-CoV-2). The inclusion of non-S protein structural proteins (nucleocapsid (N), member (M), and envelope (E)) in the target design of the vaccine aims to compensate for the T cell immune deficiency caused by Spike-only vaccine immunity. By triggering a strong cellular immune response, it synergizes with the B cell response, leading to the long-term success of the vaccine [24]. Furthermore, it has been reported that the frequencies of cytokine-secreting non-spike specific CD4+ and CD8+ T cells correlate well with those of the spike-specific T cellsfrom WT, Delta, and Omicron strains [25]. Despite partial broad-spectrum protection observed in animal studies, there are currently no broad-spectrum vaccines available on the market.

Another solution is the development of multivalent vaccines for vaccination. On one hand, it may avoid the poor production of neutralizing antibodies against subsequent variants due to the immune anchoring/antigenic guilt effect. On the other hand, recent researches have suggested that booster immunizations with bivalent vaccines including prototype and Beta strains could yield higher neutralizing antibody titers against Delta and Omicron variants than that with monovalent vaccines [26]. The phase II/III clinical trials of the Moderna bivalent mRNA vaccine consisting of the Prototype and Omicron strain as the booster showed 1.75 times higher levels of neutralizing antibodies against Omicron elicited by the bivalent one than the one induced by the monovalent one [27]. Pfizer and Moderna bivalent mRNA vaccines have been approved for emergency use by the FDA. Sinovac is developing a trivalent inactivated vaccine that includes the prototype strain as well as the Delta and Omicron variants.

For both inactivated and recombinant COVID-19 vaccines, the main target antigen is the S protein, of which the quantification is usually performed by using immunological methods such as ELISA. Due to the high similarity of S protein epitopes among different variants, it is difficult to effectively differentiate using antibodies. Therefore, the accurate quantification of each component cannot be achieved when evaluating the antigenic content of trivalent vaccines including the prototype, Delta, and Omicron strains. LC–MS/MS methods have also been reported for the quantification of S and N proteins. Zhen Long et al. and Lisa H et al. established LC–MS methods for the analysis of S and N proteins in inactivated vaccine stock solutions or patient specimens of the prototype strain, respectively [28,29]. Osnat R et al. used a similar method to evaluate the S protein content in a VSV-based vaccine [30]. Carrie P et al. developed an isotope dilution tandem mass spectrometer (IDMS) method for the quantitative detection targeting the N protein as well as the conserved sequence of the S proteins among prototype, Alpha, Beta, Gamma, and Delta strains [31]. However, studies on the detection of variant-based vaccines and multivalent vaccines have not been reported yet.

In this study, an IDMS method was developed for the quantification of S proteins among the prototype, Delta, and Omicron strains of the SARS-CoV-2 virus. This method could be used to quantify monovalent SARS-Cov-2 vaccines and it also could accurately distinguish and detect different S proteins from trivalent vaccines composed of the three strains above.

2. Methods

2.1. Selection and Synthesis of Native and Isotopically Labeled Strain Specific SARS-CoV-2 Peptides

The amino acid sequences of the S protein for the prototype, Delta, and Omicron strains of SARS-Cov-2 were digested using Biopharm Finder 4.1 software. The resulting computational peptides were compared and analyzed to identify a unique peptide sequence that could distinguish the S protein of all three strains. The selection of peptides is based on two principles: firstly, it must have strain specificity, that is, only one of the three strains has the peptide. It is best for peptides at the same location to have specificity for all three strains in order to avoid differences caused by the efficiency of trypsin digestion affected by the different S protein structures. Secondly, in order to obtain better MS response signals, it is best to select peptides that meet the general principles of MS detection peptide selection, such as the peptide length should be 7–20 amino acids. Extremely hydrophobic peptides should be avoided. Cysteine, methionine, and N-terminal glutamine amino acid residues should be avoided [32]. The final selected peptides are shown in Table 1. GenScript Biotechnology Co., Ltd. (Nanjing, China) was commissioned to synthesize both native and isotope-labeled (13C/15N Val on N terminus) peptides which were lyophilized and stored at −20 °C. The correctness of the synthesized peptide sequence was confirmed using Vanquish UHOLC-Q Exactive Plus Orbitrap Mass spectrometry.

Table 1. Strain specific peptides of the SARS-CoV-2 S protein.

Variant Strain	Peptide Name	Peptide Sequence	Precursor Ion *m/z*	Target Ion *m/z*	Collision Energy (%)
Prototype strain	E	VGGNYNYLYR	610.08 (+2)	1119.41	23
	F	**VGGNYNYLYR**	613.08 (+2)	1119.58	23
Delta (B.1.617.2)	G	VGGNYNYR	471.98 (+2)	843.33	16
	H	**VGGNYNYR**	474.56 (+2)	843.25	18
Omicron (BA.1)	I	VSGNYNYLYR	625.05 (+2)	728.58	20
	J	**VSGNYNYLYR**	628.03 (+2)	728.5	21

Underlined amino acids indicate the 13C/15N isotope labeled.

2.2. Samples

Inactivated vaccine candidates and bulks for the prototype, Delta (B.1.617.2), and Omicron (B.1.1.529) strains of the SARS-CoV-2 virus were produced by Sinovac Biotech Ltd. (Beijing, China) following a previously reported process. Specifically, each virus strain was propagated in Vero cells and inactivated using β-propiolactone. The resulting three virus bulks were purified using ultrafiltration and chromatography purification. Finally, the trivalent vaccine was prepared by mixing and diluting the three virus bulks.

2.3. Preparation of Standard Solutions and the Samples

To prepare the standard solutions, six peptides (E, G, I, F, H, and J) were accurately weighed, dissolved in acetonitrile, and diluted with 50 mM ammonium bicarbonate solution to the following concentrations: E 400 ng/mL, G 400 ng/mL, I 500 ng/mL, F 360 ng/mL, H 300 ng/mL, and J 360 ng/mL. These six peptides were then proportionally mixed and diluted to each point concentration of the standard curve as shown in Table 2, using 50 mM ammonium bicarbonate.

Table 2. Serial concentrations of standard polypeptides.

Standard Polypeptides			Concentration (ng/mL)			
Working Solution	1	2	3	4	5	6
polypeptide E	1.0	2.0	3.0	4.0	5.0	6.4
isotope-labeled peptide F	3.6	3.6	3.6	3.6	3.6	3.6
polypeptide G	0.48	0.96	1.44	1.92	2.4	2.88
isotope-labeled peptide H	3.0	3.0	3.0	3.0	3.0	3.0
polypeptide I	0.6	1.2	1.8	2.4	3	3.6
isotope-labeled peptide J	3.6	3.6	3.6	3.6	3.6	3.6

For the internal standard polypeptide solution, three isotope-labeled peptide solutions were mixed proportionally and diluted with 50 mM ammonium bicarbonate. The final mixture contained F at 7.2 ng/mL, H at 6.0 ng/mL, and J at 7.2 ng/mL.

To prepare the test samples, 0.5 mL of each sample was taken and placed into a 1.5 mL microcentrifuge tube. The tube was then vacuum centrifuged at 60 °C until it was completely dry (about 2 h). Next, 100 µL of RapiGest solution with a concentration of 0.1% (086001861, Waters Corporation, Milford, MA, USA) was added to the microcentrifuge tube. RapiGest is an anionic surfactant that helps solubilize and unfold proteins making them more amenable to cleavage without denaturing or inhibiting common proteolytic enzymes, thereby reducing the digestion time. It is prone to hydrolysis in acidic environments to produce insoluble precipitates, thus not interfering with MS detection. The contents were vigorously vortexed to ensure complete mixing and then heated at 60 °C for 30 min to denature the protein and facilitate trypsinization. After cooling to room temperature, 0.5 mL of the above mixed internal standard polypeptide solution and 5 µL of trypsin solution at a concentration of 0.5 µg/µL (V5117, Promega, Madison, WI, USA) were added to the sample. The sample was then subjected to shaking digestion at 37 °C and 1000 rpm/min for 16 h. The protein was then hydrolyzed into peptides for MS detection. Then, 10 µL of trifluoroacetic acid was added and mixed well and left at 37 °C for 30 min to stop trypsin activity, thus reducing the pH of the solution to hydrolyze RapiGest. Finally, the solution was transferred into a 1 mL volumetric flask and brought to volume by 50 mM ammonium bicarbonate solution. After centrifuging at 11,000 rpm for 10 min, RapiGest hydrolysis products and other insoluble substances were removed. The supernatant was collected as the test solution.

2.4. LC/MS/MS Instrumentation Parameters

The Vanquish UHPLC-TSQ Quantis liquid chromatography–triple quadrupole mass spectrometry (Thermo Scientific, Waltham, MA, USA) was used to separate and detect the peptides. The instrument is subjected to annual statutory testing annually to ensure its performance meets requirements. Pierce Triple Quadrupole Calibration Solution Extended Mass Range (Thermo Scientific, Waltham, MA, USA) was used for calibration. The analytical column, Shim-pack GISS-HP C18 Metal-free (2.1 × 150 mm (HSS), 3 µm (SHIMADZU) was maintained at 35 °C. The mobile phase consisted of 0.1% formic acid in acetonitrile (mobile phase A) and 0.1% formic acid in water (mobile phase B). Samples of 40 µL were injected into the analytical column and the chromatographic separation gradient was performed at a flow rate of 0.3 mL/min according to the following conditions: initial, 95% B;

1 min, 95% B; 8 min, 60% B; 8.1 min, 10% B; 10 min, 10% B; 10.1 min, 95% B; and 15 min, 95% B. The total running time was 15 min.

The TSQ system, equipped with an ESI interface operating in the positive ion MRM mode, was used to analyze the column eluent. The precursor/product pairs and collision energies for peptides were shown in Table 1. The instrument parameters were set as follows: spray voltage 3800 V, sheath gas 50 arb, auxiliary gas 15 arb, ion Transfer Tube Temperature 350 °C, and CID Gas 2 mTorr. The collision energy and tube lens of each peptide were optimized separately. Thermo Scientific Chameleon7.2 was used for instrument control and data analysis.

2.5. Data Analysis

To enhance the accuracy and precision of the quantitative method, internal standard peptides were employed using the internal standard method. Isotope-labeled peptides with the same concentration as the internal standards were added to all standard and sample solutions. The area ratio of the target peptide to the isotope-labeled peptide was subjected to linear fitting against the concentration of the target peptide to establish a standard curve. This curve was then utilized to determine the content of the peptides in the S protein. The S protein content (pmol/mL) in the samples was calculated using the following formula: $\text{S protein content (pmol/mL)} = \frac{\text{peptide content (ng/mL)}}{\text{molecular weight } \times 1000}$.

3. Results

3.1. Select of Strain Specific SARS-CoV-2 Peptides

Trypsin digestion was simulated after aligning the S proteins of three strains (prototype, Delta, and Omicron strains) and the peptide sequences after digestion were compared. We identified two sets of peptides that can be used to distinguish the three variants at the same time (Figure 1). The first group is located at positions 445–454, with the peptide sequences VGGNYNYLYR, VGGNYNYR, and VSGNYNYLYR, for the prototype, Delta, and Omicron strains, respectively. The second group is located at positions 130–147 with the peptide sequences VCEFQFCNDPFLGVYYHK, VCEFQFCNDPFLDVYYHK, and VCEFQFCNDPFL—DHK, for the three strains, respectively. The first set of distinct peptides was finally selected for quantitative studies for the following reasons: firstly, the second group of peptides contained several disulfide bonds, which could affect the stability of results. Secondly, the first group of peptides is shorter in length (the longest peptide is only 10 amino acids) compared to the second group (the longest peptide is 18 amino acids), resulting in a better mass spectrometry response. Lastly, some samples could not be detected using the second group polypeptides.

3.2. Method Validation

The method was validated for specificity, linearity, recovery, and precision.

3.2.1. Specificity

The monovalent bulks of the prototype, Delta, and Omicron strains were taken and treated according to the method described in Section 2.3 without adding any isotope-labeled peptides. The results are presented in Figure 2, which showed that only the strain-specific peptides were detected in each bulk. This demonstrated that using the strain-specific peptides can effectively distinguish the three strains and confirmed that the source of the peptides was the S protein in the corresponding strain rather than other matrix proteins, indicating specificity and anti-matrix interference ability.

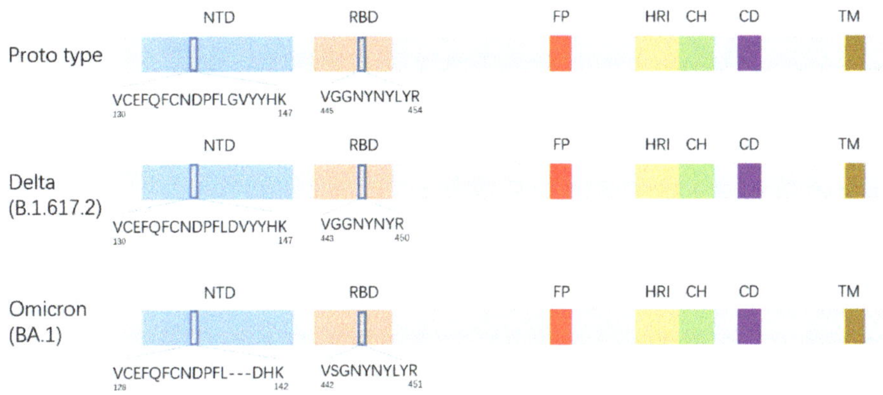

Figure 1. Schematic diagram of strain-specific SARS-CoV-2 peptide selection. There were two sets of peptides at the same location to have specificity for all three strains which were located in the RBD region at positions 445–454 and NTD at positions 130–147. The 445–454 position peptides were more suitable for MS detection, so this group of peptides was selected for method establishment.

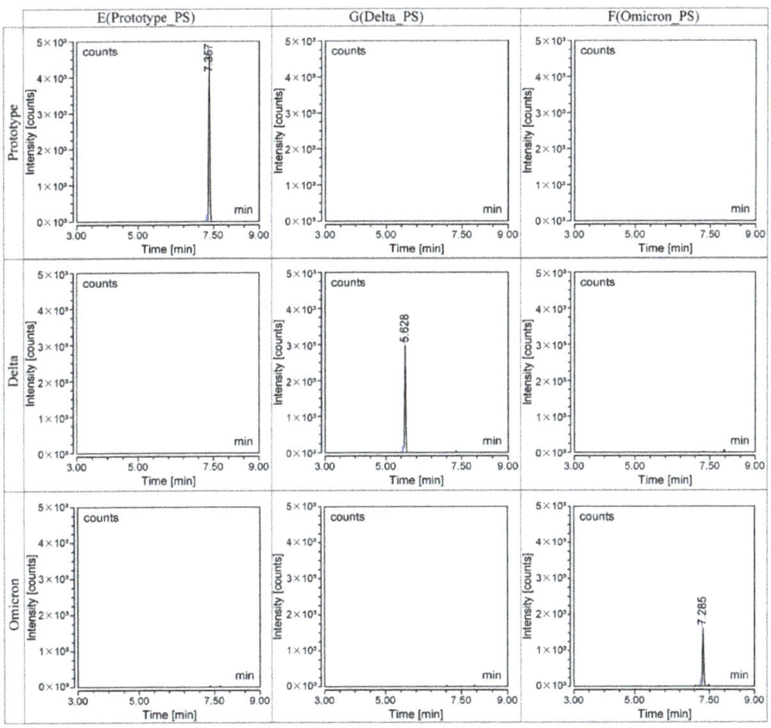

Figure 2. Specific characteristic spectrums of different variants. Monovalent bulks of the prototype, Delta, and Omicron strains were analyzed by LC–MS method. In the prototype bulk (**first line**), only peptide E can be detected. In the Delta bulk (**second line**), only peptide G can be detected. In the Omicron bulk (**third line**), only peptide I can be detected. The selected peptides have good strain specificity and can only be detected in the corresponding strains and do not exist in other strains or matrices.

3.2.2. Linearity

The standard curves were generated using different serial concentrations of the peptides. Linear regression was used to fit the peptide concentration versus the area ratio relationship. As shown in Figure 3, all three peptides showed a strong linear relationship with a regression correlation coefficient R^2 greater than 0.99.

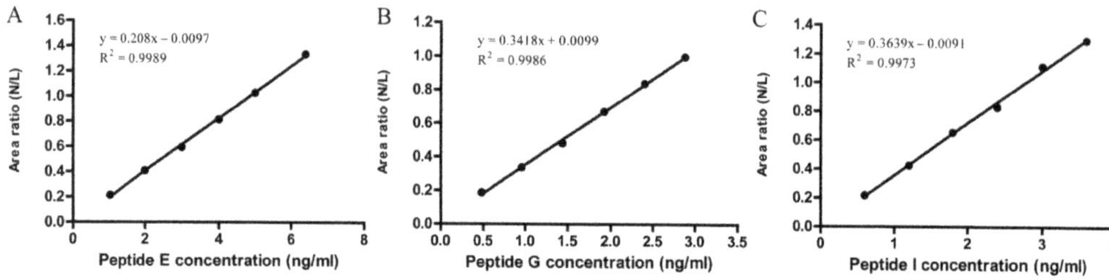

Figure 3. Standard curve for three peptide standards showing linearity. The X axis indicates the concentration of the target peptide and the Y axis indicates the area ratio of the target peptide versus the isotope-labeled peptide. The standard curve is obtained after linear fitting. The standard curves of all three viruses have good linearity with $R^2 > 0.99$. (Panel **A**) is the standard curve for the prototype strain, (Panel **B**) for the Delta strain, and (Panel **C**) for the Omicron strain.

3.2.3. Recovery

To assess the matrix effect, peptide control solutions with different concentrations were added to the trivalent vaccine samples which were processed and detected simultaneously with the samples without the added peptides. The recovery rate was calculated using the formula: $\text{recovery}(\%) = \frac{\text{sample with added peptide} - \text{sample without peptide}}{\text{theoretical peptide addition amount}} \times 100\%$. The average recovery rates of the prototype strain (peptide E), Delta strain (peptide G), and Omicron strain (peptide I) were 112.81%, 94.07%, and 87.71%, respectively. These results indicated that the method has good accuracy and the matrix has little influence on the method.

3.2.4. Precision

Repeatability was evaluated by processing a batch of trivalent vaccine samples six times, yielding RSDs of 5.5%, 5.7%, and 5.8% for peptides E, G, and H, respectively. The intermediate precision was assessed by analyzing the same batch 12 times over 2 days by 2 experimenters, yielding RSDs of 5.0%, 6.1%, and 6.8% for peptides E, G, and H, respectively. Both the repeatability and intermediate precision demonstrated that RSDs were within 10%, indicating good precision of the new method.

3.3. *Quantification of S Protein in Inactivated Vaccine Bulks and Trivalent Vaccines*

The content of S protein in vaccine bulks was quantified using the validated method. In total, 18 batches of vaccine bulks from prototype, Delta, or Omicron strains were analyzed individually. The results are shown in Table 3. Data showed that the S protein content was consistent within different batches of the same strain but varied greatly among three different strains, with the prototype bulk having the highest S protein content. The average S protein content of the Delta and Omicron bulk solutions was approximately 1/10 and 1/7 that of the prototype, respectively. Previous studies have reported that SARS-CoV-2 virus, produced in Vero cells, was prone to mutations in S1/S2 deletion which enhanced growth and stability in vitro [33]. ΔS679–688 has also beenfound in inactivated prototype strain vaccines, located at the S1/S2 junction [34]. However, no similar mutations were observed in vaccines produced with the Delta or Omicron strain. It was suggested that S1/S2 deficiency may increase the proportion of S protein, which is consistent with data observed in Table 3.

Table 3. S protein content in inactivated SARS-CoV-2 vaccine bulks and trivalent vaccines.

Samples	Lot	Spike Protein Concentration (pmol/mL)		
		Prototype Type	Delta	Omicron
Vaccine bulks of the prototype strain	1	68.5	/	/
	2	68.7	/	/
	3	67.7	/	/
	4	68.3	/	/
	5	71.3	/	/
	6	72.3	/	/
Vaccine bulks of the Delta strain	1	/	7.6	/
	2	/	7.6	/
	3	/	5.6	/
	4	/	5.8	/
	5	/	7.9	/
	6	/	5.1	/
Vaccine bulks of the Omicron strain	1	/	/	10.3
	2	/	/	11.7
	3	/	/	10.3
	4	/	/	9.2
	5	/	/	9
	6	/	/	10.3
trivalent inactivated vaccine	1	2.6	2.5	2.4
	2	2.6	2.4	2.6
	3	2.7	2.6	2.3

"/" indicate not detected.

Three batches of trivalent vaccines were produced by mixing different strain bulks, with the same S protein final concentration of 2.5 pmol/mL for three strains. The S protein content of each trivalent sample was measured and the results are shown in Table 3. The content of each strain ranged from 2.3 to 2.7 pmol/mL, in general agreement with the theoretical proportion, confirming the accuracy of the method for trivalent samples. These results further verified the specificity and accuracy of the method.

4. Discussion

The S protein is the major effective antigenic component of inactivated SARS-CoV-2 vaccine and its accurate determination is a key indicator for vaccine quality control, which is usually quantitatively detected by immunological methods such as ELISA. However, due to the high homology of S proteins among the various SARS-CoV-2 variant, methods based on antibody–antigen interactions are difficultly applied to the identification of different variants, as well as to the S protein quantification of different strains in multivalent vaccines. The results published by BIBP showed that the ELISA method for S protein of the prototype, Delta, and Omicron strains had great interference in detecting bivalent or trivalent vaccines composed by the three strains. The interference rate generated by this ELISA method displayed that the Delta/Omicron versus prototype was 6.49%, the prototype/Omicron versus is 140.7%, and the Delta/prototype versus Omicron was 184.41% in the trivalent vaccine, in which the ratio was 1:1:1 by three strains. So, it is difficult to accurately quantify the monovalent components in trivalent vaccine [35]. The ELISA method used by Sinovac had a similar situation (data unpublished). In the previous study of neuraminidase content detection of trivalent influenza vaccine, our laboratory prepared type-specific antibodies for specific peptides (15 amino acids) in N1, N2, and NB and established a Slot Blot detection method using these antibodies [36]. Although many different sites of amino acid sequences were observed in the strains between genotype, Delta, and Omicron, no specific peptides with consecutive multiple amino acids were found. Theoretically, the specific antibodies for mutant-site epitopes can largely avoid the cross-reactivity of S proteins between different strains, but so far there is no report of an immunological method that can precisely distinguish the S proteins of three strains.

Using molecular biology methods to identify SARS-CoV-2 variants was generally applied in surveillance and clinical diagnosis. Deep sequencing technology is widely utilized to identify SARS-CoV-2 variants, which can identify each mutation in the sample [37,38]. Real-time PCR assays for identifying SARS-CoV-2 variants were also reported [39,40]. While these molecular biology methods could be used to identify inactivated SARS-CoV-2 vaccines, the RNA level of vaccines may not necessarily be correlated to the S protein content.

MS methods offer high specificity, sensitivity, and accuracy for qualitative and quantitative analysis of complex biological samples. It is widely used in biomedical fields such as clinical diagnosis, drug research, and development [32,41]. It is also used in the development and quality control of HPV, Ebola, HIV, influenza, SARS-CoV-2, and other vaccines [29–31,42]. Among the many mass spectrometry quantification methods, isotope dilution tandem mass spectrometer (IDMS) has been recognized as the reference method [43]. The IDMS method is typically carried out using the following steps. The proteins that need to be quantified are enzymatically digested into peptides and specific peptides are selected for quantification. A fixed amount of synthetic 13C and 15N isotopically labeled peptides are spiked into the standards and samples as internal standards. After enzymatic digestion, samples are separated by liquid chromatography and detected by MS in multiple reaction monitoring modes. The area ratio of the target peptide and isotope-labeled peptide is linearly fitted to the concentration of the target peptide to obtain a standard curve, which was used to quantify the content of protein. Since isotopically labeled peptides are simultaneously processed with the sample for digestion and LC–MS detection, quantified by area ratio, the influence of the pretreatment and injection volume on the results can be reduced, allowing for accurate and precise quantification [31,43]. Tracie L et al. have established the IDMS method for quantifying influenza vaccine HA and other proteins, which has been applied in the evaluation of influenza vaccine strain products and the development of international standards [44]. A similar IDMS method for the quantitative detection targeting the S protein has also been established [31]. However, studies on the detection of variant-based vaccines and multivalent vaccines have not been reported yet.

Here, we established an IDMS method for quantifying S protein by targeting specific peptides that can simultaneously distinguish the three variants. This method has good specificity. In monovalent stocks, only the signal of strain-specific peptides can be detected rather than other strain or matrix proteins (Figure 2). Furthermore, the results of S protein content detection was consistent with the theoretical values in the trivalent vaccine mixed with three virus bulks (Table 3), indicating that there was no significant mutual interference among the three strains.

Although the IDMS method enables the precise and accurate quantification of S proteins, there are some limitations that require careful discussion. As an antigen, the S protein plays a critical role in stimulating the production of neutralizing antibodies that protect against infection. The use of neutralizing antibodies to establish the ELISA detection method can reflect the integrity of protein antigen structure, which is also an important reason why many vaccines use ELISA method for antigen content detection [42,45,46]. Unlike the ELISA method, MS directly detects the polypeptide after enzymatic digestion of proteins, which has the same number of moles as the protein, to calculate the protein content. Most epitopes are destroyed during enzymatic digestion of proteins, so the results obtained by the MS method may not reflect the destruction of epitopes. Therefore, for vaccine stock solutions and monovalent vaccines that can be quantified by the ELISA method, we do not recommend using this method to replace the ELISA method. Rather, as a supplementary method for better quality control of vaccines or for the preparation and assignment of standards. For the inactivated trivalent SARS-CoV-2 vaccine, from what we know, the method established in this study is currently the only accurate option for quantification.

5. Conclusions

In this study, a novel method was developed for the quantification of the S protein in monovalent or trivalent inactivated vaccines against the SARS-CoV-2 virus (prototype, Delta, and Omicron strains). Compared with the traditional methods, such as ELISA, this method can identify and quantify the S protein in different strains of the SARS-CoV-2 virus (prototype, Delta, and Omicron strains), as well as compare the S protein content in different batches of trivalent vaccines and assess the batch-to-batch stability, which can provide valuable reference points for establishing product quality control standards. The high selectivity of liquid chromatography–mass spectrometry (LC–MS) ensures minimal interference from complex sample matrices, making the method useful for the quality control of monovalent or trivalent inactivated vaccines.

Although there are many reports on SARS-CoV-2 non-spike proteins, their roles mostly remain unclear [29]. We believe that further LC–MS/MS methods could extend the quantification of these proteins when researchers grasp their characteristics deeply and understand their impact on the safety and efficacy of vaccines. In the event that a new variant emerges that cannot be distinguished by existing target peptides, accurate quantification could still be achieved by synthesizing new target peptides and isotopically labeled peptides within a few weeks, as described previously.

Author Contributions: Conceptualization, K.X. and C.L.; methodology, K.X. and H.S.; software, H.S.; validation, K.W., Y.Q. and Z.Q.; formal analysis, Y.H.; resources, Y.H. and C.L.; data curation, K.W., K.X. and H.S.; writing—original draft preparation, K.X. and H.S.; writing—review and editing, Y.H. and C.L.; visualization, K.X.; supervision, Y.Q.; project administration, C.L.; funding acquisition, C.L. All authors have read and agreed to the published version of the manuscript.

Funding: This research was funded by the Emergency Key Program of Guangzhou Laboratory (Grant no. EKPG21-30-1).

Institutional Review Board Statement: Ethics approval was not required for this research.

Informed Consent Statement: Not applicable.

Data Availability Statement: Data will be made available on request.

Conflicts of Interest: The authors declare that they have no known competing financial interest or personal relationships that could have appeared to influence the work reported in this paper.

References

1. Huang, C.; Wang, Y.; Li, X.; Ren, L.; Zhao, J.; Hu, Y.; Zhang, L.; Fan, G.; Xu, J.; Gu, X.; et al. Clinical features of patients infected with 2019 novel coronavirus in Wuhan, China. *Lancet* **2020**, *395*, 497–506. [CrossRef] [PubMed]
2. Weisblum, Y.; Schmidt, F.; Zhang, F.; DaSilva, J.; Poston, D.; Lorenzi, J.C.; Muecksch, F.; Rutkowska, M.; Hoffmann, H.H.; Michailidis, E.; et al. Escape from neutralizing antibodies by SARS-CoV-2 spike protein variants. *Elife* **2020**, *9*, e61312. [CrossRef] [PubMed]
3. Zhu, N.; Zhang, D.; Wang, W.; Li, X.; Yang, B.; Song, J.; Zhao, X.; Huang, B.; Shi, W.; Lu, R.; et al. A Novel Coronavirus from Patients with Pneumonia in China, 2019. *N. Engl. J. Med.* **2020**, *382*, 727–733. [CrossRef] [PubMed]
4. Yurkovetskiy, L.; Wang, X.; Pascal, K.E.; Tomkins-Tinch, C.; Nyalile, T.P.; Wang, Y.; Baum, A.; Diehl, W.E.; Dauphin, A.; Carbone, C.; et al. Structural and Functional Analysis of the D614G SARS-CoV-2 Spike Protein Variant. *Cell* **2020**, *183*, 739–751.e8. [CrossRef]
5. Zhou, P.; Yang, X.L.; Wang, X.G.; Hu, B.; Zhang, L.; Zhang, W.; Si, H.R.; Zhu, Y.; Li, B.; Huang, C.L.; et al. A pneumonia outbreak associated with a new coronavirus of probable bat origin. *Nature* **2020**, *579*, 270–273. [CrossRef]
6. Coronaviridae Study Group of the International Committee on Taxonomy of V. The species Severe acute respiratory syndrome-related coronavirus: Classifying 2019-nCoV and naming it SARS-CoV-2. *Nat. Microbiol.* **2020**, *5*, 536–544. [CrossRef]
7. Santacroce, L.; Charitos, I.A.; Carretta, D.M.; De Nitto, E.; Lovero, R. The human coronaviruses (HCoVs) and the molecular mechanisms of SARS-CoV-2 infection. *J. Mol. Med.* **2021**, *99*, 93–106. [CrossRef]
8. Gordon, D.E.; Jang, G.M.; Bouhaddou, M.; Xu, J.; Obernier, K.; White, K.M.; O'Meara, M.J.; Rezelj, V.V.; Guo, J.Z.; Swaney, D.L.; et al. A SARS-CoV-2 protein interaction map reveals targets for drug repurposing. *Nature* **2020**, *583*, 459–468. [CrossRef]
9. Hoffmann, M.; Kleine-Weber, H.; Schroeder, S.; Krüger, N.; Herrler, T.; Erichsen, S.; Schiergens, T.S.; Herrler, G.; Wu, N.H.; Nitsche, A.; et al. SARS-CoV-2 Cell Entry Depends on ACE2 and TMPRSS2 and Is Blocked by a Clinically Proven Protease Inhibitor. *Cell* **2020**, *181*, 271–280.e8. [CrossRef]

10. Souza, P.F.N.; Mesquita, F.P.; Amaral, J.L.; Landim, P.G.C.; Lima, K.R.P.; Costa, M.B.; Farias, I.R.; Lima, L.B.; Montenegro, R.C. The human pandemic coronaviruses on the show: The spike glycoprotein as the main actor in the coronaviruses play. *Int. J. Biol. Macromol.* **2021**, *179*, 1–19. [CrossRef]
11. Frampton, D.; Rampling, T.; Cross, A.; Bailey, H.; Heaney, J.; Byott, M.; Scott, R.; Sconza, R.; Price, J.; Margaritis, M.; et al. Genomic characteristics and clinical effect of the emergent SARS-CoV-2 B.1.1.7 lineage in London, UK: A whole-genome sequencing and hospital-based cohort study. *Lancet Infect. Dis.* **2021**, *21*, 1246–1256. [CrossRef]
12. Planas, D.; Veyer, D.; Baidaliuk, A.; Staropoli, I.; Guivel-Benhassine, F.; Rajah, M.M.; Planchais, C.; Porrot, F.; Robillard, N.; Puech, J.; et al. Reduced sensitivity of SARS-CoV-2 variant Delta to antibody neutralization. *Nature* **2021**, *596*, 276–280. [CrossRef]
13. Araf, Y.; Akter, F.; Tang, Y.D.; Fatemi, R.; Parvez, M.S.A.; Zheng, C.; Hossain, M.G. Omicron variant of SARS-CoV-2: Genomics, transmissibility, and responses to current COVID-19 vaccines. *J. Med. Virol.* **2022**, *94*, 1825–1832. [CrossRef]
14. Tian, D.; Sun, Y.; Xu, H.; Ye, Q. The emergence and epidemic characteristics of the highly mutated SARS-CoV-2 Omicron variant. *J. Med. Virol.* **2022**, *94*, 2376–2383. [CrossRef]
15. Yin, W.; Xu, Y.; Xu, P.; Cao, X.; Wu, C.; Gu, C.; He, X.; Wang, X.; Huang, S.; Yuan, Q.; et al. Structures of the Omicron spike trimer with ACE2 and an anti-Omicron antibody. *Science* **2022**, *375*, 1048–1053. [CrossRef]
16. Cele, S.; Jackson, L.; Khoury, D.S.; Khan, K.; Moyo-Gwete, T.; Tegally, H.; San, J.E.; Cromer, D.; Scheepers, C.; Amoako, D.G.; et al. Omicron extensively but incompletely escapes Pfizer BNT162b2 neutralization. *Nature* **2022**, *602*, 654–656. [CrossRef]
17. Wilhelm, A.; Widera, M.; Grikscheit, K.; Toptan, T.; Schenk, B.; Pallas, C.; Metzler, M.; Kohmer, N.; Hoehl, S.; Marschalek, R.; et al. Limited neutralisation of the SARS-CoV-2 Omicron subvariants BA.1 and BA.2 by convalescent and vaccine serum and monoclonal antibodies. *EBioMedicine* **2022**, *82*, 104158. [CrossRef]
18. Yu, X.; Wei, D.; Xu, W.; Li, Y.; Li, X.; Zhang, X.; Qu, J.; Yang, Z.; Chen, E. Reduced sensitivity of SARS-CoV-2 Omicron variant to antibody neutralization elicited by booster vaccination. *Cell Discov.* **2022**, *8*, 4. [CrossRef]
19. Su, S.; Li, W.; Jiang, S. Developing pan-β-coronavirus vaccines against emerging SARS-CoV-2 variants of concern. *Trends Immunol.* **2022**, *43*, 170–172. [CrossRef]
20. Zhou, J.; Liu, Z.; Zhang, G.; Xu, W.; Xing, L.; Lu, L.; Wang, Q.; Jiang, S. Development of variant-proof severe acute respiratory syndrome coronavirus 2, pan-sarbecovirus, and pan-beta-coronavirus vaccines. *J. Med. Virol.* **2023**, *95*, e28172. [CrossRef]
21. Tan, C.W.; Chia, W.N.; Young, B.E.; Zhu, F.; Lim, B.L.; Sia, W.R.; Thein, T.L.; Chen, M.I.C.; Leo, Y.S.; Lye, D.C.; et al. Pan-Sarbecovirus Neutralizing Antibodies in BNT162b2-Immunized SARS-CoV-1 Survivors. *N. Engl. J. Med.* **2021**, *385*, 1401–1406. [CrossRef] [PubMed]
22. Saunders, K.O.; Lee, E.; Parks, R.; Martinez, D.R.; Li, D.; Chen, H.; Edwards, R.J.; Gobeil, S.; Barr, M.; Mansouri, K.; et al. Neutralizing antibody vaccine for pandemic and pre-emergent coronaviruses. *Nature* **2021**, *594*, 553–559. [CrossRef] [PubMed]
23. Joyce, M.G.; Chen, W.H.; Sankhala, R.S.; Hajduczki, A.; Thomas, P.V.; Choe, M.; Martinez, E.J.; Chang, W.C.; Peterson, C.E.; Morrison, E.B.; et al. SARS-CoV-2 ferritin nanoparticle vaccines elicit broad SARS coronavirus immunogenicity. *Cell Rep.* **2021**, *37*, 110143. [CrossRef] [PubMed]
24. Wang, C.Y.; Peng, W.J.; Kuo, B.S.; Ho, Y.H.; Wang, M.S.; Yang, Y.T.; Chang, P.Y.; Shen, Y.H.; Hwang, K.P. Toward a pan-SARS-CoV-2 vaccine targeting conserved epitopes on spike and non-spike proteins for potent, broad and durable immune responses. *PLoS Pathog.* **2023**, *19*, e1010870. [CrossRef]
25. Huang, R.; Ying, L.; Wang, J.; Xia, J.; Zhang, Y.; Mao, H.; Zhang, R.; Zang, R.; Le, Z.; Shu, Q.; et al. Non-spike and spike-specific memory T cell responses after the third dose of inactivated COVID-19 vaccine. *Front. Immunol.* **2023**, *14*, 1139620. [CrossRef]
26. Chalkias, S.; Eder, F.; Essink, B.; Khetan, S.; Nestorova, B.; Feng, J.; Chen, X.; Chang, Y.; Zhou, H.; Montefiori, D.; et al. Safety, immunogenicity and antibody persistence of a bivalent Beta-containing booster vaccine against COVID-19: A phase 2/3 trial. *Nat. Med.* **2022**, *28*, 2388–2397. [CrossRef]
27. Hause, A.M.; Marquez, P.; Zhang, B.; Myers, T.R.; Gee, J.; Su, J.R.; Blanc, P.G.; Thomas, A.; Thompson, D.; Shimabukuro, T.T.; et al. Safety Monitoring of Bivalent COVID-19 mRNA Vaccine Booster Doses Among Persons Aged ≥12 Years—United States, August 31–October 23, 2022. *MMWR Morb. Mortal. Wkly. Rep.* **2022**, *71*, 1401–1406. [CrossRef]
28. Long, Z.; Wei, C.; Dong, X.; Li, X.; Yang, H.; Deng, H.; Ma, X.; Yin, S.; Qi, Y.; Bo, T. Simultaneous quantification of spike and nucleocapsid protein in inactivated COVID-19 vaccine bulk by liquid chromatography-tandem mass spectrometry. *J. Chromatogr. B* **2021**, *1181*, 122884. [CrossRef]
29. Cazares, L.H.; Chaerkady, R.; Samuel Weng, S.H.; Boo, C.C.; Cimbro, R.; Hsu, H.E.; Rajan, S.; Dall'Acqua, W.; Clarke, L.; Ren, K.; et al. Development of a Parallel Reaction Monitoring Mass Spectrometry Assay for the Detection of SARS-CoV-2 Spike Glycoprotein and Nucleoprotein. *Anal. Chem.* **2020**, *92*, 13813–13821. [CrossRef]
30. Rosen, O.; Jayson, A.; Dor, E.; Epstein, E.; Makovitzki, A.; Cherry, L.; Lupu, E.; Monash, A.; Borni, S.; Baruchi, T.; et al. SARS-CoV-2 spike antigen quantification by targeted mass spectrometry of a virus-based vaccine. *J. Virol. Methods* **2022**, *303*, 114498. [CrossRef]
31. Pierce-Ruiz, C.; Santana, W.I.; Sutton, W.J.H.; Fischler, D.A.; Cooper, H.C.; Marc, L.R.; Barr, J.R.; Williams, T.L. Quantification of SARS-CoV-2 spike and nucleocapsid proteins using isotope dilution tandem mass spectrometry. *Vaccine* **2021**, *39*, 5106–5115. [CrossRef]
32. Hoofnagle, A.N.; Whiteaker, J.R.; Carr, S.A.; Kuhn, E.; Liu, T.; Massoni, S.A.; Thomas, S.N.; Townsend, R.R.; Zimmerman, L.J.; Boja, E.; et al. Recommendations for the Generation, Quantification, Storage, and Handling of Peptides Used for Mass Spectrometry-Based Assays. *Clin. Chem.* **2016**, *62*, 48–69. [CrossRef]

33. Lau, S.Y.; Wang, P.; Mok, B.W.; Zhang, A.J.; Chu, H.; Lee, A.C.; Deng, S.; Chen, P.; Chan, K.H.; Song, W.; et al. Attenuated SARS-CoV-2 variants with deletions at the S1/S2 junction. *Emerg. Microbes Infect.* **2020**, *9*, 837–842. [CrossRef]
34. To, K.K.; Li, X.; Lung, D.C.; Ip, J.D.; Chan, W.M.; Chu, A.W.; Yip, C.C.Y.; Chen, J.H.; Poon, R.W.S.; Tsoi, H.W.; et al. False Coronavirus Disease 2019 Cases due to Contamination by Inactivated Virus Vaccine. *Clin. Infect. Dis.* **2022**, *74*, 1485–1488. [CrossRef]
35. Dong, Y.; Zhao, Y.X.; Zhuge, X.L.; Yang, Z.N.; Li, J.; Li, A.L. Analysis of S protein cross-interference in three different corona virus strains. *Chin. J. New Drugs* **2022**, *31*, 2136–2143.
36. Xu, K.; Li, C.; Gravel, C.; Jiang, Z.; Jaentschke, B.; Van Domselaar, G.; Li, X.; Wang, J. Universal type/subtype-specific antibodies for quantitative analyses of neuraminidase in trivalent influenza vaccines. *Sci. Rep.* **2018**, *8*, 1067. [CrossRef]
37. Izquierdo-Lara, R.; Elsinga, G.; Heijnen, L.; Munnink, B.B.O.; Schapendonk, C.M.E.; Nieuwenhuijse, D.; Kon, M.; Lu, L.; Aarestrup, F.M.; Lycett, S.; et al. Monitoring SARS-CoV-2 Circulation and Diversity through Community Wastewater Sequencing, the Netherlands and Belgium. *Emerg. Infect. Dis.* **2021**, *27*, 1405–1415. [CrossRef]
38. Nasir, J.A.; Kozak, R.A.; Aftanas, P.; Raphenya, A.R.; Smith, K.M.; Maguire, F.; Maan, H.; Alruwaili, M.; Banerjee, A.; Mbareche, H.; et al. A Comparison of Whole Genome Sequencing of SARS-CoV-2 Using Amplicon-Based Sequencing, Random Hexamers, and Bait Capture. *Viruses* **2020**, *12*, 895. [CrossRef]
39. Banada, P.; Green, R.; Banik, S.; Chopoorian, A.; Streck, D.; Jones, R.; Chakravorty, S.; Alland, D. A Simple Reverse Transcriptase PCR Melting-Temperature Assay To Rapidly Screen for Widely Circulating SARS-CoV-2 Variants. *J. Clin. Microbiol.* **2021**, *59*, e0084521. [CrossRef]
40. Dikdan, R.J.; Marras, S.A.E.; Field, A.P.; Brownlee, A.; Cironi, A.; Hill, D.A.; Tyagi, S. Multiplex PCR Assays for Identifying all Major Severe Acute Respiratory Syndrome Coronavirus 2 Variants. *J. Mol. Diagn.* **2022**, *24*, 309–319. [CrossRef] [PubMed]
41. Loos, G.; Van Schepdael, A.; Cabooter, D. Quantitative mass spectrometry methods for pharmaceutical analysis. *Philos. Trans. A Math. Phys. Eng. Sci.* **2016**, *374*, 20150366. [CrossRef] [PubMed]
42. Sharma, V.K.; Sharma, I.; Glick, J. The expanding role of mass spectrometry in the field of vaccine development. *Mass Spectrom. Rev.* **2020**, *39*, 83–104. [CrossRef] [PubMed]
43. Brun, V.; Masselon, C.; Garin, J.; Dupuis, A. Isotope dilution strategies for absolute quantitative proteomics. *J. Proteom.* **2009**, *72*, 740–749. [CrossRef] [PubMed]
44. Williams, T.L.; Pirkle, J.L.; Barr, J.R. Simultaneous quantification of hemagglutinin and neuraminidase of influenza virus using isotope dilution mass spectrometry. *Vaccine* **2012**, *30*, 2475–2482. [CrossRef]
45. Verch, T.; Trausch, J.J.; Shank-Retzlaff, M. Principles of vaccine potency assays. *Bioanalysis* **2018**, *10*, 163–180. [CrossRef]
46. Kouiavskaia, D.; Puligedda, R.D.; Dessain, S.K.; Chumakov, K. Universal ELISA for quantification of D-antigen in inactivated poliovirus vaccines. *J. Virol. Methods* **2020**, *276*, 113785. [CrossRef]

vaccines

Review

SARS-CoV-2 Neutralization Assays Used in Clinical Trials: A Narrative Review

Yeqing Sun [1,2], Weijin Huang [2], Hongyu Xiang [1,*] and Jianhui Nie [2,*]

[1] School of Life Sciences, Jilin University, Changchun 130012, China; yeqing22@mails.jlu.edu.cn
[2] Division of HIV/AIDS and Sex-Transmitted Virus Vaccines, National Institutes for Food and Drug Control, State Key Laboratory of Drug Regulatory Science, NHC Key Laboratory of Research on Quality and Standardization of Biotech Products, NMPA Key Laboratory for Quality Research and Evaluation of Biological Products, Beijing 102629, China; huangweijin@nifdc.org.cn
* Correspondence: hyxiang@jlu.edu.cn (H.X.); niejianhui@nifdc.org.cn (J.N.)

Abstract: Since the emergence of COVID-19, extensive research efforts have been undertaken to accelerate the development of multiple types of vaccines to combat the pandemic. These include inactivated, recombinant subunit, viral vector, and nucleic acid vaccines. In the development of these diverse vaccines, appropriate methods to assess vaccine immunogenicity are essential in both preclinical and clinical studies. Among the biomarkers used in vaccine evaluation, the neutralizing antibody level serves as a pivotal indicator for assessing vaccine efficacy. Neutralizing antibody detection methods can mainly be classified into three types: the conventional virus neutralization test, pseudovirus neutralization test, and surrogate virus neutralization test. Importantly, standardization of these assays is critical for their application to yield results that are comparable across different laboratories. The development and use of international or regional standards would facilitate assay standardization and facilitate comparisons of the immune responses induced by different vaccines. In this comprehensive review, we discuss the principles, advantages, limitations, and application of different SARS-CoV-2 neutralization assays in vaccine clinical trials. This will provide guidance for the development and evaluation of COVID-19 vaccines.

Keywords: SARS-CoV-2; neutralizing antibodies; clinical trials; standards; correlation of protection

1. Introduction

The severe acute respiratory syndrome coronavirus 2 (SARS-CoV-2) was first reported in Wuhan, China, in December 2019 [1], followed by the pandemic of coronavirus disease 2019 (COVID-19) [2–4]. After vaccination or viral infection, the innate/non-specific immune system is initially activated. Antigen-presenting cells, including macrophages and dendritic cells, engulf the exogenous antigens. Subsequently, antigen-presenting cells transmit antigen information to T cells and B cells, activating the adaptive/specific immune system. Activated CD4+ T cells secrete various cytokines to further stimulate immune responses, while CD8+ T cells eliminate virus-infected cells [5]. B cells are also activated to produce specific antibodies, including neutralizing antibodies (nAbs) and binding antibodies [6]. Therefore, nAbs, cellular immunity, innate immunity, and other immunological indicators are vital in assessing the effectiveness of viral vaccines [7,8]. The importance of measuring nAbs in COVID-19 lies in the fact that they can serve as correlates of protection. Quantitative detection of nAbs specific to SARS-CoV-2 facilitates monitoring of viral infections. Furthermore, nAbs can be used to assess monoclonal antibodies, convalescent sera, and vaccine effectiveness [9]. Therefore, the detection of nAbs against SARS-CoV-2 is important in clinical trial investigations.

On 5 May 2023, the World Health Organization (WHO) announced that COVID-19 no longer constituted a public health emergency of international concern [10]. However, this announcement does not signify the end of COVID-19. Research and development efforts regarding vaccines against SARS-CoV-2 are ongoing, especially for broadly protective candidate vaccines. Therefore, continuous monitoring is necessary to track emerging variants of SARS-CoV-2 and to facilitate the development of the next-generation vaccines. In this review, we searched PubMed from January 2020 to 29 April 2024, to identify all the eligible studies in order to comprehensively describe the progress in SARS-CoV-2 neutralization assays used in clinical trials.

2. Current Status of SARS-CoV-2 Vaccines

Since the first COVID-19 vaccine on 22 July 2020, as of February 2024, a total of about 13.59 billion vaccine doses have been administered worldwide [11]. The types of COVID-19 vaccines mainly include inactivated vaccines, recombinant protein subunit vaccines, viral vector vaccines, and nucleic acid vaccines (Table 1).

Table 1. Main available COVID-19 vaccines.

Vaccine Type	Name	Manufacturer	Immunogenic Detection Method	Immune Indexes	References
Inactivated vaccines	CoronaVac	Sinovac	• Neutralization potency against pseudovirus. • Neutralization potency against live SARS-CoV-2 (CPE). • RBD-specific IgG were detected using the chemiluminescence qualitative kit. • T-cell response was determined with the ELISpot method.	• Seroconversion rates of neutralizing antibodies to live SARS-CoV-2 • Seroconversion rates of binding antibodies to SARS-CoV-2 RBD-specific. • GMT of neutralizing antibodies to live SARS-CoV-2. • GMT of SARS-CoV-2 RBD-specific binding antibodies.	[12–14]
	BBIBP-CorV	Sinopharm/BIBP	• Neutralization potency against live SARS-CoV-2 (CPE).	• Seroconversion rates of neutralizing antibodies to live SARS-CoV-2 • GMT of neutralizing antibodies to live SARS-CoV-2.	[15,16]
mRNA vaccines	BNT162b2	Pfizer-BioNTech	• Neutralization potency against live SARS-CoV-2 (recombinant live virus) • Neutralization potency against pseudovirusVSV-SARS-CoV-2 spike. • SARS-CoV-2-S1–specific IgG/RBD–specific IgG direct Luminex immunoassay	• GMT of SARS-CoV-2 S1-specific binding antibodies. • GMT of neutralizing antibodies to live SARS-CoV-2.	[17–20]
	mRNA-1273	Moderna	• Neutralization potency against pseudovirus. • Neutralization potency against live SARS-CoV-2 (PRNT). • Binding antibody responses against SARS-CoV-2 S protein were assessed by ELISA. • T-cell responses against the spike protein were assessed by an intracellular cytokine–staining assay.	• GMT of SARS-CoV-2 S-specific binding antibodies. • GMT of neutralizing antibodies to pseudovirus. • GMT of neutralizing antibodies to live SARS-CoV-2.	[21,22]
	mRNA-1273.214	Moderna	• Neutralization potency against SARS-CoV-2 pseudovirus.	• GMT of neutralizing antibodies to the omicron BA.1 variant • GMT of neutralizing antibodies to the omicron BA.4 and BA.5 (BA.4/5)	[23]

Table 1. *Cont.*

Vaccine Type	Name	Manufacturer	Immunogenic Detection Method	Immune Indexes	References
Adenovirus vaccines	Ad26.COV2.S	Janssen–Cilag InternationalNV	• SARS-CoV-2 S-specific binding antibodies were measured with ELISA. • SARS-CoV-2 serum neutralizing-antibody titers in a random subgroup of samples by means of a wild-type virus microneutralization assay(wtVMA). • S-specific T-cell responses were assessed by intracellular cytokine staining.	• Seroconversion rates of neutralizing antibodies to live SARS-CoV-2 • GMC of SARS-CoV-2 S-specific binding antibodies. • GMT of neutralizing antibodies to live SARS-CoV-2.	[15,24]
	AZD1222/ Vaxzevria	AstraZeneca	• SARS-CoV-2 S-specific binding antibodies were measured with ELISA. • Neutralization potency against live SARS-CoV-2 by microneutralization assay. • Neutralization potency against live SARS-CoV-2 • T-cell response was determined with the ELISpot method.	• SARS-CoV-2 S-specific binding antibody titers • GMT of neutralizing antibodies to live SARS-CoV-2.	[25–27]
	Sputnik V	Gamaleya Nat. Center of Epidem. and Microbiol.	• SARS-CoV-2 RBD-specific IgGs were measured with ELISA. • The titer of nAbs was measured by microneutralization assay. • Antigen-specific proliferating CD4 and CD8 cells by flow cytometry	• The change from baseline in antigen-specific antibody levels. • SARS-CoV-2 S-specific binding antibody titers • Specific T-cell immunity and interferon-γ production or lymphoproliferation	[28,29]
	Ad5-nCoV/ Convidecia	CanSinoBIO	• SARS-CoV-2 RBD-specific binding antibodies were measured with ELISA. • Neutralization potency against live SARS-CoV-2 (recombinant live virus) • Neutralization potency against pseudovirusVSV-SARS-CoV-2 spike. • T-cell response was determined with the ELISpot method.	• GMT of neutralizing antibodies to live SARS-CoV-2. • GMT of neutralizing antibodies to pseudovirus. • The RBD-specific ELISA antibodies peaked. • Seroconversion rates of neutralizing antibodies to live SARS-CoV-2 • Seroconversion rates of neutralizing antibodies to pseudovirus. • Seroconversion rates of RBD-specific binding antibodies.	[30,31]

Table 1. *Cont.*

Vaccine Type	Name	Manufacturer	Immunogenic Detection Method	Immune Indexes	References
Subunit vaccines	NVX-CoV2373/ Nuvaxovid	Novavax	• IgG anti–spike protein response (ELISA) • Wild-type virus neutralization assays(MN) • Antigen-specific CD4+ T cells was determined with the intracellular cytokine staining	• ELISA anti-spike IgG geometric mean ELISA units (GMEUs) • Wild-type SARS-CoV-2 microneutralization assay at an inhibitory concentration greater than 99% (MN IC > 99%) titer responses	[32,33]
	ZF2001	Zhifei Longcom	• SARS-CoV-2 RBD-specific binding antibodies were measured with ELISA. • Neutralization potency against live SARS-CoV-2 (CPE) • T-cell responses against the spike protein were assessed by an enzyme-linked immuno-spot (ELISpot)	• Seroconversion rates of RBD-binding IgG • GMTs of RBD-binding antibodies • Seroconversion rates of neutralizing antibodies to live SARS-CoV-2. • GMTs of neutralizing antibodies to live SARS-CoV-2.	[34,35]

Cytopathic effect (CPE). Plaque reduction neutralization test (PRNT). Microneutralization (MN). Geometric mean titer (GMT). Geometric mean concentration (GMC).

The safety and efficacy of the aforementioned vaccines have been extensively validated globally [12–35]. However, the effectiveness of these vaccines may vary among different populations and in the presence of variant viruses. Because different types of vaccines have been subject to different methods during their research and development phases, it is impossible to compare them directly. Nevertheless, all vaccines that have been listed for emergency use by the WHO are proven to be highly effective in preventing severe COVID-19 illness.

As a conventional platform, inactivated vaccines have long played a critical role in the battle against infectious diseases [36]. Physical or chemical methods are used to inactivate the pathogen, rendering it non-infectious and non-pathogenic while retaining its immunogenicity. Some inactivated vaccines are formulated with appropriate adjuvants to further enhance their efficacy [37]. Two examples of inactivated COVID-19 vaccines are BBIBP-CorV from the Beijing Institute of Biological Products and CoronaVac from Sinovac Biotech. The inactivated vaccine developed by the Wuhan Institute of Biological Products, a subsidiary of the China National Pharmaceutical Group (Sinopharm), uses the SARS-CoV-2 WTV04 strain [16]. Sinovac's inactivated vaccine, CoronaVac, uses the SARS-CoV-2 CN02 strain and was officially listed on the World Health Organization's Emergency Use Listing in June 2021 [12]. Inactivated vaccines have shown good safety profiles, high immunogenicity, and minimal adverse reactions [38].

Recombinant protein subunit vaccines rely on the expression of protective antigens in prokaryotic or eukaryotic cells and are usually formulated with appropriate adjuvants. ZF2001, a recombinant protein vaccine from Chongqing Zhifei Biological Products, uses the Chinese hamster ovary (CHO) cell line to express purified SARS-CoV-2 S-RBD dimeric antigen protein and is formulated with alum as an adjuvant [34]. NVX-CoV2373, a recombinant protein vaccine from Novavax, is based on a nanoparticle containing the full-length S protein plus Matrix-M adjuvant. Subunit vaccines have high purity, stability, and safety profiles and are generally administered in a two-dose regimen [39].

Viral vector vaccines are constructed by inserting specific antigenic nucleotide sequences into viral vectors to express the antigens of interest in host cells. Ad5-nCoV, a human type 5 adenovirus vector vaccine from China, is the first viral vector vaccine to enter clinical trials [40]. This vaccine induces sustained humoral and cellular immune responses with a single-dose injection [41]. Another viral vector vaccine is AZD1222 (ChAdOx1 nCoV-19) developed by the University of Oxford/AstraZeneca, which uses a replication-deficient chimpanzee adenovirus vector to carry the SARS-CoV-2 spike (S) gene. After vaccination, the S protein expressed in vivo can trigger the immune system to generate antibodies and cellular immune responses against SARS-CoV-2 [42]. Sputnik V is one of the first COVID-19 vaccines approved for emergency use, developed by the Gamaleya Research Center in Russia, and it is the first non-Western vaccine to complete all phase III clinical trials [43]. Sputnik V uses two different adenoviral vectors (rAd26 and rAd5) to transmit the S protein gene of the virus, which can enhance the immune system's response and improve the effectiveness and duration of the vaccine [44]. Multiple studies have demonstrated that Sputnik V is highly effective in preventing both symptomatic and severe infections of COVID-19.

Nucleic acid vaccines (DNA or RNA) are designed to deliver genes that encode protective antigens into host cells and use the host translation machinery to produce the antigen protein and induce immune responses against the target pathogen [45]. mRNA vaccines have the advantage of a shorter development cycle, flexible design, and the ability to expand the range of involved antigens for an enhanced breadth of immune responses. Moreover, the exogenous mRNA itself can act as a self-adjuvant, eliminating the need for additional adjuvants [46,47]. Compared with mRNA, DNA molecules are stable and can persist in host cells for a long time, continuously generating endogenous antigens to elicit a sustained immune response [48]. However, owing to the poor delivery efficiency of DNA vaccines, relatively lower immune responses are induced in comparison with mRNA vaccines.

To address the issue of immune evasion by SARS-CoV-2 variants, a variety of approaches have been used to enhance vaccine efficacy against the circulating strains. On 31 August 2022, the U.S. Food and Drug Administration (FDA) authorized the Pfizer-BioNTech and Moderna bivalent COVID-19 vaccines for use as booster shots [49]. Moderna's bivalent vaccines, mRNA-1273.222 (targeting the original S protein and Omicron BA.4/5 variant) and mRNA-1273.214 (targeting the original S protein and Omicron variant), use a mixture of mRNAs encoding the Omicron S protein and the Wuhan-1 S protein [23,50]. Clinical trials have demonstrated that, as booster shots, the mRNA-1273.214 and mRNA-1273.222 bivalent vaccines have safety profiles similar to that of the monovalent mRNA-1273 SARS-CoV-2 vaccine and exhibit superior neutralizing antibody responses against the Omicron variant compared with mRNA-1273 used as a booster [19]. Following a similar strategy, Pfizer-BioNTech developed its bivalent SARS-CoV-2 vaccine, which encodes the S protein of both the original SARS-CoV-2 strain and the Omicron BA.4/BA.5 variant. When used as a booster, the protective efficacy against symptomatic SARS-CoV-2 infection of Pfizer-BioNTech's bivalent SARS-CoV-2 vaccine is enhanced by 28% to 56% compared with the original vaccine [51]. Various studies have demonstrated that bivalent COVID-19 vaccines can maintain immune protection against the original virus strain while triggering a specific neutralizing antibody response against the currently prevalent variant, indicating a broader spectrum of protection. However, with the continuing evolution of SARS-CoV-2 and changing antigenicity of emerging variants, the WHO, European Medicines Agency, and FDA have all recommended updating monovalent XBB component-containing COVID-19 vaccines for the autumn-winter season of 2023/2024.

3. Efficacy of SARS-CoV-2 Vaccines

All the approved vaccines on the WHO emergency use list must show a minimum efficacy of 50% or higher in clinical trials [52]. After approval, their safety and efficacy are continuously monitored. The clinical evaluation indicators for SARS-CoV-2 vaccines include the aspects of safety, immunogenicity, and effectiveness. Phase I/II trials typically focus on assessing the safety and immunogenicity of the vaccine, whereas Phase III trials primarily aim to verify its safety and effectiveness in large populations. The evaluation of vaccine effectiveness involves multiple dimensions of human immune system involvement.

Upon vaccination, the human body can generate neutralizing antibodies (nAbs) against SARS-CoV-2, which confer protective immunity [53]. These nAbs constitute only a small portion of the antibodies secreted by B cells yet possess antiviral activity. The titer of nAbs in the blood is crucial, representing a key indicator in evaluating the effectiveness of SARS-CoV-2 vaccines.

The gold standard for assessing vaccine effectiveness is the relative protection rate, which requires large-scale clinical trials. However, clinical trials are time-consuming and often struggle to keep pace with the speed of viral mutations. Currently, most studies suggest that vaccines primarily induce nAbs, leading to the possibility of using nAbs as surrogate endpoints. A surrogate endpoint refers to a biomarker that can be used as a substitute for the clinical endpoint. The detection of nAbs can effectively save time in vaccine clinical trials as compared with large-scale clinical trials [54].

4. Role of Neutralizing Antibodies

Vaccination or natural infection induces the production of cellular and humoral immunity in the body, generating antibodies through immune synergistic effects to protect against viral infections [55]. Studies have shown that different levels of antibody, including nAbs and binding antibodies, can be produced 1–2 weeks after vaccination or natural infection, with only a small portion of these antibodies being nAbs [53]. As a special type of antibody produced by B lymphocytes, nAbs can bind to the surface of viral particles, thereby blocking the viral replication cycle and preventing viral infection of cells [56]. As crucial immune markers, the mechanisms of action of nAbs generally include: (1) altering the viral surface configuration; (2) binding to virus-binding sites involved in adsorption to prevent viral attachment and cell invasion; (3) forming immune complexes with the virus, which can be easily engulfed and cleared by macrophages; and (4) activating the complement system upon binding to the surface antigens of enveloped viruses, leading to viral dissolution [57,58]. Currently, several approaches to controlling SARS-CoV-2 are being evaluated in clinical trials, including vaccines, monoclonal antibody therapy, and convalescent plasma, all of which rely on nAbs to achieve preventive or therapeutic effects [59].

SARS-CoV-2 has four structural proteins: spike (S), envelope (E), membrane (M), and nucleocapsid (N) proteins. The virus enters host cells by binding the receptor-binding domain (RBD) of the S protein to the angiotensin-converting enzyme 2 (ACE2) receptor on the surface of host cells [60]. The nAbs against SARS-CoV-2 can bind to the viral surface S protein and prevent its binding to the ACE2 receptor on host cells, thereby inhibiting viral infection [61]. Studies have shown a correlation between the level of nAbs against SARS-CoV-2 and the level of vaccine protection [62,63]. Neutralizing antibody assays can be used to determine indicators, such as the seroconversion rate of nAbs and the average antibody titer, which help assess the presence and level of nAbs in serum and evaluate the immunological efficacy of vaccines [64].

5. Detection Methods for Neutralizing Antibodies

Neutralizing antibody detection methods can be classified into three main categories: the live (conventional) virus neutralization test (cVNT), pseudovirus neutralization test (pVNT), and surrogate virus neutralization test (sVNT). Among them, the cVNT is the traditional method and is considered the gold standard for evaluating neutralizing antibody capability. The mechanism of this method involves mixing the antibodies to be tested with virus-infected cells and observing whether they can neutralize the virus and prevent infection. This method is based on live viruses, requires a high biosafety level (BSL-3), involves operational risks, and has a relatively long testing period. These factors pose challenges and limit the application of this approach in vaccine, drug, and antibody research [65] The pVNT method is an alternative to the cVNT that follows a similar mechanism. It uses a pseudovirus that can only infect once without replication. This method overcomes the limitations of traditional cVNT methods in terms of the experimental facility and equipment needed. The pVNT can be conducted in a BSL-2 laboratory, has a relatively shorter testing time, and provides greater possibilities for antibody detection, vaccine research, and drug screening. This approach can be used to combat emerging and re-emerging viruses in a more effective manner [66].

The sVNT is a non-virus neutralization antibody detection method based on molecular interactions. It does not require any viruses or cells for testing. The sVNT can be conducted in laboratories with lower biosafety level standards; it is simple and rapid, and some of those measures showed good consistency with results obtained from live virus neutralization assays (cVNT) [67].

6. Live (Conventional) Virus Neutralization Test (cVNT)

The live virus-neutralizing antibody detection methods are considered the gold standard for neutralizing antibody detection, which is a vital part of the immunogenicity analysis of the clinical trials for candidate vaccines. The main methods include the plaque reduction neutralization test (PRNT) [68] and the cytopathic effect (CPE) assay [69]. Both methods involve mixing quantified live viruses with different dilutions of serum, inoculating the mixture onto a prepared cell monolayer, and evaluating the degree of cell damage as an indicator of neutralizing antibody titer (Figure 1). Additionally, recombinant live virus detection methods with reporter genes have been employed and shown advantages in SARS-CoV-2 neutralizing antibody detection [70].

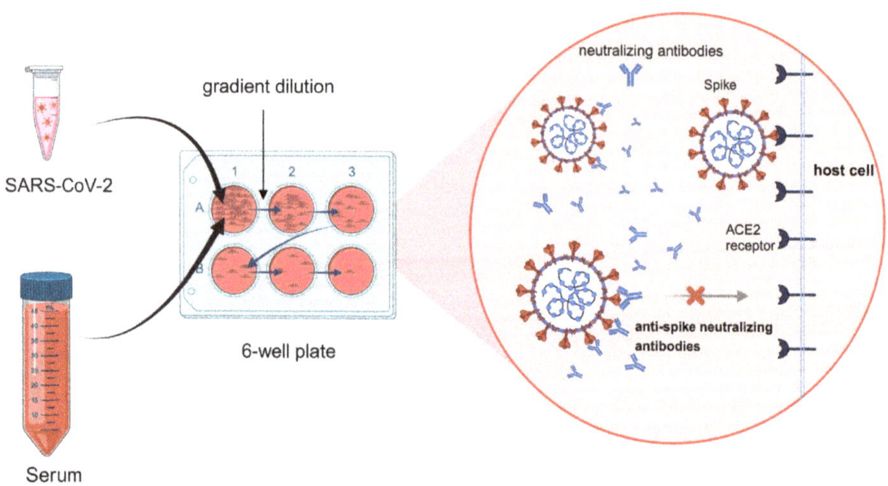

Figure 1. Mechanism of neutralizing antibody detection using live viruses. In the absence of a neutralizing antibody, the SARS-CoV-2 virus binds ACE2, followed by membrane fusion and cell entry; the release of its genetic material leads to virus propagation. The presence of neutralizing antibody blocks ACE2 interaction with the virus. (Figure created with BioRender.com accessed on 27 April 2024).

The cVNT remains the standard and classical method for evaluating nAbs in virus vaccine evaluation. It is methodologically reliable and serves as a reference for evaluating all other neutralizing antibody detection methods. However, all cVNT methods rely on the use of infectious live viruses, which restricts the testing to BSL-3/4 laboratory environments for some viruses causing severe diseases. This greatly limits the number of laboratories capable of conducting these assays. Additionally, owing to its low throughput, live virus-based neutralizing antibody detection methods are not suitable for large-scale sero-epidemics study and vaccine evaluation [71].

6.1. Cytopathic Effect (CPE)

The cytopathic effect (CPE) assay is a traditional method used to quantify viruses. CPE refers to the changes in cell morphology, structure, and function that occur when cells are infected by a virus, which ultimately leads to cell death or lysis [72]. The CPE method can be used in neutralizing antibody detection to assess the neutralizing effect of antibodies against the virus. In the laboratory, the CPE method typically involves mixing the virus with the corresponding nAbs and adding them to a host cell culture. As the virus replicates and spreads, changes in cell morphology and function can be observed. If the nAbs are effective, viral replication and spread will be inhibited, resulting in a reduction or elimination of CPE [73]. For example, in phase II clinical trials of COVID-19 inactivated vaccines such as CoronaVac and BBIBP-CorV, the CPE method was used for neutralizing antibody detection [12,16].

6.2. Plaque Reduction Neutralization Test (PRNT)

The plaque reduction neutralization test (PRNT) is considered the gold standard for measuring SARS-CoV-2 nAbs [74]. It is a highly specific method that can detect the cross-reactivity of nAbs against different viral strains or variants. PRNT is an important tool for the development and evaluation of novel antiviral drugs or vaccines in preclinical and clinical studies [75].

The PRNT is a neutralization assay based on the quantification of viral plaques formed as a result of virus-induced cytopathic effects (CPE) in cells. Briefly, cells are seeded in a multi-well plate, and the test plasma or serum samples are mixed with different dilutions of the virus solution. The mixture is then used to infect host cells. After 1 h of incubation, the culture medium is removed, and a semi-solid medium containing agarose or carboxymethylcellulose is overlayed to prevent further uncontrolled virus spread. Infected host cells undergo lysis and infect neighboring cells, resulting in a cycle of infection and lysis, ultimately forming viral plaques. These plaques can be directly observed using the naked eye or an optical microscope. Typically, 4 days are needed for SARS-CoV-2 to form visible plaques (plaque-forming units, PFU) [76]. The number of plaques is manually counted, with the assumption that each plaque arises from a single infectious virus particle. Therefore, PFU/mL represents the number of infectious virus particles per milliliter of the tested sample and corresponds to the viral infectivity level determined using titration evaluation [77,78]. If the plasma or serum sample contains nAbs, cell infection will be inhibited, resulting in a reduced number of viral plaques. As the titer of nAbs in the plasma or serum sample increases, the observed PFU decreases [79].

In clinical trials, the Moderna COVID-19 vaccine mRNA-1273 was evaluated for its neutralizing activity with the PRNT using live SARS-CoV-2 [22]. Owing to the time-intensive nature of PRNT, results of the clinical trial were only reported for the low-dose and high-dose groups on days 1 and 43.

Although PRNT is the standard method for measuring SARS-CoV-2 nAbs, it is time-consuming and unsuitable for the rapid detection of nAbs in clinical trials of emerging, highly pathogenic viral infections. Moreover, plaque identification and manual counting require experienced laboratory personnel who can recognize plaques, and different experimenters may yield variable analysis results. Additionally, it has been found that the SARS-CoV-2 Delta and Mu variants cause cell fusion without plaque formation [80], rendering the PRNT ineffective for their detection.

6.3. Recombinant Live Virus

The time-consuming and complex processes of traditional live virus-neutralizing antibody detection methods have several drawbacks, including the need for laboratory personnel with a certain level of experience in virus plaque recognition and manual counting, as well as the use of different operators, which can lead to different analysis results. To overcome these limitations, researchers have generated modified recombinant live viruses by inserting reporter genes, with a good correlation between reporter gene expression and viral replication. Additionally, recombinant viruses exhibit growth kinetics and plaque formation that are comparable to those of wild-type viruses, thereby offering a sensitive and objective alternative to neutralizing antibody assays [81–83]. For instance, researchers report a trans-complementation system that produces single-round infectious SARS-CoV-2 that can be safely used at BSL-2 laboratories for high-throughput neutralization and antiviral testing [84]. Following the emergence of SARS-CoV-2, researchers constructed a genetically stable reporter virus (mGFP Δ3678_WA1-spike) by deleting four auxiliary genes of SARS-CoV-2 and combining them with modified green fluorescent protein sequences (mGFP). This highly attenuated SARS-CoV-2 can be safely tested for nAbs in the BSL-2 laboratory [85]. In addition, researchers developed a novel single-round infection fluorescent SARS-CoV-2 virus (SFV) that can be safely used at BSL-2 laboratories for high-throughput neutralization and antiviral testing. The SFV neutralization test (SFVNT) has 100% sensitivity and specificity compared to the PRNT [86]. These recombinant viruses could successfully infect primary airway epithelial cells in culture and generate plaque morphology and growth curves similar to those of wild-type viruses. The neutralizing antibody titers detected in convalescent patient sera using this recombinant live virus system demonstrated comparable results to those of PRNT. This recombinant live virus method was also used for neutralizing antibody detection in phase I/II clinical trials of the BNT162b1 and BNT162b2 vaccines [87–90]. In this assay, inactivated test sera were incubated with the reporter virus for 1 h at 37 °C, followed by infection of Vero E6 cells pre-seeded in a 96-well plate. The neutralizing antibody titer was determined by detecting the fluorescent foci of virus infection [70].

After cell infection with recombinant live viruses containing reporter genes, fluorescence or luminescence expression can easily be detected. Visualizing the fluorescence intensity using spectrophotometric instruments reduces the time-consuming steps needed with traditional live virus-neutralizing antibody assays. Moreover, this approach offers high sensitivity and automation, making it more suitable for large-scale serological testing [91].

Traditional live virus testing primarily relies on manual interpretation for result determination and estimation, which significantly impacts the objectivity and repeatability of the outcomes. The difficulty in standardizing live virus testing methods is a key factor hindering their widespread adoption. Currently, optimizations in the interpretation of results mainly involve three approaches: virus culture combined with ELISA testing [71], virus culture coupled with quantitative real-time PCR (qRT-PCR) testing [92], and virus culture integrated with automatic cell imaging technology [93,94]. Through the optimization of live virus testing procedures, these modifications have partially enhanced the repeatability and throughput of the testing methods, making preliminary strides towards standardization.

7. Pseudovirus Neutralization Test (pVNT)

The biosafety level of SARS-CoV-2 has been classified as level 3 by the WHO. Therefore, research activities involving the live virus require a BSL-3 laboratory. Owing to the high facility and resource requirements of a BSL-3 laboratory and the risk associated with handling live strains, the use of pseudoviruses in BSL-2 laboratories has emerged as an alternative approach.

By anchoring the S protein (the receptor-binding protein of SARS-CoV-2) on the surface of the lentivirus, VSV, or other viruses, a pseudovirus mimicking SARS-CoV-2 can be constructed. This SARS-CoV-2 pseudovirus exhibits a spike (S) protein with a similar structure to that of the live virus but does not require a high-level biosafety laboratory. Additionally, reporter genes, such as GFP, red fluorescent protein, or luciferase, can be inserted in the backbone genome of the pseudovirus. Expression of the reporter gene upon cell infection correlates with the infectiousness of the viral inoculum, enabling convenient and rapid evaluation of viral infection or the neutralizing effect of antibodies. The most commonly used pseudovirus packaging systems include lentivirus (e.g., HIV) vector packaging systems, vesicular stomatitis virus (VSV) packaging systems, murine leukemia virus (MLV) packaging systems, and reverse genetic self-assembly pseudovirus systems [95,96]. Researchers have discussed various methods using SARS-CoV-2 spike-pseudo-typed viruses to assess antibody neutralization. Despite minor differences in how each virus model measures sensitivity, the results consistently correlate well with those from authentic SARS-CoV-2 neutralization assays [97], which are crucial for evaluating the effectiveness of vaccinations and the potency of therapies like convalescent plasma.

7.1. HIV Lentivirus System

Among lentivirus vector systems, HIV was the first packaging vector developed and widely used for the preparation of various pseudoviruses [98]. This is also one of the main packaging systems for SARS-CoV-2 pseudoviruses [99], typically achieved by co-transfecting two or three plasmids into cells to produce the pseudovirus. The dual-plasmid system includes one plasmid expressing the SARS-CoV-2 S protein and another plasmid expressing the packaging proteins with the envelope (E) gene deleted from the HIV backbone [100]. The HIV triple-plasmid packaging system typically consists of a packaging plasmid, a transfer plasmid containing the reporter gene, and a plasmid expressing the SARS-CoV-2 S protein [101,102]. Transfection of the plasmids into 293T cells leads to the production of pseudovirus particles with the S protein as the surface protein [103]. Compared with other viral vectors, HIV vectors have a large capacity for accommodating exogenous genes (8–9 kb) [104]. Ren et al. [105] constructed a eukaryotic expression plasmid pcDNA3.1–SARS-CoV-2 S expressing the SARS-CoV-2 S protein and co-transfected it with an HIV-1 pNL4-3.Luc.R-E- packaging plasmid containing the Fluc reporter gene in HEK293T cells, resulting in the production of a SARS-CoV-2 pseudovirus. This pseudovirus can be used for large-scale serological screening in SARS-CoV-2 epidemiological investigations and for evaluating the neutralizing activity of vaccines and therapeutic antibodies (Figure 2a). In clinical applications, the Moderna SARS-CoV-2 mRNA vaccine mRNA-1273 and its bivalent COVID-19 vaccine both used lentivirus-packaged pseudoviruses expressing the full-length S protein of SARS-CoV-2 in neutralization testing [21].

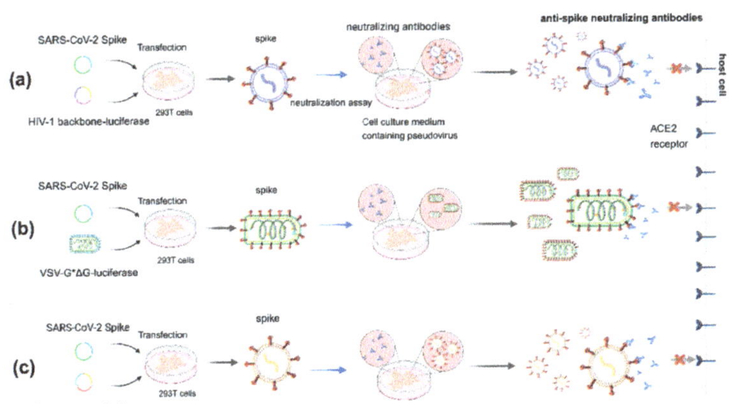

Figure 2. Schematic diagram of neutralizing antibody detection using pseudoviruses. (**a**). The 293T cells are transfected with a plasmid encoding the lentiviral backbone and a plasmid expressing envelope protein. The transfected cells produce recombined pseudoviruses. (**b**). The 293T cells are first transfected with an envelope protein expression plasmid; 24 h post-transfection, the cells are infected with VSV*ΔG encoding firefly luciferase. The transfected cells produce recombined pseudoviruses. (**c**). The 293T cells are co-transfected with an envelope protein-encoding plasmid, an MLV packaging transfer vector encoding a luciferase reporter. The transfected cells produce pseudo-typed MLV particles, like the HIV system. SARS-CoV-2 pseudoviruses are capable of binding ACE2 and entering the infected cells; non-replicative pseudovirus binds ACE2 and is internalized. Neutralizing antibody blocks ACE2 interaction with the pseudovirus. (Figure created with BioRender.com accessed on 27 April 2024).

7.2. Vesicular Stomatitis Virus (VSV) System

VSV is an enveloped negative-sense RNA virus that can infect various animals and rarely infects humans, causing only mild flu-like symptoms with infection [106]. Previous studies have shown that VSV particles do not exhibit specific selectivity for the type of viral envelope protein incorporated. When cells are co-infected with VSV and other enveloped viruses, VSV can easily become enveloped by the membrane proteins of other viruses [107]. VSV particles can bud without the presence of the G glycoprotein, and the nature of VSV can be altered by the insertion of heterologous glycoproteins. This leads to the development of recombinant VSV (rVSV), in which the gene for the G glycoprotein is replaced with a reporter gene, resulting in rVSV-ΔG, and the envelope protein is derived from the target virus under study [108,109]. Therefore, rVSV-ΔG can be used to produce single-cycle-restricted VSV pseudoviruses with any viral surface glycoprotein in a BSL-3 laboratory. After the outbreak of COVID-19, a research team synthesized the spike (S) gene sequence of SARS-CoV-2, successfully prepared SARS-CoV-2 pseudoviruses based on the VSV packaging system, and established a neutralizing antibody detection method based on SARS-CoV-2 pseudoviruses [66]. Zettl et al. [110] prepared SARS-CoV-2 pseudoviruses using the VSV*ΔG (FLuc) vector, enabling rapid quantification of SARS-CoV-19 nAbs in convalescent COVID-19 patients and vaccinated individuals under BSL-1 conditions (Figure 2b). For the COVID-19 vaccine BNT162b2, derived from German phase I/II trials, the broad-spectrum induction of antibodies by BNT162b2 vaccination was studied using VSV-based SARS-CoV-2 pVNT. The B.1.351 pseudovirus was neutralized at a lower geometric mean titer (GMT) than that of the wild-type strain; all BNT162b2 immune sera showed neutralizing activity against the 22 pseudoviruses tested [20].

This system is efficient and cost-effective, and virus production, titration, and infection assays can be completed within one week, making the system suitable for neutralizing antibody analysis or high-throughput therapeutic screening. However, an important issue with VSV-based pseudoviruses is the presence of residual VSV, which could lead to false-positive results [111].

7.3. MLV System

The HIV packaging system and VSV packaging system are the two most commonly used packaging systems for generating SARS-CoV-2 pseudoviruses. In addition to these, the Moloney MLV packaging system has been used in SARS-CoV-2 pseudovirus production. MLV is a typical simple retrovirus that is enveloped and contains positive-sense RNA, encoding three genes for the viral capsid protein, viral enzymes, and envelope protein (Gag, Pol, and Env) [112]. The process of generating pseudoviruses using the MLV system is similar to that of the HIV system, involving transfection of plasmids carrying MLV structural genes (gag and pol) and the gene encoding the heterologous viral envelope protein into cells. Pseudovirus particles carrying the heterologous viral surface protein are secreted in the cell culture medium to obtain a pseudovirus suspension [113]. Zheng et al. [114] prepared pseudovirus particles containing the SARS-CoV-2 S protein using a defective MLV vector plasmid carrying the Fluc reporter gene and a packaging plasmid encoding MLV gag/pol. This pseudovirus can be used under BSL-2 conditions for the specific measurement of neutralizing antibody titers against SARS-CoV-2 in plasma (Figure 2c).

In response to the prevalence and emergence of SARS-CoV-2, many researchers have developed various pseudovirus-based cell culture methods for neutralizing antibody detection. Compared with cVNT, pVNT does not require high-level biosafety laboratories. These methods are safer, simpler, time-saving, and have higher throughput. The inclusion of reporter genes for detection allows for objective result interpretation. Pseudoviruses serve as an alternative and conceptually validated testing approach, with high practical value in rapid clinical vaccine trials during an outbreak. However, this method also requires validation and standardization and should demonstrate comparability with live virus approaches.

8. Surrogate Virus Neutralization Test (sVNT)

Whether using a live virus or pseudovirus, there are several common issues in neutralization testing. First, the neutralizing ability of antibodies highly depends on the viral state or titer and the cell type and conditions used in the assay. If the virus and host cells are not in optimal condition for the assay, the reproducibility of the results may be poor [115]. Since the initial outbreak of COVID-19, researchers have developed molecular interaction-based non-virus neutralization antibody detection methods for rapid measurement of SARS-CoV-2 nAbs. Enzyme-linked immunosorbent assay (ELISA), which includes direct, indirect, sandwich, and competitive types, is commonly used for SARS-CoV-2 antibody detection [116]. Studies have shown that different SARS-CoV-2 proteins conjugated with horseradish peroxidase (HRP) can directly bind to the ACE2 receptor, and there is dose-dependent specific binding between ACE2 and the RBD or S1 domain but not with the nucleocapsid (N) protein [67]. Therefore, this method uses the specific binding between the SARS-CoV-2 S protein binding domain (RBD or S1) antigen and the receptor protein ACE2 to mimic virus-host cell interaction. HRP-labeled ACE2 or RBD generates a detection signal, and viral infection can be detected via a colorimetric reaction. When nAbs are present in the test sample, these block the protein interaction, resulting in a weakened colorimetric reaction (Figure 3). The test can typically be completed within a few hours and can be used for qualitative or quantitative measurement [67,117].

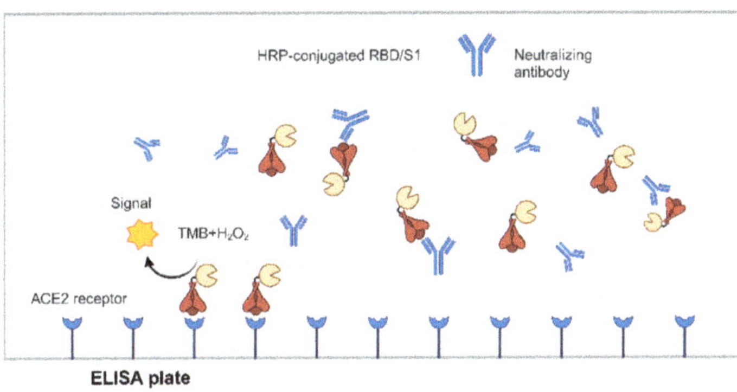

Figure 3. Principle of neutralizing antibody detection using competitive ELISA. The HRP coupling fragment of RBD or S1 binds to samples containing neutralizing antibodies. This mixture is added to an ELISA well plate that has been coated with ACE2 receptors. If the antibodies in the sample have neutralizing antibody activity, the binding between HRP-RBD/S1 and ACE2 will be broken, with cleaning to remove HRP-RBD/S1. The signal generated using a light absorption microplate reader will be lower in the presence of neutralizing antibodies. (Figure created with BioRender.com accessed on 27 March 2024).

Competitive ELISA for SARS-CoV-2 identifies epitopes, including the SARS-CoV-2 S protein, S1 domain, S2 subunit, and RBD domain of the S protein. Competitive ELISA with S1 or RBD-specific binding antibodies is commonly used for neutralizing antibody detection. Because ACE2 has the best binding affinity with RBD, the RBD–ACE2 neutralization activity of serum is nearly equivalent to the virus neutralization activity of its antibodies [67,118]. Therefore, RBD-specific binding antibodies can be used to assess the level of nAbs in individuals immunized with the COVID-19 vaccine. In clinical trials evaluating nAbs after COVID-19 vaccination, ELISA has been used to measure RBD-specific binding antibodies, and the seroconversion rate and GMT of RBD-specific binding antibodies have been reported for vaccines such as CoronaVac, Ad5-nCoV, and ZF2001 [12,30,35].

Competitive ELISA methods can be performed under standard laboratory safety conditions and offer advantages, such as high throughput and short detection time [119].

These methods have been widely used for assessing vaccine efficacy in large-scale clinical trials post-vaccination. However, whether targeting RBD or S1-specific binding antibodies, the currently established ELISA methods are directed against specific regions of the S protein rather than true nAbs. This makes it challenging to predict the neutralizing effects of antibodies out of the RBD or S1 of the virus and their protective efficacy in the host [120].

The Turn-Around Time (TAT) for different neutralizing antibody detection methods varies depending on the type of assay used. Traditional methods, like the cVNT, can take several days due to the need for virus culturing and observation of effects on cell cultures. Notably, the Recombinant live virus assay has seen significant improvement in TAT [121]. In contrast, newer methods, like high-throughput neutralizing antibody assays, which often utilize fluorescent or luminescent reporting, can provide results much more quickly, sometimes within the same day [116]. These faster methods are crucial during pandemics for the timely monitoring of immune responses. Both cVNT and pVNT can provide quantitative results, namely antibody titers, indicating to what extent serum dilution can still effectively neutralize the virus. The sVNT can detect the presence of specific antibodies and provide a relative abundance of antibodies for quantitative application. Different methods have their advantages and disadvantages. When selecting an appropriate detection method, it is necessary to decide which method to use and whether the quantitative analysis is needed, based on the specific needs of the research or diagnosis (Table 2).

Table 2. Advantages and disadvantages of different neutralizing antibody detection methods.

Method Classification	Method Name	Advantages	Disadvantages
cVNT	CPE PRNT	• The "gold standard" for neutralizing antibody detection has specificity, sensitivity, and can accurately measure the neutralizing ability of antibodies [76].	• BSL-3, low throughput, time-consuming, high condition requirements, and is influenced by subjective explanations from researchers, resulting in cumbersome steps [76,122].
	Recombinant live virus	• Reduced time consumption, dynamic monitoring, and objective results [70]. • Can be safely used at BSL-2 laboratories [84,85].	• Cumbersome and time-consuming steps [84,85].
pVNT	HIV system VSV system MLV system	• BSL-2, high throughput, less time-consuming, specific, sensitive, consistent with live virus results, accurately measure the neutralizing ability of antibodies [66]; • Lentiviral vector-based pseudoviruses differ from conventional retroviral vectors in that they have the ability to infect both dividing and non-dividing cells; • VSV systems are suitable for neutralizing antibody analysis or screening of high-content therapies [123].	• The number of envelope proteins in pseudoviruses is not directly proportional to the copy number of the core genome. The titer needs to be measured through qPCR and other methods, which loses its authenticity, especially when comparing the impact of different mutant S proteins on viral infectivity [105]; • VSV residue in the VSV system may lead to false positive results [111]; • Comparison and validation with live viruses are required.

Table 2. Cont.

Method Classification	Method Name	Advantages	Disadvantages
sVNT	ELISA	• Cell-independent, safe; high throughput, low cost, less time-consuming [119]; can be used for screening for neutralizing antibodies in the population to investigate the efficacy of protective immunity and vaccination [124].	• Does not represent actual neutralizing antibody titers, does not provide information on the neutralizing ability of antibodies and has low sensitivity for early infection [67,125].

9. Correlation of Neutralizing Antibodies (nAbs) with a Protective Effect

Both natural infection and vaccination can induce an immune response and the production of nAbs. Finding the correlation between vaccine-induced nAbs and protective efficacy is a crucial and challenging step in vaccine clinical trials [55]. Measurement of virus-neutralizing antibody titers in the blood after COVID-19 vaccination can serve as a surrogate endpoint for evaluating vaccine efficacy [126].

Clinical trial data for different vaccines and real-world data from many countries have shown a positive correlation between neutralizing antibody titers and protective efficacy [62,127,128]. A comparison of serum-neutralizing activity induced by seven major vaccines used in countries worldwide showed that, as long as serum-neutralizing activity reaches 20.2% of the mean convalescent serum-neutralizing activity, more than 50% protection is provided (95% confidence interval (CI) 14.4–28.4%). Furthermore, if the goal is to prevent a 50% chance of severe illness, a neutralizing activity of 3% of the mean convalescent serum neutralizing activity is sufficient (95% CI 0.7–13%, $p = 0.0004$) [62]. In a study conducted in Israel among 1497 healthcare workers who received two doses of vaccination, 39 breakthrough infections (0.34%) occurred. The study found that the GMT of nAbs in breakthrough infections (before infection) was 192.8, compared with 530.4 in non-infected individuals, representing only 36.35% of the latter. Additionally, among breakthrough infections, higher neutralizing antibody levels before infection were associated with lower viral loads. These findings indicate that the level of nAbs induced by COVID-19 vaccines can be used to predict post-vaccination protection [129]. Therefore, neutralizing antibody levels can be used to assess vaccine effectiveness.

Monitoring the decline of COVID-19 vaccine protection over time can be achieved using neutralizing antibody levels. Multiple studies have shown that neutralizing antibody levels increase rapidly within 14–28 days after initial/booster immunization with COVID-19 vaccines but decline within 6–8 months. However, the rate and extent of decline vary significantly among different age groups and vaccine technology platforms. Studies on neutralizing antibody titers among individuals vaccinated with inactivated COVID-19 vaccines have shown that, after receiving a second dose, the positivity rate and titers of nAbs can be maintained at a high level from 11 to 70 days post-vaccination. However, from days 70 to 332, there is a significant decrease in neutralizing antibody titers, with only a 27% positivity rate [130]. A study among 4868 healthcare workers vaccinated with the BNT162b2 vaccine found a rapid decline in neutralizing antibody titers within the first 3 months, with a lower peak level among individuals aged 65 years and above. The decline was more pronounced at 3 months post-vaccination, followed by a slower decline. The study also showed that immunosuppressed individuals had a 70% reduction in neutralizing antibody levels compared with non-immunosuppressed individuals [131,132]. Therefore, in special populations such as older or immunosuppressed individuals, the peak neutralizing antibody levels induced by COVID-19 vaccines are lower than those in younger adults, and the decline in neutralizing antibody levels over time is faster and more significant.

Another important factor contributing to the decrease in vaccine effectiveness is the emergence of SARS-CoV-2 variants. For 3 years since SARS-CoV-2 was first reported, several variants of concern have emerged, many of which are associated with increased transmissibility or some degree of immune escape. The decline in COVID-19 vaccine efficacy is significantly correlated with reduced neutralizing activity against variant strains [133–138].

10. Correlation Analysis of Testing Methods

In response to the outbreak of SARS-CoV-2, vaccine development institutions and research teams in various countries have used a range of different neutralizing antibody detection methods. However, the differences in assay formats (e.g., live virus, pseudovirus, and ELISA), target antigens (RBD, S1, S, and N), numerical readouts (absorbance optical density/relative light units/plaques/GFP%), and endpoints have led to an inability to directly compare and analyze immunological data. To address this issue, the WHO initiated a collaborative calibration with the first international standard and reference panel for anti-SARS-CoV-2 antibodies [139]. In this collaborative study, 51 laboratories evaluated the suitability of a plasma panel from 11 convalescent SARS-CoV-2 patients as an International Standard for anti-SARS-CoV-2 nAbs using 125 different methods, including ELISA, cVNT/pVNT, flow cytometry-based assays, lateral flow immunoassays, inhibition assays, and dual-antigen binding assays. For live virus assays, except for two low-titer samples and one negative sample for which titers could not be easily calculated, the geometric mean titers (GMTs) of the seven co-calibrated samples were 317.1 (PRNT), 445.3 (focus reduction neutralization test, FRNT), 93.9 (CPE), and 239.6 (microneutralization, MN). For pseudovirus assays, the total GMTs were 371.8 (PV-HIV) and 519.2 (PV-VSV). In a research project to establish the National Standard for anti-SARS-CoV-2 nAbs in China, which can be traced back to the WHO International Standard [140], six laboratories used the live virus PRNT and CPE methods, and four laboratories used the pseudovirus PV-VSV method. The co-calibration results of the two candidate standards showed that using the wild-type strain, the GMTs of samples detected using the live virus method were 133 and 194, and the GMTs of samples detected by the pseudovirus method were 641 and 1512. Using the Delta variant, the GMTs of samples 33 and 66–99 were 62 and 186 with the live virus method, and the GMTs for the pseudovirus method were 289 and 1889, respectively.

Researchers compared and analyzed the serum of hospitalized COVID-19 patients with and without acute respiratory distress syndrome (ARDS) using SARS-CoV-2 PRNT and HIV-based pVNT. The PRNT and pVNT results were comparable. However, in serum testing of patients with ARDS, the titer discrepancies observed in PRNT could not be fully resolved by pVNT, and there was no significant correlation between RBD-binding immunoglobulin G (IgG) measured via ELISA and neutralization titers. In contrast, in the serum of non-ARDS patients, there was a clear correlation between titers of RBD-binding IgG measured with ELISA and the neutralizing activity of these sera [125]. Researchers have found that the correlation between cVTN, sVNT, and pVTN varies among different virus strains, but overall shows a moderate to strong correlation. Specifically, there is a certain degree of correlation between the binding antibody levels detected by sVNT and the neutralizing antibody titers measured by cVNT [141]. Comparisons between cVNT and four alternative neutralization assays based on S-RBD or S1-specific binding (two chemiluminescence assays and two ELISA assays) revealed good agreement between cVNT and sVNT [142]. However, sVNT may only reveal a fraction of nAbs and does not measure total nAbs or neutralizing activity against epitopes outside the RBD, such as the N-terminal domain of the S protein [143]. Therefore, cVNT cannot be replaced in neutralization detection. In the summary data of the phase I/II clinical trial of the inactivated vaccine CoronaVac against SARS-CoV-2 [12], the correlation coefficients between live virus-neutralizing antibody titers and both pseudovirus-neutralizing antibody titers and RBD-specific IgG were greater than 0.8. The correlation coefficient between pseudovirus-neutralizing antibody titers and RBD-specific IgG was 0.73.

These studies indicate that, although different detection methods are based on different technical principles, they can complement each other in the determination of SARS-CoV-2 neutralizing antibodies, providing a certain degree of mutual validation. This is of great significance for evaluating the immune protection or vaccine efficacy after viral infection.

11. Standardization of Neutralizing Antibody Detection Methods

The COVID-19 pandemic triggered a global health emergency, and various vaccine development institutions worldwide continue the process of developing candidate vaccines using different platforms targeting different epitopes of SARS-CoV-2. Researchers have evaluated vaccines through in vitro neutralization assays using various infection factors, cell lines, and reporting systems. However, the diversity of testing protocols hinders direct comparisons among the different study results. Neutralizing antibody levels are important indicators of vaccine effectiveness and are critical for therapeutic and sero-epidemiological studies [116,140]. The accuracy, comparability, and reliability of different types of neutralizing antibody assays are of great importance in vaccine development, production, and application. However, owing to the use of different testing methods and laboratories, it is challenging to compare the efficacy of different assay results. Therefore, standardization of neutralizing antibody detection is crucial to ensure the comparability of testing results across different laboratories [144].

Before the availability of internationally standardized substances, qualitative detection of SARS-CoV-2 nAbs was most frequently performed. As research progresses, a clearer understanding of the relationship between levels of SARS-CoV-2 nAbs and vaccine protection is needed. To ensure consistency of SARS-CoV-2 neutralizing antibody results across multiple laboratories and testing methods, the Coalition for Epidemic Preparedness Innovations established a centralized global network in February 2020 for the evaluation and comparison of immune responses induced by candidate vaccines [145]. Using the same assay reagents and standardized protocols to assess vaccine immunogenicity, most inter-laboratory differences were eliminated, enabling head-to-head comparisons of immune responses induced by the various candidate vaccines. In July 2020, the WHO initiated the calibration of international standard substances for SARS-CoV-2. By summarizing the results of 27 neutralization assay methods, the first-generation international standard substance for SARS-CoV-2 antibodies (NIBSC code: 20/136) was released. The specified neutralizing antibody potency of the SARS-CoV-2 international standard substance was determined to be 250 IU per vial. After dissolution in 0.25 mL of distilled water, the final concentration of the preparation is 1000 IU/mL [146]. Subsequently, the SARS-CoV-2 Antibody International Reference Panel (NIBSC code: 20/268) was introduced. Using the concentration unit of 1000 IU/mL assigned in 20/136, the four samples in the international reference panel 20/268 were assigned different antibody concentrations, including neutralizing antibody and IgG concentrations against four different antigens (RBD, S1, N, and S protein) [147]. In December 2020, the WHO announced the International Standard for Anti-SARS-CoV-2 Immunoglobulin (NIBSC code: 21/340). The international standard allows precise calibration of analyses in arbitrary units, reducing inter-laboratory differences and creating a common language for reporting data. The international standard is based on human plasma collected from convalescent patients and freeze-dried in vials, with each vial containing 250 IU as the specified unit of neutralizing activity [148]. All global vaccine developers, national reference laboratories, and academic researchers should use the international standard correctly and report immunogenicity results using international standard units.

However, due to the many complex factors involved, including the characteristics of SARS-CoV-2 that are easy to mutate, the immune system's response to the virus, and the differences between experimental techniques and populations, the WHO and the international public health authorities did not announce any specific protective level of antibodies for the diseases.

The different types of antibodies detected by different methods, as well as the different sensitivities to the affinity and specificity of antibodies, can affect the accuracy of the detection results. The variation of SARS-CoV-2 may affect the sensitivity and specificity of some detection methods, especially when the antibody target changes. These factors will affect the comparability and interpretation of detection results between different methods.

12. Conclusions

In the development and evaluation of vaccines, vaccine efficacy is the most critical assessment parameter. Clinical trial data for different vaccines and real-world data from many countries have consistently shown a positive correlation between neutralizing antibody levels and vaccine protection. Therefore, one of the most important indicators for evaluating vaccine efficacy is the level of nAbs in vaccines. However, the diversity of methods and target antigens used for neutralizing antibody detection in clinical settings creates challenges in comparing results obtained using different methods. The cVNT is considered the gold standard for neutralizing antibody detection and is required in vaccine clinical trials. However, for newly emerging and highly infectious viral diseases, such as SARS-CoV-2, the use of live viruses in neutralization assays is limited by time-consuming procedures and the need for high-level biosafety laboratories. Therefore, pseudovirus-based neutralization assays are commonly used in viral vaccine evaluation. This method offers advantages, such as a high level of safety, simplicity, speed, and high throughput capability. Additionally, specific antigens can be used to detect corresponding nAbs via immune responses, such as with ELISA and other methods.

Multiple studies have shown that higher levels of neutralizing antibodies are often associated with better protection against COVID-19. With the standardization of SARS-CoV-2 neutralizing antibody detection methods and the development of standard substances, the use of nAbs as a vaccine endpoint has been rapidly advanced, promoting virological research and vaccine development for SARS-CoV-2. Researchers have attempted to resolve the differences in the specific antibody levels required for protection by a COVID-19 vaccine through the standardization of antibody titers and the use of more unified analytical methods [149]. However, due to the constantly changing strains of SARS-CoV-2, setting a clear protection threshold is challenging.

Author Contributions: Conceptualization: J.N., W.H. and Y.S.; writing—original draft preparation, J.N. and Y.S.; review and editing, J.N., H.X. and Y.S.; supervision, critical comments, and suggestions, and manuscript revision, J.N., Y.S., W.H. and H.X.; funding acquisition, J.N. All authors have read and agreed to the published version of the manuscript.

Funding: This work was supported by the National Key R and D Program of China (grant nos. 2021YFC2302500, 2021YFC2301700, and 2021YFC0863300) and the General Program of the National Natural Science Foundation of China (grant nos. 82172244 and 32070678).

Conflicts of Interest: The authors declare no conflicts of interest.

References

1. Zhou, P.; Yang, X.L.; Wang, X.G.; Hu, B.; Zhang, L.; Zhang, W.; Si, H.R.; Zhu, Y.; Li, B.; Huang, C.L.; et al. A pneumonia outbreak associated with a new coronavirus of probable bat origin. *Nature* **2020**, *579*, 270–273. [CrossRef] [PubMed]
2. Lai, C.C.; Shih, T.P.; Ko, W.C.; Tang, H.J.; Hsueh, P.R. Severe acute respiratory syndrome coronavirus 2 (SARS-CoV-2) and coronavirus disease-2019 (COVID-19): The epidemic and the challenges. *Int. J. Antimicrob. Agents* **2020**, *55*, 105924. [CrossRef] [PubMed]
3. Li, Q.; Guan, X.; Wu, P.; Wang, X.; Zhou, L.; Tong, Y.; Ren, R.; Leung, K.S.M.; Lau, E.H.Y.; Wong, J.Y.; et al. Early Transmission Dynamics in Wuhan, China, of Novel Coronavirus-Infected Pneumonia. *N. Engl. J. Med.* **2020**, *382*, 1199–1207. [CrossRef] [PubMed]
4. The species Severe acute respiratory syndrome-related coronavirus: Classifying 2019-nCoV and naming it SARS-CoV-2. *Nat. Microbiol.* **2020**, *5*, 536–544. [CrossRef] [PubMed]
5. Mateus, J.; Grifoni, A.; Tarke, A.; Sidney, J.; Ramirez, S.I.; Dan, J.M.; Burger, Z.C.; Rawlings, S.A.; Smith, D.M.; Phillips, E.; et al. Selective and cross-reactive SARS-CoV-2 T cell epitopes in unexposed humans. *Science* **2020**, *370*, 89–94. [CrossRef] [PubMed]

6. Jiang, S.; Du, L. Effect of Low-Pathogenic Human Coronavirus-Specific Antibodies on SARS-CoV-2. *Trends Immunol.* **2020**, *41*, 853–854. [CrossRef] [PubMed]
7. Stefano, G.B.; Kream, R.M. Convalescent memory T cell immunity in individuals with mild or asymptomatic SARS-CoV-2 infection may result from an evolutionarily adapted immune response to coronavirus and the 'common cold'. *Med. Sci. Monit. Int. Med. J. Exp. Clin. Res.* **2020**, *26*, e929789-1. [CrossRef] [PubMed]
8. Moss, P. The T cell immune response against SARS-CoV-2. *Nat. Immunol.* **2022**, *23*, 186–193. [CrossRef] [PubMed]
9. Gruell, H.; Vanshylla, K.; Weber, T.; Barnes, C.O.; Kreer, C.; Klein, F. Antibody-mediated neutralization of SARS-CoV-2. *Immunity* **2022**, *55*, 925–944. [CrossRef] [PubMed]
10. WHO. Statement on the Fifteenth Meeting of the IHR (2005) Emergency Committee on the COVID-19 Pandemic. Available online: https://www.who.int/news/item/05-05-2023-statement-on-the-fifteenth-meeting-of-the-international-health-regulations-(2005)-emergency-committee-regarding-the-coronavirus-disease-(COVID-19)-pandemic (accessed on 24 May 2023).
11. WHO. WHO Coronavirus (COVID-19) Dashboard. Available online: https://data.who.int/dashboards/covid19/ (accessed on 12 March 2024).
12. Zhang, Y.J.; Zeng, G.; Pan, H.X.; Li, C.G.; Hu, Y.L.; Chu, K.; Han, W.X.; Chen, Z.; Tang, R.; Yin, W.D.; et al. Safety, tolerability, and immunogenicity of an inactivated SARS-CoV-2 vaccine in healthy adults aged 18–59 years: A randomised, double-blind, placebo-controlled, phase 1/2 clinical trial. *Lancet Infect. Dis.* **2021**, *21*, 181–192. [CrossRef]
13. Jara, A.; Undurraga, E.A.; Gonzalez, C.; Paredes, F.; Fontecilla, T.; Jara, G.; Pizarro, A.; Acevedo, J.; Leo, K.; Leon, F.; et al. Effectiveness of an Inactivated SARS-CoV-2 Vaccine in Chile. *N. Engl. J. Med.* **2021**, *385*, 875–884. [CrossRef] [PubMed]
14. Tanriover, M.D.; Doganay, H.L.; Akova, M.; Guner, H.R.; Azap, A.; Akhan, S.; Kose, S.; Erdinc, F.S.; Akalin, E.H.; Tabak, O.F.; et al. Efficacy and safety of an inactivated whole-virion SARS-CoV-2 vaccine (CoronaVac): Interim results of a double-blind, randomised, placebo-controlled, phase 3 trial in Turkey. *Lancet* **2021**, *398*, 213–222. [CrossRef] [PubMed]
15. Sadoff, J.; Gray, G.; Vandebosch, A.; Cárdenas, V.; Shukarev, G.; Grinsztejn, B.; Goepfert, P.A.; Truyers, C.; Fennema, H.; Spiessens, B.; et al. Safety and Efficacy of Single-Dose Ad26.COV2.S Vaccine against COVID-19. *N. Engl. J. Med.* **2021**, *384*, 2187–2201. [CrossRef] [PubMed]
16. Xia, S.L.; Zhang, Y.T.; Wang, Y.X.; Wang, H.; Yang, Y.K.; Gao, G.F.; Tan, W.J.; Wu, G.Z.; Xu, M.; Lou, Z.Y.; et al. Safety and immunogenicity of an inactivated SARS-CoV-2 vaccine, BBIBP-CorV: A randomised, double-blind, placebo-controlled, phase 1/2 trial. *Lancet Infect. Dis.* **2021**, *21*, 39–51. [CrossRef]
17. Frenck, R.W., Jr.; Klein, N.P.; Kitchin, N.; Gurtman, A.; Absalon, J.; Lockhart, S.; Perez, J.L.; Walter, E.B.; Senders, S.; Bailey, R.; et al. Safety, Immunogenicity, and Efficacy of the BNT162b2 COVID-19 Vaccine in Adolescents. *N. Engl. J. Med.* **2021**, *385*, 239–250. [CrossRef] [PubMed]
18. Polack, F.P.; Thomas, S.J.; Kitchin, N.; Absalon, J.; Gurtman, A.; Lockhart, S.; Perez, J.L.; Pérez Marc, G.; Moreira, E.D.; Zerbini, C.; et al. Safety and Efficacy of the BNT162b2 mRNA COVID-19 Vaccine. *N. Engl. J. Med.* **2020**, *383*, 2603–2615. [CrossRef] [PubMed]
19. Walsh, E.E.; Frenck, R.W.; Falsey, A.R.; Kitchin, N.; Absalon, J.; Gurtman, A.; Lockhart, S.; Neuzil, K.; Mulligan, M.J.; Bailey, R.; et al. Safety and Immunogenicity of Two RNA-Based COVID-19 Vaccine Candidates. *N. Engl. J. Med.* **2020**, *383*, 2439–2450. [CrossRef] [PubMed]
20. Sahin, U.; Muik, A.; Vogler, I.; Derhovanessian, E.; Kranz, L.M.; Vormehr, M.; Quandt, J.; Bidmon, N.; Ulges, A.; Baum, A.; et al. BNT162b2 vaccine induces neutralizing antibodies and poly-specific T cells in humans. *Nature* **2021**, *595*, 572–577. [CrossRef]
21. Baden, L.R.; El Sahly, H.M.; Essink, B.; Kotloff, K.; Frey, S.; Novak, R.; Diemert, D.; Spector, S.A.; Rouphael, N.; Creech, C.B.; et al. Efficacy and Safety of the mRNA-1273 SARS-CoV-2 Vaccine. *N. Engl. J. Med.* **2021**, *384*, 403–416. [CrossRef]
22. Jackson, L.A.; Anderson, E.J.; Rouphael, N.G.; Roberts, P.C.; Makhene, M.; Coler, R.N.; McCullough, M.P.; Chappell, J.D.; Denison, M.R.; Stevens, L.J.; et al. An mRNA Vaccine against SARS-CoV-2-Preliminary Report. *N. Engl. J. Med.* **2020**, *383*, 1920–1931. [CrossRef]
23. Chalkias, S.; Harper, C.; Vrbicky, K.; Walsh, S.R.; Essink, B.; Brosz, A.; McGhee, N.; Tomassini, J.E.; Chen, X.; Chang, Y.; et al. A Bivalent Omicron-Containing Booster Vaccine against COVID-19. *N. Engl. J. Med.* **2022**, *387*, 1279–1291. [CrossRef] [PubMed]
24. Sadoff, J.; Le Gars, M.; Shukarev, G.; Heerwegh, D.; Truyers, C.; de Groot, A.M.; Stoop, J.; Tete, S.; Van Damme, W.; Leroux-Roels, I.; et al. Interim Results of a Phase 1-2a Trial of Ad26.COV2.S COVID-19 Vaccine. *N. Engl. J. Med.* **2021**, *384*, 1824–1835. [CrossRef] [PubMed]
25. Folegatti, P.M.; Ewer, K.J.; Aley, P.K.; Angus, B.; Becker, S.; Belij-Rammerstorfer, S.; Bellamy, D.; Bibi, S.; Bittaye, M.; Clutterbuck, E.A.; et al. Safety and immunogenicity of the ChAdOx1 nCoV-19 vaccine against SARS-CoV-2: A preliminary report of a phase 1/2, single-blind, randomised controlled trial. *Lancet* **2020**, *396*, 467–478. [CrossRef]
26. Voysey, M.; Clemens, S.A.C.; Madhi, S.A.; Weckx, L.Y.; Folegatti, P.M.; Aley, P.K.; Angus, B.; Baillie, V.L.; Barnabas, S.L.; Bhorat, Q.E.; et al. Safety and efficacy of the ChAdOx1 nCoV-19 vaccine (AZD1222) against SARS-CoV-2: An interim analysis of four randomised controlled trials in Brazil, South Africa, and the UK. *Lancet* **2021**, *397*, 99–111. [CrossRef] [PubMed]
27. Voysey, M.; Clemens, S.A.C.; Madhi, S.A.; Weckx, L.Y.; Folegatti, P.M.; Aley, P.K.; Angus, B.; Baillie, V.L.; Barnabas, S.L.; Bhorat, Q.E.; et al. Single-dose administration and the influence of the timing of the booster dose on immunogenicity and efficacy of ChAdOx1 nCoV-19 (AZD1222) vaccine: A pooled analysis of four randomised trials. *Lancet* **2021**, *397*, 881–891. [CrossRef] [PubMed]
28. Logunov, D.Y.; Dolzhikova, I.V.; Zubkova, O.V.; Tukhvatullin, A.I.; Shcheblyakov, D.V.; Dzharullaeva, A.S.; Grousova, D.M.; Erokhova, A.S.; Kovyrshina, A.V.; Botikov, A.G.; et al. Safety and immunogenicity of an rAd26 and rAd5 vector-based

heterologous prime-boost COVID-19 vaccine in two formulations: Two open, non-randomised phase 1/2 studies from Russia. *Lancet* **2020**, *396*, 887–897. [CrossRef] [PubMed]
29. Logunov, D.Y.; Dolzhikova, I.V.; Shcheblyakov, D.V.; Tukhvatulin, A.I.; Zubkova, O.V.; Dzharullaeva, A.S.; Kovyrshina, A.V.; Lubenets, N.L.; Grousova, D.M.; Erokhova, A.S.; et al. Safety and efficacy of an rAd26 and rAd5 vector-based heterologous prime-boost COVID-19 vaccine: An interim analysis of a randomised controlled phase 3 trial in Russia. *Lancet* **2021**, *397*, 671–681. [CrossRef] [PubMed]
30. Zhu, F.-C.; Li, Y.-H.; Guan, X.-H.; Hou, L.-H.; Wang, W.-J.; Li, J.-X.; Wu, S.-P.; Wang, B.-S.; Wang, Z.; Wang, L.; et al. Safety, tolerability, and immunogenicity of a recombinant adenovirus type-5 vectored COVID-19 vaccine: A dose-escalation, open-label, non-randomised, first-in-human trial. *Lancet* **2020**, *395*, 1845–1854. [CrossRef]
31. Zhu, F.-C.; Guan, X.-H.; Li, Y.-H.; Huang, J.-Y.; Jiang, T.; Hou, L.-H.; Li, J.-X.; Yang, B.-F.; Wang, L.; Wang, W.-J.; et al. Immunogenicity and safety of a recombinant adenovirus type-5-vectored COVID-19 vaccine in healthy adults aged 18 years or older: A randomised, double-blind, placebo-controlled, phase 2 trial. *Lancet* **2020**, *396*, 479–488. [CrossRef]
32. Keech, C.; Albert, G.; Cho, I.; Robertson, A.; Reed, P.; Neal, S.; Plested, J.S.; Zhu, M.; Cloney-Clark, S.; Zhou, H.; et al. Phase 1-2 Trial of a SARS-CoV-2 Recombinant Spike Protein Nanoparticle Vaccine. *N. Engl. J. Med.* **2020**, *383*, 2320–2332. [CrossRef] [PubMed]
33. Heath, P.T.; Galiza, E.P.; Baxter, D.N.; Boffito, M.; Browne, D.; Burns, F.; Chadwick, D.R.; Clark, R.; Cosgrove, C.; Galloway, J.; et al. Safety and Efficacy of NVX-CoV2373 COVID-19 Vaccine. *N. Engl. J. Med.* **2021**, *385*, 1172–1183. [CrossRef] [PubMed]
34. Dai, L.; Gao, L.; Tao, L.; Hadinegoro, S.R.; Erkin, M.; Ying, Z.; He, P.; Girsang, R.T.; Vergara, H.; Akram, J.; et al. Efficacy and Safety of the RBD-Dimer-Based COVID-19 Vaccine ZF2001 in Adults. *N. Engl. J. Med.* **2022**, *386*, 2097–2111. [CrossRef] [PubMed]
35. Yang, S.; Li, Y.; Dai, L.; Wang, J.; He, P.; Li, C.; Fang, X.; Wang, C.; Zhao, X.; Huang, E.; et al. Safety and immunogenicity of a recombinant tandem-repeat dimeric RBD-based protein subunit vaccine (ZF2001) against COVID-19 in adults: Two randomised, double-blind, placebo-controlled, phase 1 and 2 trials. *Lancet Infect. Dis.* **2021**, *21*, 1107–1119. [CrossRef] [PubMed]
36. Li, M.C.; Wang, H.; Tian, L.L.; Pang, Z.H.; Yang, Q.K.; Huang, T.Q.; Fan, J.F.; Song, L.H.; Tong, Y.G.; Fan, H.H. COVID-19 vaccine development: Milestones, lessons and prospects. *Signal Transduct. Target. Ther.* **2022**, *7*, 146. [CrossRef] [PubMed]
37. Minor, P.D. Live attenuated vaccines: Historical successes and current challenges. *Virology* **2015**, *479*, 379–392. [CrossRef] [PubMed]
38. Gao, Q.; Bao, L.; Mao, H.; Wang, L.; Xu, K.; Yang, M.; Li, Y.; Zhu, L.; Wang, N.; Lv, Z.; et al. Development of an inactivated vaccine candidate for SARS-CoV-2. *Science* **2020**, *369*, 77–81. [CrossRef] [PubMed]
39. Pollet, J.; Chen, W.H.; Strych, U. Recombinant protein vaccines, a proven approach against coronavirus pandemics. *Adv. Drug Deliv. Rev.* **2021**, *170*, 71–82. [CrossRef]
40. Wu, S.P.; Zhong, G.X.; Zhang, J.; Shuai, L.; Zhang, Z.; Wen, Z.Y.; Wang, B.S.; Zhao, Z.H.; Song, X.H.; Chen, Y.; et al. A single dose of an adenovirus-vectored vaccine provides protection against SARS-CoV-2 challenge. *Nat. Commun.* **2020**, *11*, 4081. [CrossRef]
41. Chavda, V.P.; Bezbaruah, R.; Athalye, M.; Parikh, P.K.; Chhipa, A.; Patel, S.; Apostolopoulos, V. Replicating Viral Vector-Based Vaccines for COVID-19: Potential Avenue in Vaccination Arena. *Viruses-Basel* **2022**, *14*, 759. [CrossRef]
42. Graham, S.P.; McLean, R.K.; Spencer, A.J.; Belij-Rammerstorfer, S.; Wright, D.; Ulaszewska, M.; Edwards, J.C.; Hayes, J.W.P.; Martini, V.; Thakur, N.; et al. Evaluation of the immunogenicity of prime-boost vaccination with the replication-deficient viral vectored COVID-19 vaccine candidate ChAdOx1 nCoV-19. *NPJ Vaccines* **2020**, *5*, 69. [CrossRef] [PubMed]
43. Ikegame, S.; Siddiquey, M.N.A.; Hung, C.-T.; Haas, G.; Brambilla, L.; Oguntuyo, K.Y.; Kowdle, S.; Chiu, H.-P.; Stevens, C.S.; Vilardo, A.E.; et al. Neutralizing activity of Sputnik V vaccine sera against SARS-CoV-2 variants. *Nat. Commun.* **2021**, *12*, 4598. [CrossRef] [PubMed]
44. Matveeva, O.; Ershov, A. Retrospective Cohort Study of the Effectiveness of the Sputnik V and EpiVacCorona Vaccines against the SARS-CoV-2 Delta Variant in Moscow (June–July 2021). *Vaccines* **2022**, *10*, 984. [CrossRef] [PubMed]
45. Ertl, P.F.; Thomsen, L.L. Technical issues in construction of nucleic acid vaccines. *Methods* **2003**, *31*, 199–206. [CrossRef] [PubMed]
46. Park, J.W.; Lagniton, P.N.P.; Liu, Y.; Xu, R.H. mRNA vaccines for COVID-19: What, why and how. *Int. J. Biol. Sci.* **2021**, *17*, 1446–1460. [CrossRef] [PubMed]
47. Kon, E.; Elia, U.; Peer, D. Principles for designing an optimal mRNA lipid nanoparticle vaccine. *Curr. Opin. Biotechnol.* **2022**, *73*, 329–336. [CrossRef] [PubMed]
48. Silveira, M.M.; Moreira, G.; Mendonca, M. DNA vaccines against COVID-19: Perspectives and challenges. *Life Sci.* **2021**, *267*, 118919. [CrossRef] [PubMed]
49. U.S. Food and Drug Administration. *Coronavirus (COVID-19) Update: FDA Authorizes Moderna, Pfizer-BioNTech Bivalent COVID-19 Vaccines for Use as a Booster Dose*; U.S. Food and Drug Administration: Silver Spring, MD, USA, 2022.
50. Chalkias, S.; Eder, F.; Essink, B.; Khetan, S.; Nestorova, B.; Feng, J.; Chen, X.; Chang, Y.; Zhou, H.; Montefiori, D.; et al. Safety, immunogenicity and antibody persistence of a bivalent Beta-containing booster vaccine against COVID-19: A phase 2/3 trial. *Nat. Med.* **2022**, *28*, 2388–2397. [CrossRef] [PubMed]
51. Link-Gelles, R.; Ciesla, A.A.; Fleming-Dutra, K.E.; Smith, Z.R.; Britton, A.; Wiegand, R.E.; Miller, J.D.; Accorsi, E.K.; Schrag, S.J.; Verani, J.R.; et al. Effectiveness of Bivalent mRNA Vaccines in Preventing Symptomatic SARS-CoV-2 Infection–Increasing Community Access to Testing Program, United States, September–November 2022. *Mmwr-Morb. Mortal. Wkly. Rep.* **2022**, *71*, 1526–1530. [CrossRef] [PubMed]

52. WHO. Vaccine Efficacy, Effectiveness and Protection. Available online: https://www.who.int/news-room/feature-stories/detail/vaccine-efficacy-effectiveness-and-protection (accessed on 7 May 2023).
53. Suthar, M.S.; Zimmerman, M.G.; Kauffman, R.C.; Mantus, G.; Linderman, S.L.; Hudson, W.H.; Vanderheiden, A.; Nyhoff, L.; Davis, C.W.; Adekunle, O.; et al. Rapid Generation of Neutralizing Antibody Responses in COVID-19 Patients. *Cell Rep. Med.* **2020**, *1*, 100040. [CrossRef]
54. Gilbert, P.B.; DeGruttola, V.G.; Hudgens, M.G.; Self, S.G.; Hammer, S.M.; Corey, L. What constitutes efficacy for a human immunodeficiency virus vaccine that ameliorates viremia: Issues involving surrogate end points in phase 3 trials. *J. Infect. Dis.* **2003**, *188*, 179–193. [CrossRef] [PubMed]
55. Krammer, F. A correlate of protection for SARS-CoV-2 vaccines is urgently needed. *Nat. Med.* **2021**, *27*, 1147–1148. [CrossRef] [PubMed]
56. Klasse, P.J.; Sattentau, Q.J. Occupancy and mechanism in antibody-mediated neutralization of animal viruses. *J. Gen. Virol.* **2002**, *83*, 2091–2108. [CrossRef] [PubMed]
57. Dörner, T.; Radbruch, A. Antibodies and B cell memory in viral immunity. *Immunity* **2007**, *27*, 384–392. [CrossRef] [PubMed]
58. Hua, C.K.; Ackerman, M.E. Increasing the Clinical Potential and Applications of Anti-HIV Antibodies. *Front. Immunol.* **2017**, *8*, 1655. [CrossRef] [PubMed]
59. Morales-Nunez, J.J.; Munoz-Valle, J.F.; Torres-Hernandez, P.C.; Hernandez-Bello, J. Overview of Neutralizing Antibodies and Their Potential in COVID-19. *Vaccines* **2021**, *9*, 1376. [CrossRef] [PubMed]
60. Letko, M.; Marzi, A.; Munster, V. Functional assessment of cell entry and receptor usage for SARS-CoV-2 and other lineage B betacoronaviruses. *Nat. Microbiol.* **2020**, *5*, 562–569. [CrossRef] [PubMed]
61. Stephens, D.S.; McElrath, M.J. COVID-19 and the Path to Immunity. *JAMA* **2020**, *324*, 1279–1281. [CrossRef] [PubMed]
62. Khoury, D.S.; Cromer, D.; Reynaldi, A.; Schlub, T.E.; Wheatley, A.K.; Juno, J.A.; Subbarao, K.; Kent, S.J.; Triccas, J.A.; Davenport, M.P. Neutralizing antibody levels are highly predictive of immune protection from symptomatic SARS-CoV-2 infection. *Nat. Med.* **2021**, *27*, 1205–1211. [CrossRef]
63. Earle, K.A.; Ambrosino, D.M.; Fiore-Gartland, A.; Goldblatt, D.; Gilbert, P.B.; Siber, G.R.; Dull, P.; Plotkin, S.A. Evidence for antibody as a protective correlate for COVID-19 vaccines. *Vaccine* **2021**, *39*, 4423–4428. [CrossRef]
64. Bewley, K.R.; Coombes, N.S.; Gagnon, L.; McInroy, L.; Baker, N.; Shaik, I.; St-Jean, J.R.; St-Amant, N.; Buttigieg, K.R.; Humphries, H.E.; et al. Quantification of SARS-CoV-2 neutralizing antibody by wild-type plaque reduction neutralization, microneutralization and pseudotyped virus neutralization assays. *Nat. Protoc.* **2021**, *16*, 3114–3140. [CrossRef] [PubMed]
65. Dreyfus, C.; Laursen, N.S.; Kwaks, T.; Zuijdgeest, D.; Khayat, R.; Ekiert, D.C.; Lee, J.H.; Metlagel, Z.; Bujny, M.V.; Jongeneelen, M.; et al. Highly Conserved Protective Epitopes on Influenza B Viruses. *Science* **2012**, *337*, 1343–1348. [CrossRef] [PubMed]
66. Nie, J.H.; Li, Q.Q.; Wu, J.J.; Zhao, C.Y.; Hao, H.; Liu, H.; Zhang, L.; Nie, L.L.; Qin, H.Y.; Wang, M.; et al. Establishment and validation of a pseudovirus neutralization assay for SARS-CoV-2. *Emerg. Microbes Infect.* **2020**, *9*, 680–686. [CrossRef] [PubMed]
67. Tan, C.W.; Chia, W.N.; Qin, X.; Liu, P.; Chen, M.I.; Tiu, C.; Hu, Z.; Chen, V.C.; Young, B.E.; Sia, W.R.; et al. A SARS-CoV-2 surrogate virus neutralization test based on antibody-mediated blockage of ACE2-spike protein-protein interaction. *Nat. Biotechnol.* **2020**, *38*, 1073–1078. [CrossRef] [PubMed]
68. Case, J.B.; Rothlauf, P.W.; Chen, R.E.; Liu, Z.; Zhao, H.; Kim, A.S.; Bloyet, L.M.; Zeng, Q.; Tahan, S.; Droit, L.; et al. Neutralizing Antibody and Soluble ACE2 Inhibition of a Replication-Competent VSV-SARS-CoV-2 and a Clinical Isolate of SARS-CoV-2. *Cell Host Microbe* **2020**, *28*, 475–485.e475. [CrossRef] [PubMed]
69. Kaminskyy, V.; Zhivotovsky, B. To kill or be killed: How viruses interact with the cell death machinery. *J. Intern Med.* **2010**, *267*, 473–482. [CrossRef] [PubMed]
70. Muruato, A.E.; Fontes-Garfias, C.R.; Ren, P.; Garcia-Blanco, M.A.; Menachery, V.D.; Xie, X.P.; Shi, P.Y. A high-throughput neutralizing antibody assay for COVID-19 diagnosis and vaccine evaluation. *Nat. Commun.* **2020**, *11*, 4059. [CrossRef] [PubMed]
71. Frische, A.; Brooks, P.T.; Gybel-Brask, M.; Sækmose, S.G.; Jensen, B.A.; Mikkelsen, S.; Bruun, M.T.; Boding, L.; Strandh, C.P.; Jørgensen, C.S.; et al. Optimization and evaluation of a live virus SARS-CoV-2 neutralization assay. *PLoS ONE* **2022**, *17*, e0272298. [CrossRef] [PubMed]
72. Heaton, N.S. Revisiting the concept of a cytopathic viral infection. *PLoS Pathog.* **2017**, *13*, e1006409. [CrossRef]
73. Traggiai, E.; Becker, S.; Subbarao, K.; Kolesnikova, L.; Uematsu, Y.; Gismondo, M.R.; Murphy, B.R.; Rappuoli, R.; Lanzavecchia, A. An efficient method to make human monoclonal antibodies from memory B cells: Potent neutralization of SARS coronavirus. *Nat. Med.* **2004**, *10*, 871–875. [CrossRef]
74. Perera, R.A.; Mok, C.K.; Tsang, O.T.; Lv, H.; Ko, R.L.; Wu, N.C.; Yuan, M.; Leung, W.S.; Chan, J.M.; Chik, T.S.; et al. Serological assays for severe acute respiratory syndrome coronavirus 2 (SARS-CoV-2), March 2020. *Euro. Surveill.* **2020**, *25*, 2000421. [CrossRef] [PubMed]
75. Coates, H.V.; Alling, D.W.; Chanock, R.M. An antigenic analysis of respiratory syncytial virus isolates by a plaque reduction neutralization test. *Am. J. Epidemiol.* **1966**, *83*, 299–313. [CrossRef] [PubMed]
76. Lau, E.H.Y.; Tsang, O.T.Y.; Hui, D.S.C.; Kwan, M.Y.W.; Chan, W.H.; Chiu, S.S.; Ko, R.L.W.; Chan, K.H.; Cheng, S.M.S.; Perera, R.; et al. Neutralizing antibody titres in SARS-CoV-2 infections. *Nat. Commun.* **2021**, *12*, 63. [CrossRef]
77. Kisch, A.L.; Johnson, K.M. A plaque assay for respiratory syncytial virus. Proceedings of the Society for Experimental Biology and Medicine. *Soc. Exp. Biol. Med.* **1963**, *112*, 583–589. [CrossRef] [PubMed]

78. Dulbecco, R. Production of Plaques in Monolayer Tissue Cultures by Single Particles of an Animal Virus. *Proc. Natl. Acad. Sci. USA* **1952**, *38*, 747–752. [CrossRef] [PubMed]
79. Septisetyani, E.P.; Prasetyaningrum, P.W.; Anam, K.; Santoso, A. SARS-CoV-2 Antibody Neutralization Assay Platforms Based on Epitopes Sources: Live Virus, Pseudovirus, and Recombinant S Glycoprotein RBD. *Immune Netw.* **2021**, *21*, e39. [CrossRef] [PubMed]
80. Mlcochova, P.; Kemp, S.A.; Dhar, M.S.; Papa, G.; Meng, B.; Ferreira, I.; Datir, R.; Collier, D.A.; Albecka, A.; Singh, S.; et al. SARS-CoV-2 B.1.617.2 Delta variant replication and immune evasion. *Nature* **2021**, *599*, 114–119. [CrossRef] [PubMed]
81. Fujino, M.; Yoshida, N.; Kimura, K.; Zhou, J.; Motegi, Y.; Komase, K.; Nakayama, T. Development of a new neutralization test for measles virus. *J. Virol. Methods* **2007**, *142*, 15–20. [CrossRef] [PubMed]
82. Rostad, C.A.; Currier, M.C.; Moore, M.L. Fluorescent and Bioluminescent Reporter Myxoviruses. *Viruses* **2016**, *8*, 214. [CrossRef]
83. Zhang, F.Q.; Perez-Martin, E.; Juleff, N.; Charleston, B.; Seago, J. A replication-competent foot-and-mouth disease virus expressing a luciferase reporter. *J. Virol. Methods* **2017**, *247*, 38–44. [CrossRef]
84. Zhang, X.; Liu, Y.; Liu, J.; Bailey, A.L.; Plante, K.S.; Plante, J.A.; Zou, J.; Xia, H.; Bopp, N.E.; Aguilar, P.V.; et al. A trans-complementation system for SARS-CoV-2 recapitulates authentic viral replication without virulence. *Cell* **2021**, *184*, 2229–2238.e13. [CrossRef] [PubMed]
85. Zou, J.; Kurhade, C.; Chang, H.C.; Hu, Y.; Meza, J.A.; Beaver, D.; Trinh, K.; Omlid, J.; Elghetany, B.; Desai, R.; et al. An Integrated Research-Clinical BSL-2 Platform for a Live SARS-CoV-2 Neutralization Assay. *Viruses* **2023**, *15*, 1855. [CrossRef] [PubMed]
86. Zou, J.; Xia, H.; Shi, P.Y.; Xie, X.; Ren, P. A Single-Round Infection Fluorescent SARS-CoV-2 Neutralization Test for COVID-19 Serological Testing at a Biosafety Level-2 Laboratory. *Viruses* **2022**, *14*, 1211. [CrossRef] [PubMed]
87. Liu, J.; Liu, Y.; Xia, H.; Zou, J.; Weaver, S.C.; Swanson, K.A.; Cai, H.; Cutler, M.; Cooper, D.; Muik, A.; et al. BNT162b2-elicited neutralization of B.1.617 and other SARS-CoV-2 variants. *Nature* **2021**, *596*, 273–275. [CrossRef] [PubMed]
88. Liu, Y.; Liu, J.; Xia, H.; Zhang, X.; Zou, J.; Fontes-Garfias, C.R.; Weaver, S.C.; Swanson, K.A.; Cai, H.; Sarkar, R.; et al. BNT162b2-Elicited Neutralization against New SARS-CoV-2 Spike Variants. *N. Engl. J. Med.* **2021**, *385*, 472–474. [CrossRef] [PubMed]
89. Falsey, A.R.; Frenck, R.W., Jr.; Walsh, E.E.; Kitchin, N.; Absalon, J.; Gurtman, A.; Lockhart, S.; Bailey, R.; Swanson, K.A.; Xu, X.; et al. SARS-CoV-2 Neutralization with BNT162b2 Vaccine Dose 3. *N. Engl. J. Med.* **2021**, *385*, 1627–1629. [CrossRef] [PubMed]
90. Muñoz, F.M.; Sher, L.D.; Sabharwal, C.; Gurtman, A.; Xu, X.; Kitchin, N.; Lockhart, S.; Riesenberg, R.; Sexter, J.M.; Czajka, H.; et al. Evaluation of BNT162b2 COVID-19 Vaccine in Children Younger than 5 Years of Age. *N. Engl. J. Med.* **2023**, *388*, 621–634. [CrossRef] [PubMed]
91. Chiem, K.; Vasquez, D.M.; Park, J.G.; Platt, R.N.; Anderson, T.; Walter, M.R.; Kobie, J.J.; Ye, C.; Martínez-Sobrido, L.; Chamari, K.; et al. Generation and Characterization of recombinant SARS-CoV-2 expressing reporter genes. *J. Virol.* **2021**, *95*, e02209-20. [CrossRef] [PubMed]
92. Lichtenegger, S.; Saiger, S.; Hardt, M.; Kulnik, S.; Wagner, G.E.; Kleinhappl, B.; Assig, K.; Zauner, A.; Ober, M.; Kimpel, J.; et al. Development of a Rapid Live SARS-CoV-2 Neutralization Assay Based on a qPCR Readout. *J. Clin. Microbiol.* **2022**, *60*, e0037622. [CrossRef]
93. Liu, X.; Li, Y.; Wang, Z.; Cao, S.; Huang, W.; Yuan, L.; Huang, Y.J.; Zheng, Y.; Chen, J.; Ying, B.; et al. Safety and superior immunogenicity of heterologous boosting with an RBD-based SARS-CoV-2 mRNA vaccine in Chinese adults. *Cell Res.* **2022**, *32*, 777–780. [CrossRef]
94. Wu, S.; Huang, J.; Zhang, Z.; Wu, J.; Zhang, J.; Hu, H.; Zhu, T.; Zhang, J.; Luo, L.; Fan, P.; et al. Safety, tolerability, and immunogenicity of an aerosolised adenovirus type-5 vector-based COVID-19 vaccine (Ad5-nCoV) in adults: Preliminary report of an open-label and randomised phase 1 clinical trial. *Lancet Infect. Dis.* **2021**, *21*, 1654–1664. [CrossRef] [PubMed]
95. Ansarah-Sobrinho, C.; Nelson, S.; Jost, C.A.; Whitehead, S.S.; Pierson, T.C. Temperature-dependent production of pseudoinfectious dengue reporter virus particles by complementation. *Virology* **2008**, *381*, 67–74. [CrossRef] [PubMed]
96. Chen, P.; Song, Z.; Qi, Y.; Feng, X.; Xu, N.; Sun, Y.; Wu, X.; Yao, X.; Mao, Q.; Li, X.; et al. Molecular determinants of enterovirus 71 viral entry: Cleft around GLN-172 on VP1 protein interacts with variable region on scavenge receptor B 2. *J. Biol. Chem.* **2012**, *287*, 6406–6420. [CrossRef] [PubMed]
97. Schmidt, F.; Weisblum, Y.; Muecksch, F.; Hoffmann, H.H.; Michailidis, E.; Lorenzi, J.C.C.; Mendoza, P.; Rutkowska, M.; Bednarski, E.; Gaebler, C.; et al. Measuring SARS-CoV-2 neutralizing antibody activity using pseudotyped and chimeric viruses. *J. Exp. Med.* **2020**, *217*, e20201181. [CrossRef] [PubMed]
98. Lillico, S.; Vasey, D.; King, T.; Whitelaw, B. Lentiviral transgenesis in livestock. *Transgenic. Res.* **2011**, *20*, 441–442. [CrossRef] [PubMed]
99. Banga Ndzouboukou, J.L.; Zhang, Y.D.; Fan, X.L. Recent Developments in SARS-CoV-2 Neutralizing Antibody Detection Methods. *Curr. Med. Sci.* **2021**, *41*, 1052–1064. [CrossRef] [PubMed]
100. Ledgerwood, J.E.; Coates, E.E.; Yamshchikov, G.; Saunders, J.G.; Holman, L.; Enama, M.E.; DeZure, A.; Lynch, R.M.; Gordon, I.; Plummer, S.; et al. Safety, pharmacokinetics and neutralization of the broadly neutralizing HIV-1 human monoclonal antibody VRC01 in healthy adults. *Clin. Exp. Immunol.* **2015**, *182*, 289–301. [CrossRef] [PubMed]
101. Hu, J.; Gao, Q.; He, C.; Huang, A.; Tang, N.; Wang, K. Development of cell-based pseudovirus entry assay to identify potential viral entry inhibitors and neutralizing antibodies against SARS-CoV-2. *Genes Dis.* **2020**, *7*, 551–557. [CrossRef] [PubMed]

102. Huang, S.W.; Tai, C.H.; Hsu, Y.M.; Cheng, D.; Hung, S.J.; Chai, K.M.; Wang, Y.F.; Wang, J.R. Assessing the application of a pseudovirus system for emerging SARS-CoV-2 and re-emerging avian influenza virus H5 subtypes in vaccine development. *Biomed. J.* **2020**, *43*, 375–387. [CrossRef]
103. Crawford, K.H.D.; Eguia, R.; Dingens, A.S.; Loes, A.N.; Malone, K.D.; Wolf, C.R.; Chu, H.Y.; Tortorici, M.A.; Veesler, D.; Murphy, M.; et al. Protocol and Reagents for Pseudotyping Lentiviral Particles with SARS-CoV-2 Spike Protein for Neutralization Assays. *Viruses* **2020**, *12*, 513. [CrossRef]
104. Kumar, M.; Keller, B.; Makalou, N.; Sutton, R.E. Systematic determination of the packaging limit of lentiviral vectors. *Hum. Gene Ther.* **2001**, *12*, 1893–1905. [CrossRef] [PubMed]
105. Yang, R.; Huang, B.; A, R.; Li, W.; Wang, W.; Deng, Y.; Tan, W. Development and effectiveness of pseudotyped SARS-CoV-2 system as determined by neutralizing efficiency and entry inhibition test in vitro. *Biosaf. Health* **2020**, *2*, 226–231. [CrossRef] [PubMed]
106. Lichty, B.D.; Power, A.T.; Stojdl, D.F.; Bell, J.C. Vesicular stomatitis virus: Re-inventing the bullet. *Trends Mol. Med.* **2004**, *10*, 210–216. [CrossRef] [PubMed]
107. Fukushi, S.; Watanabe, R.; Taguchi, F. Pseudotyped vesicular stomatitis virus for analysis of virus entry mediated by SARS coronavirus spike proteins. *Methods Mol. Biol.* **2008**, *454*, 331–338. [CrossRef] [PubMed]
108. Whitt, M.A. Generation of VSV pseudotypes using recombinant ΔG-VSV for studies on virus entry, identification of entry inhibitors, and immune responses to vaccines. *J. Virol. Methods* **2010**, *169*, 365–374. [CrossRef] [PubMed]
109. Witte, O.N.; Baltimore, D. Mechanism of formation of pseudotypes between vesicular stomatitis virus and murine leukemia virus. *Cell* **1977**, *11*, 505–511. [CrossRef] [PubMed]
110. Zettl, F.; Meister, T.L.; Vollmer, T.; Fischer, B.; Steinmann, J.; Krawczyk, A.; V'Kovski, P.; Todt, D.; Steinmann, E.; Pfaender, S.; et al. Rapid Quantification of SARS-CoV-2-Neutralizing Antibodies Using Propagation-Defective Vesicular Stomatitis Virus Pseudotypes. *Vaccines* **2020**, *8*, 386. [CrossRef]
111. Li, Q.; Liu, Q.; Huang, W.; Li, X.; Wang, Y. Current status on the development of pseudoviruses for enveloped viruses. *Rev. Med. Virol.* **2018**, *28*, e1963. [CrossRef] [PubMed]
112. Fan, H. Leukemogenesis by Moloney murine leukemia virus: A multistep process. *Trends Microbiol.* **1997**, *5*, 74–82. [CrossRef]
113. Millet, J.K.; Whittaker, G.R. Host cell entry of Middle East respiratory syndrome coronavirus after two-step, furin-mediated activation of the spike protein. *Proc. Natl. Acad. Sci. USA* **2014**, *111*, 15214–15219. [CrossRef]
114. Zheng, Y.; Larragoite, E.T.; Williams, E.S.C.P.; Lama, J.; Cisneros, I.; Delgado, J.C.; Slev, P.; Rychert, J.; Innis, E.A.; Coiras, M.; et al. Neutralization assay with SARS-CoV-1 and SARS-CoV-2 spike pseudotyped murine leukemia virions. *Virol. J.* **2021**, *18*, 1. [CrossRef] [PubMed]
115. Mukherjee, S.; Dowd, K.A.; Manhart, C.J.; Ledgerwood, J.E.; Durbin, A.P.; Whitehead, S.S.; Pierson, T.C. Mechanism and Significance of Cell Type-Dependent Neutralization of Flaviviruses. *J. Virol.* **2014**, *88*, 7210–7220. [CrossRef] [PubMed]
116. Liu, K.-T.; Han, Y.-J.; Wu, G.-H.; Huang, K.-Y.A.; Huang, P.-N. Overview of Neutralization Assays and International Standard for Detecting SARS-CoV-2 Neutralizing Antibody. *Viruses* **2022**, *14*, 1560. [CrossRef]
117. Byrnes, J.R.; Zhou, X.X.; Lui, I.; Elledge, S.K.; Glasgow, J.E.; Lim, S.A.; Loudermilk, R.P.; Chiu, C.Y.; Wang, T.T.; Wilson, M.R.; et al. Competitive SARS-CoV-2 Serology Reveals Most Antibodies Targeting the Spike Receptor-Binding Domain Compete for ACE2 Binding. *Msphere* **2020**, *5*, 10–128. [CrossRef] [PubMed]
118. Kostin, N.N.; Bobik, T.V.; Skryabin, G.A.; Simonova, M.A.; Knorre, V.D.; Abrikosova, V.A.; Mokrushina, Y.A.; Smirnov, I.V.; Aleshenko, N.L.; Kruglova, N.A.; et al. An ELISA Platform for the Quantitative Analysis of SARS-CoV-2 RBD-neutralizing Antibodies As an Alternative to Monitoring of the Virus-Neutralizing Activity. *Acta Naturae* **2022**, *14*, 109–119. [CrossRef] [PubMed]
119. Kohmer, N.; Ruhl, C.; Ciesek, S.; Rabenau, H.F. Utility of Different Surrogate Enzyme-Linked Immunosorbent Assays (sELISAs) for Detection of SARS-CoV-2 Neutralizing Antibodies. *J. Clin. Med.* **2021**, *10*, 2128. [CrossRef] [PubMed]
120. Girl, P.; Zwirglmaier, K.; von Buttlar, H.; Wölfel, R.; Müller, K. Evaluation of Two Rapid Lateral Flow Tests and Two Surrogate ELISAs for the Detection of SARS-CoV-2 Specific Neutralizing Antibodies. *Front. Med.* **2022**, *9*, 820151. [CrossRef] [PubMed]
121. Shi, A.C.; Ren, P. SARS-CoV-2 serology testing: Progress and challenges. *J. Immunol. Methods* **2021**, *494*, 113060. [CrossRef] [PubMed]
122. Xie, X.; Muruato, A.E.; Zhang, X.; Lokugamage, K.G.; Fontes-Garfias, C.R.; Zou, J.; Liu, J.; Ren, P.; Balakrishnan, M.; Cihlar, T.; et al. A nanoluciferase SARS-CoV-2 for rapid neutralization testing and screening of anti-infective drugs for COVID-19. *Nat. Commun.* **2020**, *11*, 5214. [CrossRef] [PubMed]
123. Condor Capcha, J.M.; Lambert, G.; Dykxhoorn, D.M.; Salerno, A.G.; Hare, J.M.; Whitt, M.A.; Pahwa, S.; Jayaweera, D.T.; Shehadeh, L.A. Generation of SARS-CoV-2 Spike Pseudotyped Virus for Viral Entry and Neutralization Assays: A 1-Week Protocol. *Front. Cardiovasc. Med.* **2020**, *7*, 618651. [CrossRef] [PubMed]
124. von Rhein, C.; Scholz, T.; Henss, L.; Kronstein-Wiedemann, R.; Schwarz, T.; Rodionov, R.N.; Corman, V.M.; Tonn, T.; Schnierle, B.S. Comparison of potency assays to assess SARS-CoV-2 neutralizing antibody capacity in COVID-19 convalescent plasma. *J. Virol. Methods* **2021**, *288*, 114031. [CrossRef] [PubMed]
125. Herrlein, M.-L.; Hein, S.; Zahn, T.; Mhedhbi, I.; Raupach, J.; Husria, Y.; Benz, N.I.; Eisert, J.; Bender, D.; Haberger, V.; et al. Comparative Investigation of Methods for Analysis of SARS-CoV-2-Spike-Specific Antisera. *Viruses* **2022**, *14*, 410. [CrossRef] [PubMed]

126. Liu, J.; Mao, Q.; Wu, X.; He, Q.; Bian, L.; Bai, Y.; Wang, Z.; Wang, Q.; Zhang, J.; Liang, Z.; et al. Considerations for the Feasibility of Neutralizing Antibodies as a Surrogate Endpoint for COVID-19 Vaccines. *Front. Immunol.* **2022**, *13*, 814365. [CrossRef] [PubMed]
127. Gilbert, P.B.; Montefiori, D.C.; McDermott, A.B.; Fong, Y.; Benkeser, D.; Deng, W.; Zhou, H.; Houchens, C.R.; Martins, K.; Jayashankar, L.; et al. Immune correlates analysis of the mRNA-1273 COVID-19 vaccine efficacy clinical trial. *Science* **2022**, *375*, 43–50. [CrossRef] [PubMed]
128. Lim, W.W.; Mak, L.; Leung, G.M.; Cowling, B.J.; Peiris, M. Comparative immunogenicity of mRNA and inactivated vaccines against COVID-19. *Lancet Microbe* **2021**, *2*, e423. [CrossRef] [PubMed]
129. Bergwerk, M.; Gonen, T.; Lustig, Y.; Amit, S.; Lipsitch, M.; Cohen, C.; Mandelboim, M.; Levin, E.G.; Rubin, C.; Indenbaum, V.; et al. COVID-19 Breakthrough Infections in Vaccinated Health Care Workers. *N. Engl. J. Med.* **2021**, *385*, 1474–1484. [CrossRef] [PubMed]
130. Zhang, H.; Jia, Y.; Ji, Y.; Cong, X.; Liu, Y.; Yang, R.; Kong, X.; Shi, Y.; Zhu, L.; Wang, Z.; et al. Inactivated vaccines against SARS-CoV-2: Neutralizing antibody titers in vaccine recipients. *Front. Microbiol.* **2022**, *13*, 816778. [CrossRef]
131. Levin, E.G.; Lustig, Y.; Cohen, C.; Fluss, R.; Indenbaum, V.; Amit, S.; Doolman, R.; Asraf, K.; Mendelson, E.; Ziv, A.; et al. Waning Immune Humoral Response to BNT162b2 COVID-19 Vaccine over 6 Months. *N. Engl. J. Med.* **2021**, *385*, e84. [CrossRef]
132. Israel, A.; Shenhar, Y.; Green, I.; Merzon, E.; Golan-Cohen, A.; Schäffer, A.A.; Ruppin, E.; Vinker, S.; Magen, E. Large-Scale Study of Antibody Titer Decay following BNT162b2 mRNA Vaccine or SARS-CoV-2 Infection. *Vaccines* **2021**, *10*, 64. [CrossRef] [PubMed]
133. Chen, R.E.; Zhang, X.; Case, J.B.; Winkler, E.S.; Liu, Y.; VanBlargan, L.A.; Liu, J.; Errico, J.M.; Xie, X.; Suryadevara, N.; et al. Resistance of SARS-CoV-2 variants to neutralization by monoclonal and serum-derived polyclonal antibodies. *Nat. Med.* **2021**, *27*, 717–726. [CrossRef] [PubMed]
134. Shen, X.; Tang, H.; McDanal, C.; Wagh, K.; Fischer, W.; Theiler, J.; Yoon, H.; Li, D.; Haynes, B.F.; Sanders, K.O.; et al. SARS-CoV-2 variant B.1.1.7 is susceptible to neutralizing antibodies elicited by ancestral spike vaccines. *Cell Host Microbe* **2021**, *29*, 529–539.e3. [CrossRef] [PubMed]
135. Edara, V.V.; Norwood, C.; Floyd, K.; Lai, L.; Davis-Gardner, M.E.; Hudson, W.H.; Mantus, G.; Nyhoff, L.E.; Adelman, M.W.; Fineman, R.; et al. Infection- and vaccine-induced antibody binding and neutralization of the B.1.351 SARS-CoV-2 variant. *Cell Host Microbe* **2021**, *29*, 516–521.e3. [CrossRef] [PubMed]
136. Rössler, A.; Riepler, L.; Bante, D.; von Laer, D.; Kimpel, J. SARS-CoV-2 Omicron Variant Neutralization in Serum from Vaccinated and Convalescent Persons. *N. Engl. J. Med.* **2022**, *386*, 698–700. [CrossRef] [PubMed]
137. Altarawneh, H.N.; Chemaitelly, H.; Ayoub, H.H.; Hasan, M.R.; Coyle, P.; Yassine, H.M.; Al-Khatib, H.A.; Smatti, M.K.; Al-Kanaani, Z.; Al-Kuwari, E.; et al. Protective Effect of Previous SARS-CoV-2 Infection against Omicron BA.4 and BA.5 Subvariants. *N. Engl. J. Med.* **2022**, *387*, 1620–1621. [CrossRef] [PubMed]
138. Callaway, E. New Omicron-specific vaccines offer similar protection to existing boosters. *Nature* **2022**, *609*, 232–233. [CrossRef]
139. Knezevic, I.; Mattiuzzo, G.; Page, M.; Minor, P.; Griffiths, E.; Nuebling, M.; Moorthy, V. WHO International Standard for evaluation of the antibody response to COVID-19 vaccines: Call for urgent action by the scientific community. *Lancet Microbe* **2022**, *3*, e235–e240. [CrossRef] [PubMed]
140. Guan, L.; Mao, Q.; Tan, D.; Liu, J.; Zhang, X.; Li, L.; Liu, M.; Wang, Z.; Cheng, F.; Cui, B.; et al. Establishment of national standard for anti-SARS-Cov-2 neutralizing antibody in China: The first National Standard calibration traceability to the WHO International Standard. *Front. Immunol.* **2023**, *14*, 1107639. [CrossRef]
141. Lerdsamran, H.; Anusorntanawat, R.; Sangsiriwut, K.; Sawadpongpan, S.; Prasertsopon, J.; Thinpan, N.; Intalapaporn, P.; Techasuwanna, R.; Okada, P.; Puthavathana, P. Higher correlation between neutralizing antibodies and surrogate neutralizing or binding antibodies in COVID-19 patients than vaccine recipients. *PLoS ONE* **2024**, *19*, e0298033. [CrossRef]
142. Pieri, M.; Infantino, M.; Manfredi, M.; Nuccetelli, M.; Grossi, V.; Lari, B.; Tomassetti, F.; Sarubbi, S.; Russo, E.; Amedei, A.; et al. Performance evaluation of four surrogate Virus Neutralization Tests (sVNTs) in comparison to the in vivo gold standard test. *FBL* **2022**, *27*, 74. [CrossRef] [PubMed]
143. Chi, X.; Yan, R.; Zhang, J.; Zhang, G.; Zhang, Y.; Hao, M.; Zhang, Z.; Fan, P.; Dong, Y.; Yang, Y.; et al. A neutralizing human antibody binds to the N-terminal domain of the Spike protein of SARS-CoV-2. *Science* **2020**, *369*, 650–655. [CrossRef] [PubMed]
144. Chmielewska, A.M.; Czarnota, A.; Bieńkowska-Szewczyk, K.; Grzyb, K. Immune response against SARS-CoV-2 variants: The role of neutralization assays. *NPJ Vaccines* **2021**, *6*, 142. [CrossRef] [PubMed]
145. CEPI. CEPI Establishes Global Network of Laboratories to Centralise Assessment of COVID-19 Vaccine Candidates. Available online: https://cepi.net/news_cepi/cepi-establishes-global-network-of-laboratories-to-centralise-assessment-of-COVID-19-vaccine-candidates/ (accessed on 19 May 2023).
146. WHO. First WHO International Standard for Anti-SARS-CoV-2 Immunoglobulin (Human) NIBSC Code: 20/136 Instructions for Use. Available online: https://www.nibsc.org/documents/ifu/20-136.pdf (accessed on 21 May 2023).
147. WHO. First WHO International Reference Panel for Anti-SARS-CoV-2 Immunoglubulin NIBSC Code: 20/268 Instructions for Use. Available online: https://www.nibsc.org/documents/ifu/20-268.pdf (accessed on 19 May 2023).

148. Kristiansen, P.A.; Page, M.; Bernasconi, V.; Mattiuzzo, G.; Dull, P.; Makar, K.; Plotkin, S.; Knezevic, I. WHO International Standard for anti-SARS-CoV-2 immunoglobulin. *Lancet* **2021**, *397*, 1347–1348. [CrossRef]
149. Khoury, D.S.; Schlub, T.E.; Cromer, D.; Steain, M.; Fong, Y.; Gilbert, P.B.; Subbarao, K.; Triccas, J.A.; Kent, S.J.; Davenport, M.P. Correlates of Protection, Thresholds of Protection, and Immunobridging among Persons with SARS-CoV-2 Infection. *Emerg. Infect. Dis.* **2023**, *29*, 381–388. [CrossRef]

vaccines

MDPI

Review

A Clinical Insight on New Discovered Molecules and Repurposed Drugs for the Treatment of COVID-19

Surojit Banerjee [1], Debadri Banerjee [1], Anupama Singh [1], Sumit Kumar [2], Deep Pooja [3], Veerma Ram [1], Hitesh Kulhari [4,*] and Vikas Anand Saharan [1,*]

1 School of Pharmaceutical Sciences and Technology, Sardar Bhagwan Singh University, Balawala, Dehradun 248001, India
2 Department of Pharmaceutical Sciences, Central University of Haryana, Jant-Pali, Mahendergarh 123031, India
3 School of Pharmacy, National Forensic Sciences University, Sector 9, Gandhinagar 382007, India
4 School of Nano Sciences, Central University of Gujarat, Sector 30, Gandhinagar 382030, India
* Correspondence: hitesh.kulhari@cug.ac.in (H.K.); vikas.pharmaceutics@gmail.com (V.A.S.)

Abstract: Severe acute respiratory syndrome coronavirus 2 (SARS-CoV-2) began churning out incredulous terror in December 2019. Within several months from its first detection in Wuhan, SARS-CoV-2 spread to the rest of the world through droplet infection, making it a pandemic situation and a healthcare emergency across the globe. The available treatment of COVID-19 was only symptomatic as the disease was new and no approved drug or vaccine was available. Another challenge with COVID-19 was the continuous mutation of the SARS-CoV-2 virus. Some repurposed drugs, such as hydroxychloroquine, chloroquine, and remdesivir, received emergency use authorization in various countries, but their clinical use is compromised with either severe and fatal adverse effects or non-availability of sufficient clinical data. Molnupiravir was the first molecule approved for the treatment of COVID-19, followed by Paxlovid™, monoclonal antibodies (MAbs), and others. New molecules have variable therapeutic efficacy against different variants or strains of SARS-CoV-2, which require further investigations. The aim of this review is to provide in-depth information on new molecules and repurposed drugs with emphasis on their general description, mechanism of action (MOA), correlates of protection, dose and dosage form, route of administration, clinical trials, regulatory approval, and marketing authorizations.

Keywords: SARS-CoV-2; antiviral; monoclonal antibody; cytokine blockers; correlates of protection

1. Introduction

Novel coronavirus disease 2019 (COVID-19), also known as severe acute respiratory syndrome coronavirus 2 (SARS-CoV-2), has become a grave issue in today's world [1]. Initially, SARS-CoV-2 infection was limited to the Republic of China, but with time, its infection spread across the entire globe. Although the peak of COVID-19 is on the decline after a long time, still cases of infection are increasing due to the emergence of new strains, resulting in changes in symptoms and treatment strategies [2]. A grave challenge with COVID-19 is the continuous mutation of SARS-CoV-2 virus leading to several variants (B.1.1.7, B.1.351, P.1, B.1.617.2, B.1.1.529 etc.) spread all over the world [3]. To control the situation, drug regulatory agencies (such as the United States Food and drug Administration (USFDA) and the European Medicines Agency (EMA)) have given emergency use approval (EUA) to several novel molecules and pre-existing drugs for controlling the pandemic. Research on COVID-19 vaccines and drugs is still ongoing to find the best cure. Different vaccines, such as ChAdOx1, BNT162 (especially BNT162b2), Covaxin, mRNA-1273, and AD26.COV2.S, have been found effective against different strains of SARS-CoV-2 [3]. However, the effect of vaccines on boosting immunity varies for different strains and their variant. Research funding agencies are aggressively collaborating with pharmaceutical companies and research institutions for the development of vaccines and drugs for the treatment of COVID-19.

Vis-a-vis a vaccine development program, pharmaceutical companies are also aggressively working to discover new antivirus molecules against COVID-19. The quest for finding a treatment started with repurposing efforts on approved drugs. Some pre-existing drugs have shown good therapeutic efficacy against SARS-CoV-2 [1]. Repurposing of drugs is advantageous due to the reduced cost and speeding up of the approval process from regulatory authorities. However, several failures have been observed as in the case of remdesivir, chloroquine, and hydroxychloroquine [4,5]. These drugs either showed adverse effects at higher doses or were effective only in combination with other drugs [6–9]. EMA has authorized some of these drugs for their emergency use and for the conduct of clinical trials [10–12].

However, like developed vaccines, new molecules and repurposed drugs work by boosting the human immune system [2]. These new molecules and repurposed drugs are directly or indirectly associated with the inhibition of the mitogen-activated $CD3^+$, $CD4^+$, and $CD8^+$ T cell proliferation. Regulatory T cells are also controlled by these repurposed drugs and new molecules. In the absence of other immune cells, Wharton's jelly mesenchymal stem cells inhibit the proliferation of alloreactive $CD4^+$ and $CD8^+$ T cells. Wharton's jelly mesenchymal stem cells also inhibit the proliferation of B cells and T-cell-secreted interferon-γ. Different monoclonal antibodies (MAbs) (such as sarilumab, tocilizumab, and baricitinib), repurposed drugs (such as dexamethasone, hydroxychloroquine, and naproxen), and convalescent plasma inhibit IL-6, IL-10 (responsible for T-cell and B-cell stimulation) and serve as anti-inflammatory agents and anticytokines. They are involved in diminishing C-reactive protein and ferritin, inflammatory markers responsible for COVID-19 disease. This review critically investigates various pharmaceutical formulations, their mechanism of action, correlates of protection, dose and dosage form, route of administration, clinical trials, regulatory approval, and marketing authorizations of repurposed drugs and new molecules.

2. New Discovered Molecules

New molecules/drugs are discovered by humans for the therapeutic management of new diseases or to address existing health challenges. New molecules/drugs require extensive clinical investigations, followed by approval from drug regulatory agencies for marketing. Several new molecules have been discovered to combat SARS-CoV-2. The newly discovered molecules have been tested in silico via molecular docking studies to assess their interaction with the target receptor on SARS-CoV-2. These new molecules are either antivirals or immunomodulators [13,14]. It has been observed in different studies that these molecules can block SARS-CoV-2 in their conventional way. For example, molnupiravir blocks RNA-dependent RNA polymerase (RdRp) to stop the replication of SARS-CoV-2. Novel MAbs (both alone and in combination) have also been investigated in clinical trials. Drug regulatory authorities have approved several new drugs via the EUA route. Some of these new drugs or molecules are now available commercially for the treatment of COVID-19.

On the other hand, research on new drugs, both monotherapy and combination therapy, is also ongoing. In 2021, the World Health Organization (WHO) recommended EUA to two new drugs, baricitinib and sotrovimab, and later, full approval was granted in 2022 [10].

2.1. *Molnupiravir*

2.1.1. General Description

Molnupiravir (EIDD-2801/MK-4482) is a ribonucleoside prodrug of N-hydroxycytidine [11–13]. It was developed by Drug Innovation Ventures at Emory University and later procured by Ridgeback Biotherapeutics in collaboration with Merck, US. It is a prescription antiviral drug and is available as pills, inhaled powder, liquid, and intravenous solutions.

2.1.2. Mechanism of Action

This novel antiviral agent blocks RdRp of the SARS-CoV-2 virus to inhibit the viral genome transcription and replication [14–18]. It was reported that molnupiravir becomes cleaved in plasma and forms β-D-N4-hydroxycytidine (NHC). NHC is then distributed and metabolized into NHC triphosphate in the cytoplasm of the liver cells, which acts as the substrate of RdRp. NHC causes mutation in the viral RNA with the help of RdRp and changes the genome sequence, leading to the inhibition of genome transcription and replication for SARS-CoV-2.

2.1.3. Clinical Trial, Route of Administration, Dose, and Dosage Form

In preclinical trials, molnupiravir showed effect on the blocking replication of SARS-CoV-2 in mice models [19]. It was found safe and efficacious in ferrets at 2.3 and 7 mg/kg of body weight when administered BID orally [16]. In a different study, the administration of molnupiravir in a Syrian hamster showed blocking of the viral replication of SARS-CoV-2 [20].

The phase 1 randomized, double-blind, and placebo controlled trial (NCT04392219), in 130 adult human volunteers divided into four groups, for molnupiravir was conducted by the developer, Ridgeback Biotherapeutics, at the Covance Leeds Clinical Research Unit, UK [14,21–23]. The first group received a single oral-dose capsules of 50 to 1600 mg in fasted state, the second group was administered two oral-dose capsules of 200 mg in fed or fasted state, and the third group received twice daily an oral-dose capsule of the drug or placebo. It was observed that molnupiravir was more tolerable than placebo, and adverse events were also lesser than placebo. Phase 2 (NCT04405570) randomized, double-blind, and placebo-controlled clinical trial, in 204 adults and older adults divided into four groups, was conducted by Ridgeback and Merck at different locations in the US [14,21,22,24]. The groups were administered with a 200, 400, or 800 mg BID oral capsule for 5 days. Placebo was taken as oral capsule BID for 5 days. Viral replication was observed in 1.9% participants treated with molnupiravir and 16.5% participants who received placebo. Moreover, 400 and 800 mg BID doses were found effective in reducing viral replication with no significant adverse effects. In another phase 2/3 randomized, double-blind, and placebo-controlled trial (NCT04575597), in 1850 adults and older adults at 200, 400, or 800 mg BID oral dose and placebo every 12 h for 5 days, was conducted by Merck at 173 different locations in the US, Brazil, Chile, Argentina, Canada, Colombia, Egypt, France, Germany, Mexico, and so on [11,12,25]. It was observed that molnupiravir was effective in reducing death, 6.8% (95% CI) for the drug and 9.7% (95% CI) for the placebo, in hospitalized patients. Adverse events were alleviated with the drug (30.4%) when compared with the placebo (33%). Unfortunately, this trial was terminated by the data safety monitoring board (DSMB) probably because of the doubtful beneficial data of the drug as compared with the placebo.

2.1.4. Regulatory Approval and Marketing Authorization

Molnupiravir was approved by the United Kingdom Medicines and Healthcare Products Regulatory Agency (MHRA) in November 2021. MSD (Merck Sharp & Dohme) markets this drug as 200 mg hard capsules in the UK as Lagevrio® [26,27]. EMA (November 2021) gave EUA to it and recommends its administration to patients who do not need supplemental oxygen [28]. EUA was also granted by USFDA (December 2021) for molnupiravir [29]. Sun Pharma, Cipla and Torrent market it as Molxvir®, Cipmolnu®, and Molnutor®, respectively, in India after obtaining DCGI (Drug Controller General of India) approval [30–32].

2.2. Paxlovid™

2.2.1. General Description

Paxlovid™ is a combination of two drugs, nirmatrelvir and ritonavir, developed by Pfizer [33]. It is a protease inhibitor, which is intended for oral administration.

2.2.2. Mechanism of Action

Nirmatrelvir blocks the SARS-CoV-23-CL protease in the proteolysis phase, leading to the inhibition of the replication of the virus. It is administered with a lower dose of ritonavir, which is a strong CYP3A4 inhibitor and HIV protease inhibitor [34]. Therefore, this combination results in the synergistic inhibition of SARS-CoV-2 viral protease [33].

2.2.3. Clinical Trial, Route of Administration, Dose, and Dosage Form

The phase 1 open-label, randomized, single-dose, and crossover clinical trial (NCT05263921) for Paxlovid™ was conducted in 12 adults (divided in different groups) [35]. In some groups, participants were administered oral commercial tablets in fasted state, whereas in other groups, participants received oral powder. Nirmatrelvir (150 mg)/ritonavir (100 mg) in combination was orally administered in fasted state with water, with applesauce, or with vanilla pudding to the participants. An EPIC-HR (evaluation of protease inhibition for COVID-19 in high-risk patients) study, EPIC-SR (evaluation of protease inhibition for COVID-19 in standard-risk patients) study, and EPIC-PEP (evaluation of protease inhibition for COVID-19 in postexposure prophylaxis) studies have been conducted to assess the clinical benefits of Paxlovid™ [33]. The EPIC-HR study (NCT04960202) was a phase 2/3 randomized and quadruple-blinded trial, conducted across Europe, Africa, North and South America, and Asia in 2246 adults and older adults, was divided into different studies; that is, in that trial, the participants were given the drug combination and placebo through the oral route as a tablet and/or capsule. According to the interim analysis result of Pfizer, the drug combination showed about 89% risk of hospitalization, and chances of death were decreased when compared with the placebo group. Adverse events were less with the drug combination rather than in the placebo, that is, 1.7% and 6.6%, respectively. Final analysis showed that the therapeutic efficacy was maintained in 1379 patients with a percentage point difference of −5.81 (95% CI, −7.78 to −3.84; $p < 0.001$) [36].

2.2.4. Regulatory Approval and Marketing Authorization

Paxlovid™ received its EUA from EMA in December 2021 [37], and EUA from USFDA was obtained in December 2021 [33]. In India, Torrent Pharma and Aurobindo Pharma have been licensed to commercialize this product [38].

2.3. *Baricitinib*

2.3.1. General Description

Baricitinib is an orally active immunomodulator, anti-inflammatory, and anticancer agent [39,40].

2.3.2. Mechanism of Action

Baricitinib selectively blocks JAK 1/2 and activates the JAK-STAT (signal transducers and activators of transcription) signaling pathway. It also inhibits numb-associated kinase (NAK) family enzymes, which include BMP-2-inducible kinase (BIKE), serine/threonine-protein kinase 16 (STK16), adaptor-associated kinase 1 (AAK1), and cyclin G-associated kinase (GAK). This results in the blocking of the AP-2 scaffolding protein, which is responsible for the entry and propagation of SARS-CoV-2 [41]. On the other hand, baricitinib also acts as anticytokines and reduces the inflammatory markers of COVID-19 disease, which includes C-reactive protein (CRP), IL-6, and ferritin.

2.3.3. Clinical Trial, Route of Administration, Dose, and Dosage Form

Eli Lilly conducted and completed the phase 3 (NCT04421027) trial, a randomized, double-blinded, and parallel-assigned study (COV-BARRIER), on baricitinib in 1525 adults and older adults at 96 different locations in the US, Argentina, Republic of Korea, Brazil, Mexico, Germany, India, Japan, Italy, Russian Federation, Puerto Rico, Spain, and UK to estimate the therapeutic efficacy of baricitinib in hospitalized COVID-19 patients [42]. Participants were administered 4 mg baricitinib once daily through the oral route against the standard of care. As a result, the chance of mortality was found to be reduced by 38.2%, and one death was found averted per 20 participants. It was also observed that in the baricitinib-treated group, the mortality was 10% in 60 days ($n = 79$; 95% CI; $p = 0.0050$), and in the placebo group, it was 15% ($n = 116$; 95% CI; $p = 0.0050$). Serious adverse events were seen to be slightly lower in the case of the baricitinib-treated group than in the placebo group, that is, 15% and 18%, respectively.

2.3.4. Regulatory Approval and Marketing Authorization

Recently, Eli Lilly received EUA from USFDA (October 2022) for their product Olumiant® (Baricitinib) [43]. EMA has started (in 2021) and is continuing the evaluation of Olumiant® (baricitinib) in the treatment of COVID-19 [44].

2.4. Wharton's Jelly Mesenchymal Stem Cells (WJ-MSC)

2.4.1. General Description

WJ is a gelatinous tissue found in the umbilical cord. It contains myofibroblast-like stromal cells [45]. MSCs are stromal cells, having the capability of self-renewal and multilineage proliferation [46]. MSCs are isolated from the umbilical cord, bone marrow, endometrial polyps, and so on. However, WJ is considered to be the pivotal source of MSCs [47]. MSCs are also considered medicinal signaling cells, which are therapeutically effective stomatal cells. They can be classified in different types of cells, including myocytes, osteoblasts, chondrocytes, and adipocytes. MSCs are found in the cord cells, amniotic fluid adipose tissue, bone marrow, and molar cells. It had proved its effectiveness previously for the treatment of autoimmune diseases, such as Crohn's disease, systemic lupus erythematosus (SLE), multiple sclerosis, osteoarthritis, and graft versus host disease [48]. Apart from this, it also has immunomodulatory effects and antimicrobial effects. WJ-MSCs are obtained from the cord tissue of newborns and cultured to intensity for MSCs. Then these are placed in saline solution (25 mL) containing 0.5% human serum albumin [49,50].

2.4.2. Mechanism of Action

WJ-MSC curbs mitogen-induced T-cell responses rather than other MSCs. Along with this, WJ-MSC blocks the multiplication of mitogen-activated $CD3^+$, $CD4^+$, and $CD8^+$ T cells [51–53]. Apart from peripheral blood mononuclear cells (PBMC) and interferon-γ (IFN_γ), WJ-MSC can also mediate T_{reg} cells (regulatory T cells). This helps in the polarization procedure of monocytes/macrophages toward an anti-inflammatory phenotype (type 2) and blocks the separation into the phenotype (type 1) and DCs. More specifically, the MSC-secreted interleukin 1 receptor antagonist (IL1-RA) helps in the polarization procedure of macrophages toward phenotype (type 2). Along with these, proinflammatory Th1 is switched to anti-inflammatory Th2 cells, and the profiles of cytokines are modified. It enhances IL-4 secretion by Th2 cells, increases the production of T_{reg}, and blocks the multiplication, stimulation, and cytotoxicity of natural killer cells (NK cells). It directly blocks the multiplication of alloreactive $CD4^+$ and $CD8^+$ T cells in the absence of other immune cells, which are stimulated by the SC-derived galectin-1. Effector $CD4^+$ T cells can be easily converted to $Foxp3^+$ T_{regs} by B_{regs} (regulatory B cells), which produce IL-10. Along with T cells, it also blocks the multiplication of B cells and T-cell-secreted IFN_γ. In this way, the upregulation of genes involved in the process of phagocytosis in macrophages is increased along with the downregulation of inflammatory cytokines via macrophages.

2.4.3. Clinical Trial, Route of Administration, Dose, and Dosage Form

Assistance Publique–Hôpitaux de Paris had already completed a phase 1/2, randomized, triple-blinded, and parallel-assigned clinical trial (NCT04333368) in 47 adults and older adults [54]. The intervention was administered through the central venous line for 1 h by tubing and at a dose of 1 mg/kg. The placebo group was given 0.9% NaCl through the IV infusion route. The difference in ratio of partial pressure of oxygen to fractional inspired oxygen (PaO_2/FiO_2) was not significant between the intervention and placebo group

(95% CI) [55]. Adverse events were observed in 28.6% participants from the intervention group and 25% patients from the placebo group. Efficacy of WJ-MSC was also evaluated for the treatment of COVID-19 related acute respiratory distress syndrome (ARDS) [56]. However, sponsors did not disclose results for this clinical trial (NCT04625738).

2.4.4. Regulatory Approval and Marketing Authorization

WJ-MSC is not approved by USFDA and EMA so far for use against COVID-19.

2.5. Convalescent Plasma (CP)

2.5.1. General Description

CP is the plasma that is collected from patients who have survived from any viral disease [57]. CP contains antibodies against a virus. COVID-19 survivors have antibodies against SARS-CoV-2 in their blood, and these antibodies have the potential to treat COVID-19 patients. Collecting CP from patients surviving from COVID-19 infection and administering it to patients suffering from this infection may boost the immunity of COVID-19 patients. Although CP therapy is not approved yet, USFDA proclaimed that CP therapy can be used for the treatment of critically ill COVID-19 patients.

2.5.2. Mechanism of Action

In COVID-19 patients, some issues were observed, such as infiltration of inflammatory cells and cytokine storm in the alveoli of the lungs, which further produces ARDS [58,59]. Moreover, the amount of lymphocytes and cytokine levels are also reduced; on the other hand, the levels of IL-6, TNF-α, IL-10, colony-stimulating factor, and granulocyte macrophage are enhanced significantly. CP administration in COVID-19 patients enhances the lymphocyte amount and oxygen saturation and restores liver function. Therefore, CP therapy can boost immunogenicity and reduce the inflammation of lungs. However, this therapy also treats different hemorrhagic fevers induced by Ebola, influenza A (H5N1), SARS, Middle East respiratory syndrome (MERS), and other viruses [60].

2.5.3. Clinical Trial, Route of Administration, Dose, and Dosage Form

An open-label study (NCT04321421) was conducted by Foundation IRCCS San Matteo Hospital in collaboration with Ospedale Carlo Poma Asst Mantova, Ospedale Maggiore Lodi, and Ospedale Asst Cremona in 49 adults and older adults having moderate to severe ARDS, which was lasting <10 days, at Catherine Klersy, Italy [61]. Participants were administered 250–300 mL CP three times over 5 days through IV infusion. However, in this study, the developers were uncertain whether CP was beneficial against COVID-19 or not [62,63]. A different study was conducted by Mayo Clinic (NCT04338360) in adults and older adults at four places in the US, namely, Arizona, Florida, Minnesota, and Wisconsin [64]. Serious adverse events were significantly reduced (<1%), and the mortality rate, up to 30 days of study, was 25.2% (95% CI) [65]. The study was successful, and the developers obtained marketing approval.

2.5.4. Regulatory Approval and Marketing Authorization

Though USFDA has not approved CP therapy as yet, an IND application for it proclaims that CP can be used effectively in emergency cases and in critically ill patients [66].

2.6. Sarilumab

2.6.1. General Description

Sarilumab is a potent MAb (IgG_1) that inhibits IL-6 receptors [67,68]. It is approved by FDA and EMA for the treatment of rheumatoid arthritis through SC administration. In COVID-19 infection, this agent is thought to be a potent agent because of the evidence of its in vitro and in vivo therapeutic efficacy.

2.6.2. Mechanism of Action

IL-6 is considered to be the pleiotropic cytokine that is responsible for the stimulation of T cells and B cells [67–71]. It is noteworthy that hepatocytes are also stimulated by IL-6 and CRP, which secrete fibrinogen and serum amyloid A (SAA). IL-6 also plays a pivotal role in the process of pathological inflammation. Sarilumab binds to the IL-6 receptor (both soluble IL-6 receptor, namely, sIL-6Rα, and membrane-bound IL-6 receptor, namely, mIL-6Rα). It blocks the gp130 and signal transducer and signaling transcription protein 3 (STAT3) pathway and cis- and trans- signaling pathways (though in vitro). Complexes of IL-6 and mIL-6Rα inducing trans-signaling pathways are also inhibited in those respective cells, which express specifically gp130 and mIL-6Rα.

2.6.3. Clinical Trial, Route of Administration, Dose, and Dosage Form

A randomized, double-blind, and placebo-controlled phase 2/3 clinical trial (NCT04315298) was conducted at 62 locations in the US by Regeneron Pharmaceuticals in collaboration with Sanofi in 1912 adult or older adult patients having confirmed COVID-19 infection and hospitalized with pneumonia or multisystem organ dysfunctions [72]. The trial was supported by USFDA and the Biomedical Advanced Research and Development Authority (BARDA) [73,74]. In the phase 2 clinical trial, the first two experimental groups were administered 200 and 400 mg of sarilumab through the IV route [75]. The next four groups were divided into three cohorts, which received 200 mg (Cohort 1), 400 mg (Cohort 1), 800 mg (Cohort 2), and 800 mg (Cohort 3) of the drug through the IV route. It was observed (especially in phase 3) that the percentage of improvement was 43.2% for the interventional group and 35.5% for the placebo group. Another phase 3, randomized, quadruple-blinded, parallel-mode (NCT04327388) trial, by Sanofi in collaboration with Regeneron Pharmaceuticals, in 420 adults or older adults was conducted at 47 different locations in Brazil, Argentina, Chile, Canada, Germany, France, Japan, Israel, Russian Federation, Italy, and Spain [76]. The participants were divided into three groups, of which the first experimental group was administered a single dose of 200 mg sarilumab through IV injection once a day on the first day. The second group was administered a single dose of 400 mg of sarilumab through IV injection once a day. The comparator group was provided placebo through IV injection once a day on the first day. However, no significant difference in therapeutic activity was observed between groups receiving 200 mg sarilumab (n = 159 (38%)), 400 mg sarilumab (n = 173 (42%)), and the placebo (n = 84 (24%)) [77]. Adverse events were 70% with 400 mg sarilumab, 65% with 200 mg sarilumab, and 65% in the placebo group.

2.7. *Tocilizumab*

2.7.1. General Description

Tocilizumab is a recombinant MAb that contains human and murine components [78,79]. Antigen-binding domains of the murine antihuman IL-6R antibody are grafted to human IgG_1 scaffolding. It is effective against inflammation and autoimmune diseases by blocking IL-6R. This MAb is thought to be a potential agent to combat against COVID-19. Tocilizumab is already available in the market as Actemra® and ACTPen™. Actemra® is single-dose manual injection of tocilizumab, and ACTPen™ is a single-dose autoinjector containing tocilizumab [80,81]. Actemra® and ACTPen™ can be administered via the subcutaneous and IV infusion route. Hoffmann-La Roche marketed these products for the treatment of arthritis [82].

2.7.2. Mechanism of Action

Like sarilumab, tocilizumab competitively blocks IL-6R receptors, both sIL-6R and mIL-6R, resulting in the inhibition of signal transduction, leading to the reduction in inflammation [78]. B cells and T cells generate IL-6, which is a proinflammatory cytokine and helps in the formation of an antibody and differentiation of cytotoxic T cells and is responsible for the stimulation of hepatocytes, CRP, fibrinogen, and SAA secretion along with the diminishing of the differentiation of T_{reg} cells. It is conventionally used for the treatment of rheumatoid arthritis, systemic juvenile idiopathic arthritis, giant cell arteritis, polyarticular juvenile idiopathic arthritis, cytokine release syndrome, and so on.

2.7.3. Clinical Trial, Route of Administration, Dose, and Dosage Form

Another phase 3, randomized, double-blinded, parallel-mode, and multicenter clinical trial (NCT04320615) was conducted by Hoffmann-La Roche in 452 adult or older adult patients at 62 different locations in the U.S., Canada, Denmark, Germany, France, Italy, Netherlands, Spain, and UK [83]. Participants were administered 8 mg/kg of this drug (maximum up to 800 mg each dose), and one more dose was only administered in case of worse condition or lack of improvement through IV infusion. It was observed that 24.5% of the participants from the interventional group and 25% of the participants from the placebo group died by the 60th day [84]. Adverse events were seen in 24.1% of the participants (interventional group) and 29.4% of the participants (placebo group). Result indicates that no beneficial outcomes were obtained in this study.

2.7.4. Regulatory Approval and Marketing Authorization

Actemra® (tocilizumab), manufactured by Genentech (member of the Roche Group) is approved by USFDA in December 2022 for severe COVID-19 patients [85].

2.8. Bevacizumab

2.8.1. General Description

Like tocilizumab, bevacizumab is also a humanized IgG_1 MAb that has the ability to bind with vascular endothelial growth factor A (VEGF), resulting in the apoptosis of tumor cells [86]. It is conventionally used for the treatment of various types of cancers, such as cervical cancer, colorectal cancer, non-small cell lung cancer, fallopian tube cancer, ovarian cancer, peritoneal cancer, and renal cell carcinoma. It is already a FDA-approved agent for the treatment of cancer [87,88]. Currently, it is being evaluated for its potential efficacy against COVID-19 infection for which many clinical trials are being continued. Bevacizumab is already available in the market as Avastin®, Mvasi®, and Zirabev®. Avastin® injections are used for the treatment of diseases such as colorectal cancer and non-small cell lung cancer [86,87].

2.8.2. Mechanism of Action

Bevacizumab binds to VEGF, a well-known proangiogenic growth factor expressed by tumor cells, and blocks VEGF receptors of lung tissues, leading to an increase in vascular permeability [87–89]. Therefore, oxygen perfusion and anti-inflammatory effects increase.

2.8.3. Clinical Trial, Route of Administration, Dose and Dosage Form

Another phase 2 open-label trial (NCT04275414), in 27 adults or older adults having confirmed COVID-19 infection, was conducted by Qilu Hospital of Shandong University in collaboration with Renmin Hospital of Wuhan University and Moriggia-Pelascini Gravedona Hospital, Gravedona, Italy [90]. Patients were administered with 500 mg of bevacizumab along with 100 mL normal saline through IV drip for a minimum of 90 min and onwards. Body temperature was normalized in 93% patients and improvement in PaO_2/FiO_2 ratios was observed in 92% patients who received both bevacizumab and the standard of care [89].

2.8.4. Regulatory Approval and Marketing Authorization

Till date, bevacizumab is not approved for the treatment of COVID-19.

3. Comparative Analysis of Newly Discovered Molecules

There are several differences among new molecules in different aspects, such as dose, dosage forms, duration of administration, efficacy, and adverse events. Molnupiravir was found effective at higher doses when compared with others (400 and 800 mg BID oral dose) [14,21,22,24]. Although only 6.8% death or hospitalization case was reported for molnupiravir, the mutation of the patient's DNA is a significant adverse effect [91]. On the other hand, only a negligible difference was observed between molnupiravir and placebo for other adverse events. A recent study showed that on day 5, 18.42% of patients treated with molnupiravir survived from COVID-19 (especially caused by the omicron variant), while in the placebo group, none survived [92]. Adverse events were reported in 3.9% of COVID-19 patients who were treated with molnupiravir. Unlike molnupiravir, Paxlovid™ did not cause any mutation in host DNA [35]. Molnupiravir inhibits RdRp to block the viral transcription, while Paxlovid™ (especially nirmatrelvir) is a protease inhibitor and acts on the 3-chymotrypsin-like cysteine protease enzyme (M^{pro}) [91]. Pharmacokinetic parameters of nirmatrelvir increases when administered with ritonavir due to CYP3A4 inhibition, leading to a reduction in the metabolism of nirmatrelvir [93]. In an EPIC-PEP study, molnupiravir showed a superior efficacy of 89% reduction in hospitalization or death cases [33]. On the other hand, 1.7% of patients from the Paxlovid™ group showed adverse events, while 6.6% of patients reported adverse effects from placebo.

Besides other novel molecules, novel MAbs have also shown beneficial effects in COVID-19. Baricitinib showed a higher mortality rate (38.2%) when compared with molnupiravir (6.8%) and Paxlovid™ (11%) [42]. More trials on baricitinib are required to be conducted in a wide population pool to have a true picture on adverse events. It may furnish a more clear result. Moreover, 15% of patients from the baricitinib group reported adverse effects, whereas adverse effects for molnupiravir and Paxlovid™ were reported by 3.9% and 1.7% of patients, respectively. Sarilumab is administered through the IV route, while other drugs require oral administration. However, improvements were observed in 43.2% of patients with sarilumab and 35.5% of patients reported improvement with placebo administration. These results clearly indicate that molnupiravir and Paxlovid™ have superior efficacy than sarilumab. Moreover, higher adverse effects were observed with sarilumab (70% of patients from 400 mg and 65% of patients from 200 mg) in comparison with molnupiravir and Paxlovid™ [77]. In the clinical trial, 24.5% of patients treated with tocilizumab were found dead on day 60 [84]. This result was better than sarilumab (56.8% patients), but worse than molnupiravir (6.8% of patients), Paxlovid™ (11% of patients), and baricitinib (38.2% of patients). Besides, adverse events were observed in 24.1% of patients with tocilizumab. Contrary to this, fewer patients reported adverse effects when treated with molnupiravir (3.9% of patients), Paxlovid™ (1.7% of patients), and baricitinib (15% of patients). The condition of 92% of patients improved with the IV administration of bevacizumab [89]. Bevacizumab normalized the body temperature of patients within a very short time. Unfortunately, adverse events observed for bevacizumab were high, almost 38%.

WJ-MSC and CP therapy were used in hospitalized patients and emergency conditions, but clinical trials have not yet ascertained their therapeutic efficacy. WJ-MSCs work on mitogen-induced T cells, and CP works on the cytokine storm to provide protection for patients from COVID-19 [50–53,58,59]. However, both therapies require IV administration. For WJ-MSC therapy, adverse events were seen in 28.6% of patients, and for CP therapy, it was less than 1% [55,65]. On the basis of adverse events, it can be proclaimed that CP therapy is well tolerated by patients. This is probably because CP has been obtained from patients who have survived COVID-19. Figures 1 and 2 describe the percentage of survival rates and adverse events, respectively, for these new molecules. Molnupiravir and Paxlovid™ are better in terms of survival rate and adverse events.

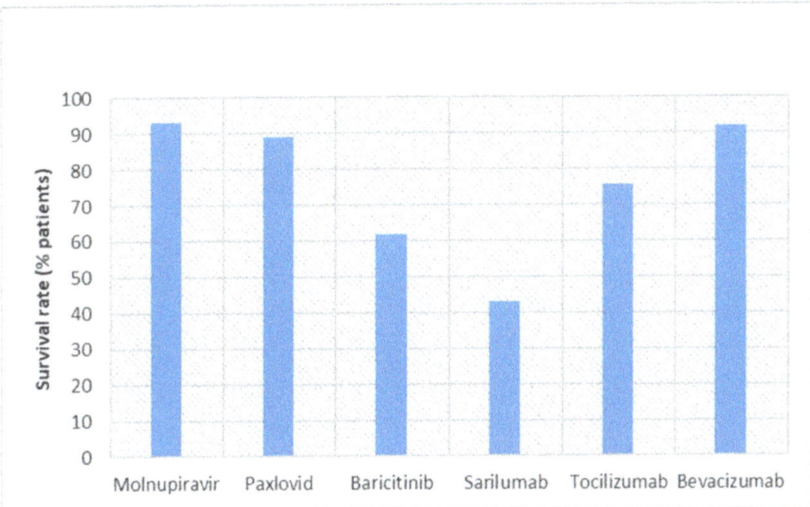

Figure 1. Comparison of the survival rates of COVID-19 patients in clinical trials conducted with new discovered molecules.

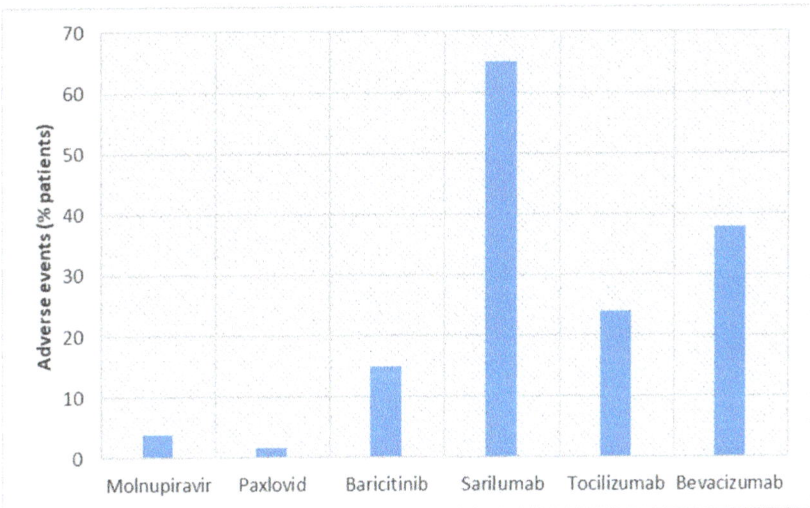

Figure 2. Comparison of adverse events observed in COVID-19 patients in clinical trials conducted with new discovered molecules.

Drugs such as sarilumab and tocilizumab target cytokines such as IL-6 for reducing inflammation. Some drugs inhibit RNA replication either by blocking RdRp (such as molnupiravir) or 3CL (such as Paxlovid™). The rest of the drugs block the entry of the virus either by the JAK-STAT pathway (like sarilumab and baricitinib) or by blocking the NAK family (such as baricitinib). However, the blocking of inflammation can lead to the alleviation of COVID-19 symptoms (such as pneumonia). Therefore, targeting cytokines for blocking inflammation may be a therapeutic management of COVID-19. MOAs of new molecules are illustrated in Figure 3.

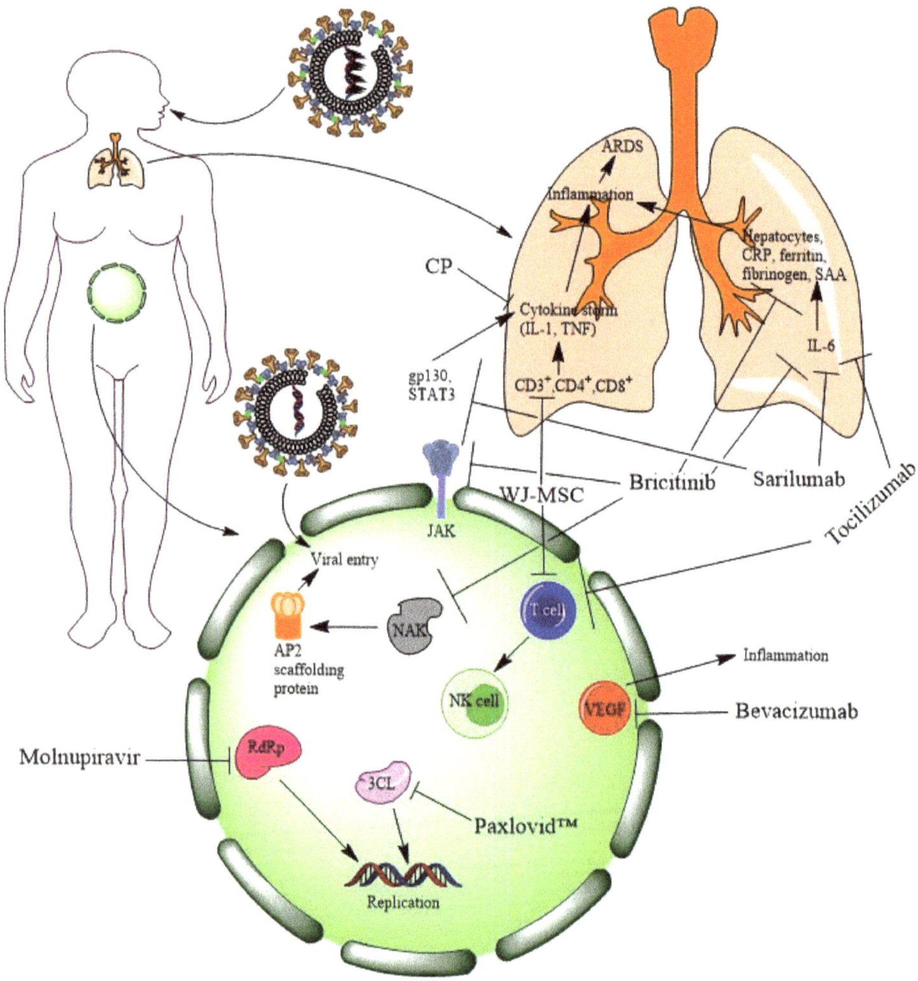

Figure 3. At a glance MOAs of newly discovered molecules for COVID-19 (figure is generated by ChemDraw Professional 15.1).

Table 1 summarizes some clinical trials conducted on new molecules and materials for the treatment of COVID-19.

Table 1. Some clinical trials on new molecules and materials for the treatment of COVID-19 (source: clinicaltrials.gov; accessed on 24 January 2023).

Clinicaltrials.gov Identifier	Nature of Clinical Trial	Selection Criteria of Volunteer	No. of Volunteers	Sponsor(s)	Drug Product(s)/Treatment	Phase	Final Outcome
NCT04392219	Randomized, double blind, and placebo controlled	Adults and older adults	130	Ridgeback Biotherapeutics at Covance Leeds Clinical Research Unit, UK	Molnupiravir	Phase 1	Molnupiravir was more tolerable than placebo, and adverse events were also lesser than placebo
NCT04405570	Randomized, double blind, and placebo controlled	Adults and older adults	204	Ridgeback and Merck	Molnupiravir	Phase 2	400 and 800 mg dose bid was effective with no side effects
NCT04575597	Randomized, double-blind, and placebo-controlled trial	Adults and older adults	1850	Merck	Molnupiravir	Phase 2/3	Effective in reducing death, 6.8% (95% CI) for the drug and 9.7% (95% CI) for placebo
NCT04960202	Randomized and quadruple-blinded trial	Adults and older adults	2246	Pfizer	Paxlovid	Phase 2/3	Less death risk and adverse effects; therapeutic efficacy maintained with percentage point difference of −5.81
NCT05011513	Randomized, quadruple blinded, and parallel sequenced	Adults and older adults	1140		Paxlovid		Terminated due to lower number of deaths and hospitalized standard-risk patients
NCT04421027	Randomized, double blinded, and parallel assigned	Adults and older adults	1525	Eli Lilly	Baricitinib	Phase 3	Mortality was reduced by 38.2%
NCT04333368	Randomized, triple blinded, and parallel assigned	Adults and older adults	47	Assistance Publique–Hôpitaux de Paris	Wharton's jelly mesenchymal stem cells	Phase 1/2	Adverse effects were reported in 28.6% and 25% of patients from the intervention and placebo group, respectively
NCT04321421	Open-label study	Adults and older adults, having moderate to severe ARDS	49	Foundation IRCCS San Matteo Hospital	Convalescent plasma		Result showed uncertainty about beneficial effect of CP
NCT04315298	Randomized, double blind, and placebo controlled	Adult or older adult patients having confirmed COVID-19 infection	1912	Regeneron Pharmaceuticals in collaboration with Sanofi	Sarilumab	Phase 2/3	The percentage of improvement was 43.2% and 35.5% in the intervention and placebo group, respectively
NCT04327388	Randomized, quadruple blinded, parallel mode	Adults or older adults	420	Sanofi in collaboration with Regeneron Pharmaceuticals	Sarilumab	Phase 3	Adverse effects were reported by 70%, 65%, and 65% of patients with 400 and 200 mg sarilumab and placebo, respectively
NCT04320615	Randomized, double blinded, parallel mode, and multicenter	Adult or older adult patients	452	Hoffmann-La Roche	Tocilizumab	Phase 3	No beneficial outcomes were obtained
NCT04275414	Open label	Adult or older adults having confirmed COVID-19 infection	27	Qilu Hospital of Shandong University	Bevacizumab	Phase 2	Body temperature was normalized in 93% of patients and improvement in PaO_2/FiO_2 ratios

4. Repurposed Drugs

Several repurposed drugs, which had received approval from drug regulatory authorities for the treatment of diseases other than COVID-19, have been tested clinically against COVID-19. Several clinical trials on pre-existing drugs for the treatment of COVID-19 are still ongoing.

4.1. Dexamethasone

4.1.1. General Description

Dexamethasone is a corticosteroid that blocks the release of inflammation causing compounds [94]. For many years, it has been widely used for the treatment of hematologic, endocrine, respiratory, dermatologic, allergic, collagen, gastrointestinal, ophthalmic, neoplastic, rheumatic, edematous, and other conditions, more specifically breathing disorders, ulcerative colitis, psoriasis, systemic lupus erythematosus, and so on [94]. Generally, it cannot be used in patients having fungal infection as it weakens the immune system of the body. It has potent anti-inflammatory activity, which facilitates it for assessing through different clinical trials in COVID-19 patients.

4.1.2. Mechanism of Action

Dexamethasone blocks the demargination and apoptosis of neutrophil and phospholipase A2, resulting in the reduction in the production of the derivatives of arachidonic acid [94,95]. It also blocks nuclear factor κB (NFκB) and proinflammatory cytokines (IL-1, IL-6, IL-2, IL-8, TNF, VEGF, IFN_{γ}, and prostaglandin), which are responsible for the transcription of inflammation and also helps in the activation of anti-inflammatory genes.

4.1.3. Clinical Trial, Route of Administration, Dose, and Dosage Form

A phase 3, randomized, open-label, parallel-mode (NCT04327401) clinical trial of dexamethasone was conducted in 290 patients at 21 places of Brazil by the Hospital Israelita Albert Einstein, Hospital do Coracao, Brazilian Research in Intensive Care Network and Ache Laboratorios Farmaceuticos S.A. [96]. Patients were administered dexamethasone 20 mg once in a day for 5 days, followed by 10 mg once in a day for 5 days through an intravenous route. Unfortunately, the trial was terminated by the data monitoring committee for the unaccepted recovery results in trial. Dr. Negrin University Hospital in collaboration with the Li Ka Shing Knowledge Institute, Consorcio Centro de Investigación Biomédica en Red, M.P., conducted a phase 4, randomized, open-label, parallel-mode (NCT04325061) clinical trial on dexamethasone in 200 adult or older adult patients at 24 different places in Spain [97]. Patients received 20 mg dexamethasone dose daily up to 5 days randomly and, after that, 10 mg dose daily up to 5 days randomly through the IV route. This trial was also terminated due to lack of participant enrollment.

4.1.4. Regulatory Approval

Dexamethasone is not approved till date for the treatment of COVID-19. However, WHO recommended dexamethasone for the treatment of severe COVID-19 patients in October 2020 [98].

4.2. Naproxen

4.2.1. General Description

Naproxen is an over-the-counter (OTC) nonsteroidal anti-inflammatory drug (NSAID) indicated for the treatment of acute pain and pain involved in rheumatoid arthritis and other diseases, such as headache, tendonitis, muscle pain, menstrual pain, dental pain, swelling, and joint stiffness [99,100]. It is also available in the market alone and in combination with esomeprazole (for gastric ulcers) or with sumatriptan (for migraine).

4.2.2. Mechanism of Action

Naproxen is a nonselective cyclooxygenase (COX) inhibitor that inhibits arachidonate (ester form of arachidonic acid), leading to its anti-inflammatory and analgesic activities [100,101]. Naproxen blocks the nucleoprotein of SARS-CoV-2, resulting in blocking the RNA replication of the virus [102].

4.2.3. Clinical Trial, Route of Administration, Dose, and Dosage Form

Assistance Publique—Hôpitaux de Paris conducted a phase 3, open-label, randomized, parallel model (NCT04325633) clinical trial on naproxen in 584 adult or older adult patients [103]. Naproxen was orally administered at 250 mg twice a day along with 30 mg of lansoprazole. This treatment was compared with the standard of care. The location of this trial was not recorded in the clinical trial record of the U.S. However, this trial was terminated due to insufficient recruitment of participants.

4.2.4. Regulatory Approval

No regulatory authorities have approved nor recommended naproxen till date for the treatment of COVID-19.

4.3. Remdesivir

4.3.1. General Description

Remdesivir is a 1′-cyano-substituted nucleotide analogue and phosphoramidate prodrug. It is well known for its broad-spectrum effect against Ebola, SARS-CoV, MERS-CoV, Nipah virus, and Hendra virus [104,105]. An in vitro study of this drug proved it as a therapeutically potent agent against SARS-CoV-2 and safe for humans.

4.3.2. Mechanism of Action

Remdesivir acts as a nucleoside analog and is a RdRp blocker that attacks the replication cycle of a viral genome [105–108]. CoVs utilize RdRp for running their replication process of RNA-dependent genomes. The drug is metabolized in the human body, and the metabolite GS-441524 blocks RdRp, leading to a gradual stopping of RNA formation.

4.3.3. Clinical Trial, Route of Administration, Dose, and Dosage Form

A phase 3, randomized, open-label, parallel-mode clinical trial (NCT04292899) was conducted in 6000 hospitalized child, adult, or older adult patients by Gilead Sciences at 179 places in the US, France, Japan, China, Hong Kong, Italy, Republic of Korea, Taiwan, Germany, Netherlands, Singapore, Sweden, Spain, Switzerland, and UK [109]. In this trial, the patients were divided into three groups. The first group received 200 mg drug for 5 days and subsequently 100 mg drug for up to 5 days through IV infusion. The second group was administered with 200 mg drug for 10 days and then 100 mg drug for up to 10 days through IV infusion. The third group received 200 mg drug for 5 or 10 days and then 100 mg drug for up to 5 or 10 days through IV infusion. After 14 days, 74.4% of patients from interventional groups and 59% of patients from noninterventional groups survived (95% CI; $p < 0.001$) [110]. The National Institute of Allergy and Infectious Diseases (NIAID) conducted a phase 3, randomized, double-blinded, and parallel-mode (NCT04280705) clinical trial in 572 hospitalized adult or older adult patients at 68 different places [111]. In this study, 200 mg of remdesivir was administered to patients through the IV route along with a maintenance dose of 100 mg once a day during hospitalization for 10 days, and its efficacy was compared with placebo against COVID-19 infection. It was observed that the median recovery time was 10 days and 15 days for the intervention and placebo group, respectively [112]. Serious adverse events were seen in 24.6% of patients in the interventional group and 31.6% of patients treated with placebo.

4.3.4. Regulatory Approval

Veklury® (remdesivir) IV infusion, developed by Gilead Sciences, obtained EUA from USFDA in October 2020 and obtained expedited approval from USFDA on January 2022 [113,114]. Veklury® is approved for the treatment of hospitalized adult and pediatric patients (at least having 40 kg body weight) who are at high risk of COVID-19. EMA granted full approval for Veklury® (remdesivir) IV infusion in December 2021 [115].

4.4. Hydroxychloroquine

4.4.1. General Description

Hydroxychloroquine (chemically 4-aminoquinoline) is a well-known derivative of chloroquine [116,117]. Apart from its antimalarial activity, it has proven effect as anti-inflammatory, antiautophagy, and immunosuppressive activities for the treatment of systemic lupus erythematosus and rheumatoid arthritis. It is also used as a chemotherapeutic drug against malarial parasites.

4.4.2. Mechanism of Action

Hydroxychloroquine blocks sialic acid receptors at the upper respiratory pathway and prevents the glycosylation of the angiotensin converting enzyme 2 (ACE2) receptor [116–119]. Prevention of ACE2 receptor glycosylation results in the failure of the viral spike protein binding to the ACE2 receptor. Therefore, the entry of the COVID-19 virus in humans was restricted. Hydroxychloroquine also blocks antigen processing and lowers the T-cell level in the human body. This results in the prevention of cytokine storm.

4.4.3. Clinical Trial, Route of Administration, Dose, and Dosage Form

Several clinical trials, NCT04341493, NCT04342156, NCT04341441, NCT04325893, and NCT04329923, were terminated due to reasons such as continuous decreasing number of participants, increasing side effects, or lack of beneficial results. A phase 3, randomized, quadruple-blinded, parallel-mode clinical trial (NCT04332991) was conducted in 510 adult or older adult patients at 43 locations in the US by Massachusetts General Hospital in collaboration with the National Heart, Lung, and Blood Institute (NHLBI) [120,121]. Participants were divided into the experimental group and comparator group, of which the experimental group was administered 400 mg hydroxychloroquine sulfate BID on the first day and then 200 mg of this drug BID for the rest of 4 days through the oral route (tablet), whereas the comparator group was administered placebo through the oral route [122]. By day 28, it was observed that 10.4% of participants from the interventional group and 10.6% of participants from the placebo group had died [123]. The result of the trial indicates that there were no significant differences between the interventional group and the placebo group.

4.4.4. Regulatory Approval

Hydroxychloroquine is approved for treating emergency cases of COVID-19 infection and for the prophylaxis of healthcare professionals by FDA and EMA [116,124,125]. USFDA granted EUA to hydroxychloroquine for the emergency treatment of COVID-19 [126,127]. EMA released news on 1 April 2020 stating that chloroquine and hydroxychloroquine can be used for the emergency condition of a patient and in clinical trials [128].

4.5. Favipiravir

4.5.1. General Description

The drug favipiravir (also known as Avigan or Favilavir) is a derivative of pyrazine and is developed by Fujifilm Toyama Chemical Co., Ltd. of Tokyo, Japan [129,130]. It was approved for the treatment of influenza A and B as it attacks RdRp, required for the replication cycle and transcription process of the viral genome. Its therapeutic efficacy against Lassa virus and Ebola virus is under investigation.

4.5.2. Mechanism of Action

Favipiravir is a prodrug that is biotransformed into its active metabolite favipiravir ribofuranosyl-5′-triphosphate (favipiravir-RTP). Favipiravir-RTP interacts with RdRp and blocks the replication cycle of the virus genome [78,118,131]. Till now, the interaction mechanism is not very clear, but a hypothesis regarding it claims that this active metabolite may link to the conserved domains of polymerase, or it may be added in a nascent viral RNA, for which nucleotides cannot be added, and the process of transcription and replication is blocked. To accomplish the incorporation, this active metabolite has to compete with the purine bases or nucleotides.

4.5.3. Clinical Trial, Route of Administration, Dose, and Dosage Form

A phase 2, randomized, open-label, parallel-mode clinical trial (NCT04346628) was conducted in 120 adult or older adult patients by Stanford University [132]. Participants were split into two sections, of which the experimental section received 1800 mg favipiravir for the first day along with 800 mg twice a day for up to 10 days through the oral route. However, no significant differences (95% CI) were observed between the experimental group and the comparator group for mortality, symptom resolution, and transition as well as accumulation of mutation in the viral genome [133]. Another phase 3, randomized, open-label, parallel-mode clinical trial (NCT04349241) was conducted by Ain Shams University in 100 adult or older adult patients at the Faculty of Medicine, Ain Shams University Research Institute–Clinical Research Center, Cairo, Egypt [134]. The experimental group was administered 1600 mg of favipiravir BID every 12 h on the 1st day and 600 mg of this drug BID every 12 h as the maintenance dose from the 2nd day to the 10th day through the oral route, and the comparator group received the standard of care according to the national Egyptian guideline, that is, 75 mg of oseltamivir every 12 h from the 5th day to the 10th day and 400 mg of hydroxychloroquine every 12 h on the 1st day and 200 mg of this drug every 12 h daily from the 2nd day to the 10th day through the oral route. Among all groups, the favipiravir group showed potential results, where two consecutive SARS-CoV-2-negative reports were observed after the treatment and body temperature was also normalized [135,136].

4.5.4. Regulatory Approval

DCGI (Drug Controller General of India) granted EUA to favipiravir for the treatment of severe COVID-19 patients on July 2020 [137]. It is not approved by USFDA and EMA till date.

4.6. Umifenovir (Arbidol)

4.6.1. General Description

Umifenovir is a derivative of the indole moiety. In Russia, it is an approved drug to treat influenza A and B [78,138,139]. This antiviral drug is effective against viruses such as the Lassa virus, coxsackie virus B5, herpes simplex virus, hepatitis B and C, Zika virus, Ebola virus, and chikungunya virus.

4.6.2. Mechanism of Action

Umifenovir can interact with amino acid residues such as tryptophan and tyrosine and thereby aids in preventing the entry of viruses in the human body [140]. It also interacts with aromatic residues present in glycoproteins of a virus (S protein of SARS-CoV-2). The glycoprotein is generally responsible for the recognition for cells and the process of fusion with the plasma membrane. Through the clathrin-mediated exocytosis, the process and trafficking in the cells are blocked. In enveloped viruses, such as SARS-CoV-2, it can also directly bind with the lipid envelope, leading to the stabilization of the plasma membrane, which prevents the entry of the virus in the human body.

4.6.3. Clinical Trial, Route of Administration, Dose, and Dosage Form

Another phase 4, randomized, open-label, parallel-mode clinical trial (NCT04252885) was conducted by Guangzhou Eighth People's Hospital in 125 adult or older adult patients at Guangzhou Eighth People's Hospital, Guangdong, China [141]. Participants were divided into three groups. One group was administered 200 mg of lopinavir oral tablet with 50 mg of ritonavir oral tablet twice a day for 7 to 14 days along with the standard treatment. Another group received 200 mg of umifenovir (i.e., two tablets of 100 mg of arbidol) thrice in a day for 7 to 14 days along with the standard treatment. The third group was not provided any intervention. On the seventh day, 23.5% of participants from the lopinavir/ritonavir group, 8.6% of participants from the arbidol group, and 11.8% of participants from the control group showed deterioration from moderate to severe condition ($p = 0.206$) [142]. Adverse events were observed in 35.3% of patients treated with lopinavir/ritonavir and 14.3% of patients treated with arbidol. Adverse events were not observed in patients who received placebo. Result indicates that arbidol monotherapy could be beneficial for mild, moderate, and severe COVID-19 patients.

4.6.4. Regulatory Approval

Umifenovir is not approved by USFDA and EMA for the treatment of COVID-19 till date.

4.7. *Azvudine*

4.7.1. General Description

Azvudine (2′-deoxy-2′-β-fluoro-4′-azidocytidine), also known as FNC, is a prodrug that is biotransformed to its active form, FNC triphosphate. Azvudine is the first oral antiviral drug approved in China by SFDA [143]. It is a nucleoside analogue, having broad spectrum antiviral efficacy against RNA viruses such as Ebola, HIV, HBV, HCV, and so on [144]. It contains 4′-modified nucleoside, which is essential for drug resistance and antiviral effect [143]. It is notable that azvudine, manufactured by Zhengzhou Granlen PharmaTech Ltd. (a division of Granlen LLC, Carlsbad, CA, USA), received approval from the China Food and Drug Administration (SFDA) in April 2013 as an investigational new drug for the treatment of HIV [145]. Azvudine, manufactured by Genuine Biotech Limited, Pingdingshan, Henan Province, Central China, is available in the Chinese market as 1 mg oral tablet for the treatment of COVID-19 [146].

4.7.2. Mechanism of Action

FNC works against SARS-CoV-2 by blocking RdRp [144,147]. The mechanism or action is divided into two steps, antiviral action in the thymus (one of the vital immune organs, where T lymphocytes are born) of the human chest and boosting immunity against the virus in whole body. After distribution, FNC is phosphorylated three times in the cytoplasm by deoxycytidine kinase and forms the active product FNC triphosphate, which inhibits RdRp of SARS-CoV-2 in the thymus part. However, after inhibition of RdRp, the FNC triphosphate elevates IL-4, IL-10, and IL-13 levels in the thymus to combat cytokine storm. FNC can also control the expression of P-glycoprotein (Pgp), multidrug resistance-associated protein 2 (MRP2), breast cancer resistance protein (BCRP), and so on [148].

4.7.3. Clinical Trial, Route of Administration, Dose, and Dosage Form

A randomized, open-label, controlled pilot study (phase 0, ChiCTR2000029853) on FNC was conducted in 20 mild to moderate COVID-19 patients at People's Hospital of Guangshan County, Guangshan County, He'nan, China [147]. Twelve patients (60%) were provided routine treatment. Among these 12 patients, 6 patients were administered placebo, and the other 6 received FNC (5 mg per day and 5 oral tablets QD). After that, some patients were provided antiviral treatment, some received antibiotic treatment, some patients received adjuvant treatment, and some were administered traditional Chinese medicine. As a result, the placebo group showed adverse events with 30% incidence ($p = 0.06$), while the FNC group did not show any adverse event. After 5 days of treatment with FNC, 100% of patients survived from COVID-19 pneumonia, while in the case of the placebo group, 100% survival was observed after 8 days. FNC was also found effective in those patients who received other treatments (antiviral, antibiotic treatment, etc.) [149].

4.7.4. Regulatory Approval

In July 2022, azvudine obtained conditional approval from the National Medical Products Administration, China [143]. On the other hand, azvudine obtained approval from the National Health Commission and National Administration of Traditional Chinese Medicine, China, and the National Healthcare Security Administration, China, for incorporating it in the Diagnosis and Treatment Program for Novel Coronavirus Pneumonia (Ninth Edition) and a medical reimbursement list as a potential drug for the treatment of COVID-19 in August 2022.

5. Comparative Analysis of Repurposed Drugs

Repurposed drugs have been symptomatically used for treating inflammation, pneumonia, and other symptoms observed in COVID-19 patients. Dexamethasone has been used due to its efficacy against inflammation [94]. According to one study, dexamethasone has reduced the risk of death by 1/3 in patients kept on ventilation and by 1/5 among patients requiring oxygen [150,151]. Despite having advantages, such as low cost, prolonged action, and so on, dexamethasone reduced immunity in COVID-19 patients. Dexamethasone also causes other side effects, such as anxiety and disturbance in sleep hormonal imbalance. It also causes some withdrawal symptoms, such as myalgia, arthralgia, vomiting, and loss of weight headache. In an open-level, randomized clinical trial (NCT04381936; RECOVERY trial), 6 mg of dexamethasone (QD up to 10 days) was administered to hospitalized COVID-19 patients through the oral or IV route [152]. In that trial, 2104 patients received dexamethasone and 4321 patients received standard care. It was observed that 22.9% of patients from the dexamethasone group and 25.7% of patients from the standard of care group died within 28 days (95% CI, $p < 0.001$). Results indicated a marginal reduction in death rate. Another study reported that patients treated with dexamethasone showed 12.7% adverse events [153].

Naproxen, a nonselective COX inhibitor, such as dexamethasone, also inhibits inflammation to combat COVID-19 [100,101]. However, in a randomized placebo controlled, double-blinded clinical trial, naproxen oral capsule (500 mg BID) showed potential efficacy in treating breathing issue and cough in COVID-19 patients [154]. It also elevated mean the corpuscular volume and helped to normalize blood pressure. In overall aspects, naproxen has proven itself more beneficial than dexamethasone. In general, treatment with NSAIDs may cause gastrointestinal issues, but naproxen did not show any serious adverse event. Naproxen is therefore better tolerated when compared with dexamethasone.

1′-Cyano-substituted nucleotide analogue remdesivir has shown both positive and negative effects in COVID-19 patients [104,105]. Like molnupiravir, remdesivir blocks RdRp to inhibit RNA formation in SARS-CoV-2 [105–107]. In a clinical trial (NCT04292899), it was observed that the survival rate of patients treated with iv infusion of remdesivir (IV infusion) was 74.4% (95% CI; $p < 0.001$) [110]. For the IV route, the death rate was recorded at 7.6% (95% CI; $p = 0.001$). Results gave a clear indication that treatment though the IV route for remdesivir was better than the infusion route. Despite having a good survival rate, remdesivir also showed different adverse effects, such as reaction at the infusion site, kidney injuries, anemia, multiorgan failure, pyrexia, hyperglycemia, hypokalemia, torsade de point, hypotension, cardiac arrest, bradycardia, and complete heart block= [155]. Another study indicated that remdesivir could induce cytotoxicity on cardiomyocytes [156]. However, WHO provided a conditional approval for the administration of remdesivir in hospitalized patients having severe COVID-19 [5]. Moreover, in the another study, remdesivir showed 24.6% serious adverse events [112]. Remdesivir (7.6% patients) showed more beneficial effects than dexamethasone (22.9% patients) in reducing the death rate, but adverse events for dexamethasone (12.7% of patients) were lesser than remdesivir (24.6% of patients). Like remdesivir, favipiravir also blocks RdRp to combat COVID-19 [78,118,131]. Due to its good efficacy in inhibiting viral proliferation, favipiravir protects the lungs from fibrosis and inflammatory damage [157]. In this aspect favipiravir possess more beneficial effects than remdesivir. Remdesivir may also protect the lung but has serious adverse events, leading to its cautious use for the treatment of COVID-19. Most of the clinical trials of favipiravir have been conducted in combination other drugs such as oseltamivir, darunavir, ritonavir, and hydroxychloroquine. This is probably due to the synergistic or additive effect in various symptoms of COVID-19. It has also been observed that a high dose of favipiravir may lead to serious adverse events such as elevated hepatic enzymes, tachycardia, and vomiting [158,159]. A study reported serious adverse events in 10.36% of COVID-19 patients when treated with favipiravir [160]. A different study revealed a 46.2% fatality rate for favipiravir [161]. In a nutshell, no beneficial effect has been observed in COVID-19 patients treated with favipiravir.

The antimalarial drug hydroxychloroquine reduces cytokines to give symptomatic relief in inflammation to COVID-19 patients [116,117]. In one study, 10.4% of COVID-19 patients died when they received 200–400 mg BID oral hydroxychloroquine treatment [123]. No significant difference in death rate between the hydroxychloroquine group and the placebo group (10.6%) was observed in the clinical trial. Serious adverse events from hydroxychloroquine were reported in 30% of COVID-19 patients when treated with hydroxychloroquine [162]. Hydroxychloroquine showed the highest adverse effects in COVID-19 patients when compared with other repurposed drugs. Both remdesivir and hydroxychloroquine were recommended conditionally for treating COVID-19, but reports of serious adverse events had compromised their clinical use. Later, regulatory authorities had restricted the use of both of these drugs in treating COVID-19. ICMR (India) had recommended the use of chloroquine or hydroxychloroquine for prophylaxis of COVID-19. For showing the desired therapeutic efficacy in COVID-19 patients, high doses of chloroquine or hydroxychloroquine is required, but high doses may cause serious adverse effects [6,9]. Overdose of chloroquine can cause muscle weakness, mood swings, nausea, skin irritation, headache, bleeding from the nose, pale skin, swelling, cramps, and hearing problems. On the other hand, an overdose of hydroxychloroquine can cause cramps, drowsiness, headaches, nausea, irreversible blindness, lowering of blood glucose, depression, reduced appetite, and heart failure. An overdose for both cases can cause death. Therefore, it is highly recommended not to take these drugs without the advice of a physician. A Chinese study reported hydroxychloroquine to be ineffective against COVID-19. A different study reported hydroxychloroquine to be effective when it is used in combination with azithromycin [163].

Umifenovir (arbidol) has been used for treating diseases caused by several viruses such as hepatitis B and C, Zika, and influenza [78,138,139]. Adverse events were found in 14.3% of COVID-19 patients treated with umifenovir, and therefore, it may be regarded as superior to remdesivir (24.6% of patients) and hydroxychloroquine (30% of patients) but inferior when compared with favipiravir (10.36% of patients) [142]. Notably, the dose of umifenovir is lower in comparison with other repurposed drugs. Similarity was found in the case of the route of administration, oral route. A recent study revealed the mortality rate for the umifenovir group and the control group, that is, 33.3% and 38%, respectively [164]. These results indicated no statistically significant difference in mortality rate between umifenovir and the control group. Furthermore, the mortality rate of umifenovir (33.3% of patients) was more than the death rate observed for remdesivir (7.6% of patients) and hydroxychloroquine (10.4% of patients). However, another study revealed that umifenovir has potential in reducing the mortality rate [165,166]. The first approved (SFDA) oral antiviral drug, azvudine, works like favipiravir [143]. As an RdRp inhibitor, both favipiravir and azvudine can be compared. According to the trial results, as per the survival rate of COVID-19 patients and adverse events, azvudine was superior to favipiravir [147]. Azvudine seems better among all repurposed drugs discussed here as it has the highest survival rate (100%) and adverse events are negligible. However, more clinical trials and more participants are required to prove its efficacy. Figures 4 and 5 illustrate the survival rates and adverse events, respectively, for the repurposed drugs.

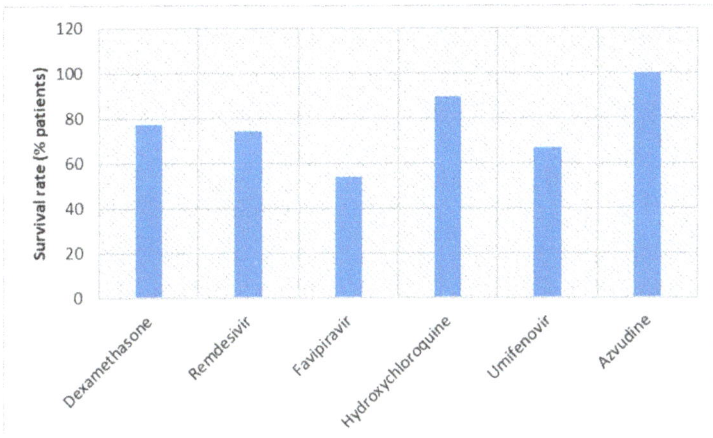

Figure 4. Comparison of survival rates of COVID-19 patients in clinical trials conducted with repurposed drugs.

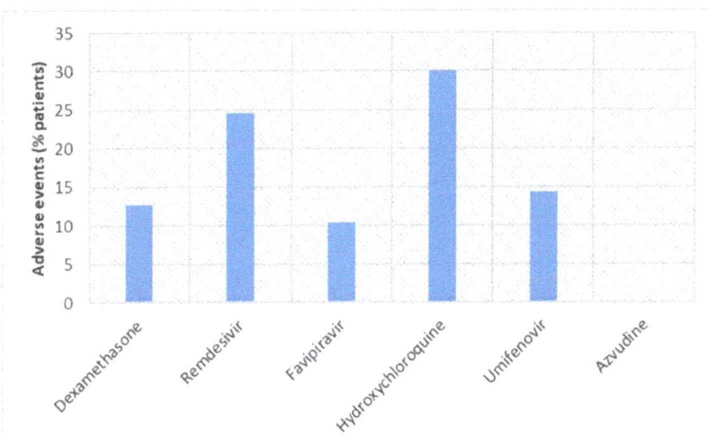

Figure 5. Comparison of adverse events observed in COVID-19 patients in clinical trials conducted with repurposed drugs.

Some of the clinical trials on pre-existing drugs are listed in Table 2. The main therapeutic management approach for repurposed drugs was to block cytokines and therefor inhibit inflammation. To achieve this target, repurposed drugs may either directly block (like dexamethasone) cytokines or indirectly by blocking RdRp (like favipiravir). NSAIDs (such as naproxen) target COX to inhibit the cytokines' mediated inflammation. Some of the repurposed drugs also target the replication of viral RNA (such as azvudine). Apart from that, a few drugs tried to stop the virus by blocking the ACE2 receptor or its glycosylation process (such as hydroxychloroquine). Some repurposed drugs (such as umifenovir) also work on amino acids and S-glycoprotein to prevent entry of SARS-CoV-2 in the human body. Blocking inflammation or viral RNA replication may inhibit SARS-CoV-2. MOAs of repurposed drugs are illustrated in Figure 6.

Table 2. Clinical trials of the some repurposed drugs for the treatment of COVID-19 (collected from clinicaltrials.gov; accessed on 24 January 2023).

Clinicaltrials.gov Identifier	Nature of Clinical Trial	Selection Criteria of Volunteer	No. of Volunteers	Sponsor(s)	Drug Product(s)/Treatment	Phase	Final Outcome
NCT04327401	Randomized, open label, parallel mode	Adult or older adults	290	Hospital Israelita Albert Einstein, Hospital do Coracao, Brazilian Research in Intensive Care Network and Ache Laboratorios Farmaceuticos S.A.	Dexamethasone	Phase 3	Terminated by the data monitoring committee for the unaccepted recovery results
NCT04325633	Open label, randomized, parallel model	Adult or older adult patients	584	Assistance Publique—Hôpitaux de Paris	Naproxen	Phase 3	Terminated due to insufficient recruitment of participants
NCT04292899	Randomized, open label, parallel	Hospitalized child, adult, or older adults	6000	Gilead Sciences	Remdesivir	Phase 3	After 14 days. 74.4% and 59% of patients survived from the intervention and noninterventional group.
NCT04280705	Randomized, double blinded, and parallel mode	Hospitalized adult or older adult patients	572	National Institute of Allergy and Infectious Diseases (NIAID)	Remdesivir	Phase 3	Adverse events were seen in 24.6% and 31.6% of patients in the interventional and placebo groups, respectively
NCT04332991	Randomized, quadruple blinded, parallel	Adult or older adult patients	510	Massachusetts General Hospital in collaboration with National Heart, Lung, and Blood Institute (NHLBI)	Hydroxychloroquine	Phase 3	No significant differences between the interventional group and the placebo group was found
NCT04346628	Randomized, open label, parallel mode	Adult or older adult patients	120	Stanford University	Favipiravir	Phase 2	No significant differences (95% CI) were observed between groups in mortality and symptom resolution
NCT04349241	Randomized, open label, parallel	Adult or older adult patients	100	Faculty of Medicine, Ain Shams University Research Institute- Clinical Research Center	Favipiravir	Phase 3	Two consecutive SARS-CoV-2-negative reports were observed after the treatment
NCT04252885	Randomized open label, parallel	Adult or older adult patients	125	Guangzhou Eighth People's Hospital	Umifenovir	Phase 4	Arbidol monotherapy could be beneficial for mild, moderate, and severe COVID-19 patients

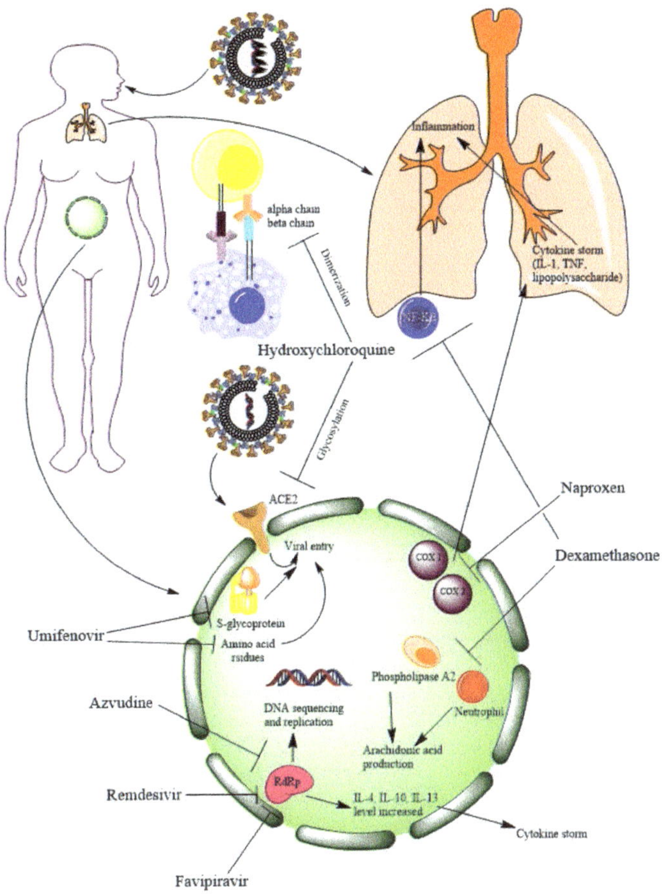

Figure 6. At a glance MOAs of repurposed drugs (figure is generated by ChemDraw Professional 15.1).

Different drugs are there in the market, which are repurposed for the treatment of COVID-19. They are approved by regulatory authorities such as USFDA, EMA, and DCGI. Some of them are listed in Table 3 along with their approved brand names.

Table 3. Approved repurposed drugs with their approved brand names.

Drug	Formulation	Approved by and Type of Approval	Date of Approval	Marketed Name	References
		New Drugs			
Baricitinib	Oral tablets	USFDA (EUA)	October 2022	Olumiant®	[43]
Casirivimab and Imdevimab	Injectable solution	FDA (EUA), EMA (full approval)	November 2020 (USFDA), February 2021 (EMA for REGN-COV2), October 2021 (EMA for Ronapreve)	REGEN-COV® (USFDA), REGN-COV2 (EMA), Ronapreve (EMA)	[167–169]
Molnupiravir	Oral capsules	USFDA (EUA), MHRA, EMA (EUA)	December 2021 (USFDA), November 2021 (MHRA, EMA)	Lagevrio™	[26–29]
Nirmatrelvir and Ritonavir	Oral tablets	USFDA (EUA), EMA (EUA)	December 2021 (USFDA, EMA)	Paxlovid™	[33,37]
Tocilizumab	Injectable solution	USFDA (full approval)	December 2022	Actemra®	[85]
		Repurposed Drugs			
Favipiravir	Tablets	DCGI (EUA)	July 2020	Fabiflu®	[137]
Hydroxychloroquine	Oral formulation	USFDA (EUA), EMA (EUA)	March 2020April 2020		[124–127]
Remdesivir	Injection, solution	USFDA (EUA), EMA (EUA)	October 2020December 2021	Veklury®	[113–115]
Tixagevimab and Cilgavimab	Injection, solution	USFDA (EUA)	December 2021	Evusheld™	[170]

6. Conclusions

Clinical studies on new discovered molecules and repurposed drugs have proven effectiveness and safety for the treatment of COVID-19. The first approved antiviral drug, molnupiravir, has shown good therapeutic efficacy, but the risk of mutation and high dose may limit its wider usage in treating COVID-19. Paxlovid™, the combination of nirmatrelvir and ritonavir, has shown good therapeutic efficacy due to the synergistic effect of nirmatrelvir and ritonavir. Ritonavir improves the pharmacokinetics of nirmatrelvir, leading to advantage in pharmacodynamics. Several studies have revealed acceptable human toleration for new discovered molecules such as Paxlovid™ and molnupiravir. MAbs have also shown potential results in clinical studies. Among all MAbs, bevacizumab has been observed better based on the survival rate of COVID-19 patients. Other MAbs (such as sarilumab, baricitinib, and tocilizumab) have scored lower due to higher mortality rate or serious adverse events. Almost all repurposed drugs, discussed here, have serious adverse events, which are dose dependent. Different clinical trials have addressed this issue to find their best dose, but still the quest to find the best dose seems unresolved. Some drugs have been initially recommended for the treatment of COVID-19, such as remdesivir and hydroxychloroquine. Unfortunately, these drugs were later found very much harmful due to their adverse effects, leading to restrictions imposed on them. Notably, most of the route of administration for most of COVID-19 drugs is oral, and very few drugs have been tested clinically for other routes, such as IV or infusion. Several drugs (such as favipiravir and umifenovir) are recommended to be administered in combination with other drugs for their synergistic effects. More clinical studies on monotherapy are required for these drugs (especially umifenovir). As per a single clinical study, it is quite impressive that azvudine had 100% survival rate and negligible adverse events. More clinical studies are required to ascertain the efficacy and safety of azvudine.

This review has some limitations, which can be addressed in future reviews. Two neutralizing antibodies, imdevimab and casirivimab, are not highlighted in this work due to lack of data on clinical trials. These antibodies can be explored in future reviews. A comparison on efficacy of the repurposed drugs against each variant of SARS-CoV-2 is not presented in this review due to lack of availability of data on clinical trials. Various clinical trials are still ongoing or for some others results are not yet available in the public domain. This limitation allows a future scope of reviews. Besides, this review has furnished details of only one study where combination drugs are used (Paxlovid™). More information on combinations of drugs used against COVID-19 can be introduced in future works.

Further work may be conducted to determine the efficacy of each drug against each variant of SARS-CoV-2. The risk of mutations in humans for other RdRp inhibitors should be determined. Adverse events from new discovered molecules seem lower when compared with repurposed drugs. However, further studies are required to prove this statement. In future studies, some new discovered drugs, such as molnupiravir and Paxlovid™, should be compared with repurposed drugs.

Author Contributions: All authors have contribution in the design, data assimilation, interpretation, analysis, and revision of this article. All authors have read and agreed to the published version of the manuscript.

Funding: Surojit Banerjee acknowledges the kind support of DST-INSPIRE Fellowship from Department of Science and Technology, Ministry of Science and Technology, Govt. of India (Application Reference No. DST/INSPIRE/03/2021/002695; IF No. IF200218) for pursuing Ph.D.

Institutional Review Board Statement: Not applicable.

Informed Consent Statement: Not applicable.

Data Availability Statement: Not applicable.

Acknowledgments: Authors are also highly grateful to Sonali Rawat for reading and improving the English language.

Conflicts of Interest: The authors declare no conflict of interest.

References

1. Gao, K.; Nguyen, D.D.; Chen, J.; Wang, R.; Wei, G.-W. Repositioning of 8565 Existing Drugs for COVID-19. *J. Phys. Chem. Lett.* **2020**, *11*, 5373–5382. [CrossRef] [PubMed]
2. Hu, B.; Guo, H.; Zhou, P.; Shi, Z.-L. Characteristics of SARS-CoV-2 and COVID-19. *Nat. Rev. Microbiol.* **2021**, *19*, 141–154. [CrossRef] [PubMed]
3. Banerjee, S.; Banerjee, D.; Singh, A.; Saharan, V.A. A Comprehensive Investigation Regarding the Differentiation of the Procurable COVID-19 Vaccines. *AAPS PharmSciTech* **2022**, *23*, 95. [CrossRef] [PubMed]
4. Cohen, J. The 'Very, Very Bad Look' of Remdesivir, the First FDA-Approved COVID-19 Drug. *Science*, 28 MOctoberay 2020. Available online: https://www.science.org/content/article/very-very-bad-look-remdesivir-first-fda-approved-Covid-19-drug(accessed on 31 October 2022).
5. World Health Organization. WHO Recommends against the Use of Remdesivir in COVID-19 Patients. Available online: https://www.who.int/news-room/feature-stories/detail/who-recommends-against-the-use-of-remdesivir-in-Covid-19-patients (accessed on 31 October 2022).
6. Acharjee, S. COVID-19: The Bitter Truth about Using Hydroxychloroquine as a Preventive Drug. *India Today*, 25 May 2020. Available online: https://www.indiatoday.in/india-today-insight/story/covid-19-the-bitter-truth-about-using-hydroxychloroquine-as-a-preventive-drug-1659116-2020-03-24(accessed on 15 July 2022).
7. Kaul, R. Stay away from Indiscriminate Use of Anti-Malaria Drug: Experts on COVID-19. *Hindustan Times*, 25 March 2020. Available online: https://www.hindustantimes.com/india-news/stay-away-from-indiscriminate-use-of-anti-malaria-drug-experts-on-covid-19/story-UY2zFNCcWivkdw7zToWHcI.html(accessed on 19 December 2022).
8. India News. Assam Doctor Who Took Anti-Malaria Drug amid COVID-19 Outbreak Dies. *Hindustan Times*, 30 March 2020. Available online: https://www.hindustantimes.com/india-news/assam-doctor-who-took-anti-malaria-drug-amid-covid-19-outbreak-dies/story-mHqu2dLT8gKxLohCk9JkeP.html(accessed on 19 December 2022).
9. Terry, M. Amidst Controversy, FDA Issues EUA Allowing Anti-Malaria Drugs to Be Used to Treat COVID-19. Biospace. *Biospace*, 30 March 2020. Available online: https://www.biospace.com/article/fda-allows-anti-malaria-drug-chloroquine-to-be-used-to-treat-covid-19/(accessed on 19 December 2022).
10. World Health Organization. WHO Recommends Two New Drugs to Treat COVID-19. Available online: https://www.who.int/news/item/14-01-2022-who-recommends-two-new-drugs-to-treat-Covid-19 (accessed on 31 October 2022).
11. Jayk Bernal, A.; Gomes da Silva, M.M.; Musungaie, D.B.; Kovalchuk, E.; Gonzalez, A.; Delos Reyes, V.; Martín-Quirós, A.; Caraco, Y.; Williams-Diaz, A.; Brown, M.L.; et al. Molnupiravir for Oral Treatment of COVID-19 in Nonhospitalized Patients. *N. Engl. J. Med.* **2022**, *386*, 509–520. [CrossRef]
12. Butler, C.C.; Hobbs, F.D.R.; Gbinigie, O.A.; Rahman, N.M.; Hayward, G.; Richards, D.B.; Dorward, J.; Lowe, D.M.; Standing, J.F.; Breuer, J.; et al. Molnupiravir plus Usual Care versus Usual Care Alone as Early Treatment for Adults with COVID-19 at Increased Risk of Adverse Outcomes (PANORAMIC): An Open-Label, Platform-Adaptive Randomised Controlled Trial. *Lancet* **2022**, *401*, 281–293. [CrossRef]
13. Kozlov, M. Merck's COVID Pill Loses Its Lustre: What That Means for the Pandemic. *Nature*, 13 December 2021. Available online: https://www.nature.com/articles/d41586-021-03667-0(accessed on 3 March 2022).
14. Singh, A.K.; Singh, A.; Singh, R.; Misra, A. Molnupiravir in COVID-19: A Systematic Review of Literature. *Diabetes Metab. Syndr. Clin. Res. Rev.* **2021**, *15*, 102329. [CrossRef]
15. Kabinger, F.; Stiller, C.; Schmitzová, J.; Dienemann, C.; Kokic, G.; Hillen, H.S.; Höbartner, C.; Cramer, P. Mechanism of Molnupiravir-Induced SARS-CoV-2 Mutagenesis. *Nat. Struct. Mol. Biol.* **2021**, *28*, 740–746. [CrossRef]
16. Toots, M.; Yoon, J.-J.; Hart, M.; Natchus, M.G.; Painter, G.R.; Plemper, R.K. Quantitative Efficacy Paradigms of the Influenza Clinical Drug Candidate EIDD-2801 in the Ferret Model. *Transl. Res.* **2020**, *218*, 16–28. [CrossRef]
17. Pourkarim, F.; Pourtaghi-Anvarian, S.; Rezaee, H. Molnupiravir: A New Candidate for COVID-19 Treatment. *Pharmacol. Res. Perspect.* **2022**, *10*, e00909. [CrossRef]
18. Khiali, S.; Khani, E.; B Rouy, S.; Entezari-Maleki, T. Comprehensive Review on Molnupiravir in COVID-19: A Novel Promising Antiviral to Combat the Pandemic. *Future Microbiol.* **2022**, *17*, 377–391. [CrossRef]
19. Wahl, A.; Gralinski, L.E.; Johnson, C.E.; Yao, W.; Kovarova, M.; Dinnon, K.H.; Liu, H.; Madden, V.J.; Krzystek, H.M.; De, C.; et al. SARS-CoV-2 Infection is Effectively Treated and Prevented by EIDD-2801. *Nature* **2021**, *591*, 451–457. [CrossRef]
20. Rosenke, K.; Hansen, F.; Schwarz, B.; Feldmann, F.; Haddock, E.; Rosenke, R.; Barbian, K.; Meade-White, K.; Okumura, A.; Leventhal, S.; et al. Orally Delivered MK-4482 Inhibits SARS-CoV-2 Replication in the Syrian Hamster Model. *Nat. Commun.* **2021**, *12*, 2295. [CrossRef]
21. Painter, W.P.; Holman, W.; Bush, J.A.; Almazedi, F.; Malik, H.; Eraut, N.C.J.E.; Morin, M.J.; Szewczyk, L.J.; Painter, G.R. Human Safety, Tolerability, and Pharmacokinetics of Molnupiravir, a Novel Broad-Spectrum Oral Antiviral Agent with Activity against SARS-CoV-2. *Antimicrob. Agents Chemother.* **2021**, *65*, e02428-20. [CrossRef]
22. Khoo, S.H.; Fitzgerald, R.; Fletcher, T.; Ewings, S.; Jaki, T.; Lyon, R.; Downs, N.; Walker, L.; Tansley-Hancock, O.; Greenhalf, W.; et al. Optimal Dose and Safety of Molnupiravir in Patients with Early SARS-CoV-2: A Phase I, Open-Label, Dose-Escalating, Randomized Controlled Study. *J. Antimicrob. Chemother.* **2021**, *76*, 3286–3295. [CrossRef]
23. Lee, C.-C.; Hsieh, C.-C.; Ko, W.-C. Molnupiravir—A Novel Oral Anti-SARS-CoV-2 Agent. *Antibiotics* **2021**, *10*, 1294. [CrossRef]

24. Fischer, W.; Eron, J.J.; Holman, W.; Cohen, M.S.; Fang, L.; Szewczyk, L.J.; Sheahan, T.P.; Baric, R.; Mollan, K.R.; Wolfe, C.R.; et al. Molnupiravir, an Oral Antiviral Treatment for COVID-19. *medRxiv* **2021**. [CrossRef]
25. Mahase, E. COVID-19: Molnupiravir Reduces Risk of Hospital Admission or Death by 50% in Patients at Risk, MSD Reports. *BMJ* **2021**, *375*, n2422. [CrossRef]
26. Merck & Co., Inc. Merck and Ridgeback's Molnupiravir, an Oral COVID-19 Antiviral Medicine, Receives First Authorization in the World. Available online: https://www.merck.com/news/merck-and-ridgebacks-molnupiravir-an-oral-covid-19-antiviral-medicine-receives-first-authorization-in-the-world/ (accessed on 19 December 2022).
27. EMC. Lagevrio 200 Mg Hard Capsules. Available online: https://www.medicines.org.uk/emc/product/13044/smpc#gref (accessed on 20 November 2022).
28. European Medicines Agency. EMA Issues Advice on Use of Lagevrio (Molnupiravir) for the Treatment of COVID-19. Available online: https://www.ema.europa.eu/en/news/ema-issues-advice-use-lagevrio-molnupiravir-treatment-covid-19 (accessed on 20 November 2022).
29. US Food and Drug Administration. Coronavirus (COVID-19) Update: FDA Authorizes Additional Oral Antiviral for Treatment of COVID-19 in Certain Adults. Available online: https://www.fda.gov/news-events/press-announcements/coronavirus-covid-19-update-fda-authorizes-additional-oral-antiviral-treatment-covid-19-certain (accessed on 19 December 2022).
30. Sun Pharma. Sun Pharma Geared to Introduce Molnupiravir, Licensed from MSD and Ridgeback, under the Brand Name Molxvir®in India. Available online: https://sunpharma.com/wp-content/uploads/2021/11/Press-Release-Sun-Pharma-gearing-up-to-introduce-molnupiravir-under-the-brand-name-Molxvir%C2%AE-in-India.pdf (accessed on 19 December 2022).
31. Torrent Pharma. Torrent Pharma to Launch Molnupiravir under the Brand Name Molnutor®in India. Available online: https://torrentpharma.com/pdf/investors/Press_Release_Molnupiravir_Final.pdf (accessed on 19 December 2022).
32. Cipla. Cipla Receives Emergency Use Authorisation (EUA) to Launch Oral Antiviral Drug Cipmolnu®(Molnupiravir 200 Mg) in India for Treatment of Adult Patients with COVID-19, with SpO2 > 93% and Who Have High Risk of Progression of the Disease Including Hospitaliza. Available online: https://www.gov.uk/government/news/first-oral-antiviral-for-covid-19-lagevrio-molnupiravir-approved-by-mhra (accessed on 19 December 2022).
33. Pfizer. Pfizer's Novel COVID-19 Oral Antiviral Treatment Candidate Reduced Risk of Hospitalization or Death by 89% in Interim Analysis of Phase 2/3 EPIC-HR Study. Available online: https://www.pfizer.com/news/press-release/press-release-detail/pfizers-novel-covid-19-oral-antiviral-treatment-candidate (accessed on 3 April 2022).
34. National Institute of Health. Ritonavir-Boosted Nirmatrelvir (Paxlovid). Available online: https://www.covid19treatmentguidelines.nih.gov/therapies/antiviral-therapy/ritonavir-boosted-nirmatrelvir--paxlovid-/ (accessed on 25 March 2022).
35. U.S. National Library of Medicine. Relative Bioavailability Study of PF-07321332/Ritonavir Oral Powder Relative to the Commercial Tablets in Healthy Participants. Available online: https://clinicaltrials.gov/ct2/show/NCT05263921?term=covid+19&cond=nirmatrelvir&draw=2&rank=3 (accessed on 25 March 2022).
36. Hammond, J.; Leister-Tebbe, H.; Gardner, A.; Abreu, P.; Bao, W.; Wisemandle, W.; Baniecki, M.; Hendrick, V.M.; Damle, B.; Simón-Campos, A.; et al. Oral Nirmatrelvir for High-Risk, Nonhospitalized Adults with COVID-19. *N. Engl. J. Med.* **2022**, *386*, 1397–1408. [CrossRef]
37. European Medicines Agency. Paxlovid. Available online: https://www.ema.europa.eu/en/medicines/human/EPAR/paxlovid (accessed on 3 April 2022).
38. Pilla, V. Torrent & Aurobindo Receive Licenses to Sell Pfizer's COVID Pill Paxlovid in India, Low Income Countries. Available online: https://economictimes.indiatimes.com/industry/healthcare/biotech/pharmaceuticals/torrent-aurobindo-receive-licenses-to-sell-pfizers-covid-pill-paxlovid-in-india-low-income-countries/articleshow/90297321.cms (accessed on 3 April 2022).
39. National Center for Biotechnology Information. PubChem Compound Summary for CID 44205240, Baricitinib. Available online: https://pubchem.ncbi.nlm.nih.gov/compound/Baricitinib (accessed on 3 April 2022).
40. Richardson, P.; Griffin, I.; Tucker, C.; Smith, D.; Oechsle, O.; Phelan, A.; Rawling, M.; Savory, E.; Stebbing, J. Baricitinib as Potential Treatment for 2019-nCoV Acute Respiratory Disease. *Lancet* **2020**, *395*, e30–e31. [CrossRef]
41. Stebbing, J.; Krishnan, V.; De Bono, S.; Ottaviani, S.; Casalini, G.; Richardson, P.J.; Monteil, V.; Lauschke, V.M.; Mirazimi, A.; Youhanna, S.; et al. Mechanism of Baricitinib Supports Artificial Intelligence-Predicted Testing in COVID-19 Patients. *EMBO Mol. Med.* **2020**, *12*, e12697. [CrossRef]
42. Marconi, V.C.; Ramanan, A.V.; de Bono, S.; E Kartman, C.; Krishnan, V.; Liao, R.; Piruzeli, M.L.B.; Goldman, J.D.; Alatorre-Alexander, J.; Pellegrini, R.D.C.; et al. Efficacy and Safety of Baricitinib for the Treatment of Hospitalised Adults with COVID-19 (COV-BARRIER): A Randomised, Double-Blind, Parallel-Group, Placebo-Controlled Phase 3 Trial. *Lancet Respir. Med.* **2021**, *9*, 1407–1418. [CrossRef] [PubMed]
43. Eli Lilly. Baricitinib Receives Emergency Use Authorization from the FDA for the Treatment of Hospitalized Patients with COVID-19. Available online: https://investor.lilly.com/news-releases/news-release-details/baricitinib-receives-emergency-use-authorization-fda-treatment (accessed on 19 December 2022).
44. BerGenBio ASA. BerGenBio and Oslo University Hospital Announce the AXL Inhibitor Bemcentinib Will Be Studied in the EU Funded EU-SolidAct Trial in Hospitalised COVID-19 Patients. *Cission PR Newswire*, 27 January 2022. Available online: https://www.prnewswire.com/news-releases/bergenbio-and-oslo-university-hospital-announce-the-axl-inhibitor-

bemcentinib-will-be-studied-in-the-eu-funded-eu-solidact-trial-in-hospitalised-covid-19-patients-301469371.html(accessed on 4 November 2022).
45. Kim, N.; Cho, S.-G. Clinical Applications of Mesenchymal Stem Cells. *Korean J. Intern. Med.* **2013**, *28*, 387–402. [CrossRef] [PubMed]
46. Ding, D.-C.; Shyu, W.-C.; Lin, S.-Z. Mesenchymal Stem Cells. *Cell Transplant.* **2011**, *20*, 5–14. [CrossRef] [PubMed]
47. La Rocca, G. Editorial Connecting the Dots: The Promises of Wharton's Jelly Mesenchymal Stem Cells for Tissue Repair and Regeneration. *Open Tissue Eng. Regen. Med. J.* **2011**, *4*, 3–5. [CrossRef]
48. Carvello, M.; Lightner, A.; Yamamoto, T.; Kotze, P.G.; Spinelli, A. Mesenchymal Stem Cells for Perianal Crohn's Disease. *Cells* **2019**, *8*, 764. [CrossRef]
49. Saleh, M.; Vaezi, A.A.; Aliannejad, R.; Sohrabpour, A.A.; Kiaei, S.Z.F.; Shadnoush, M.; Siavashi, V.; Aghaghazvini, L.; Khoundabi, B.; Abdoli, S.; et al. Cell Therapy in Patients with COVID-19 Using Wharton's Jelly Mesenchymal Stem Cells: A Phase 1 Clinical Trial. *Stem Cell Res. Ther.* **2021**, *12*, 410. [CrossRef]
50. Golchin, A.; Seyedjafari, E.; Ardeshirylajimi, A. Mesenchymal Stem Cell Therapy for COVID-19: Present or Future. *Stem Cell Rev. Rep.* **2020**, *16*, 427–433. [CrossRef]
51. Khanh, V.C.; Fukushige, M.; Chang, Y.H.; Hoang, N.N.; Yamashita, T.; Obata-Yasuoka, M.; Hamada, H.; Osaka, M.; Hiramatsu, Y.; Ohneda, O. Wharton's Jelly Mesenchymal Stem Cell-Derived Extracellular Vesicles Reduce SARS-CoV2-Induced Inflammatory Cytokines under High Glucose and Uremic Toxin Conditions. *Stem Cells Dev.* **2021**, *30*, 758–772. [CrossRef]
52. Paladino, F.V.; Rodrigues, J.D.M.; da Silva, A.; Goldberg, A.C. The Immunomodulatory Potential of Wharton's Jelly Mesenchymal Stem/Stromal Cells. *Stem Cells Int.* **2019**, *2019*, 3548917. [CrossRef]
53. Weiss, A.R.R.; Dahlke, M.H. Immunomodulation by Mesenchymal Stem Cells (MSCs): Mechanisms of Action of Living, Apoptotic, and Dead MSCs. *Front. Immunol.* **2019**, *10*, 1191. [CrossRef]
54. Mebarki, M.; Abadie, C.; Larghero, J.; Cras, A. Human Umbilical Cord-Derived Mesenchymal Stem/Stromal Cells: A Promising Candidate for the Development of Advanced Therapy Medicinal Products. *Stem Cell Res. Ther.* **2021**, *12*, 152. [CrossRef]
55. Monsel, A.; Hauw-Berlemont, C.; Mebarki, M.; Heming, N.; Mayaux, J.; Tchoumba, O.N.; Diehl, J.-L.; Demoule, A.; Annane, D.; Marois, C.; et al. Treatment of COVID-19-Associated ARDS with Mesenchymal Stromal Cells: A Multicenter Randomized Double-Blind Trial. *Crit. Care* **2022**, *26*, 48. [CrossRef]
56. US National Library of Medicine. Efficacy of Infusions of MSC From Wharton Jelly in the SARS-CoV-2 (COVID-19) Related Acute Respiratory Distress Syndrome (MSC-COVID19). Available online: https://clinicaltrials.gov/ct2/show/NCT04625738?term=Wharton%27s+Jelly+Mesenchymal+stem+cells&cond=COVID-19&draw=2&rank=2 (accessed on 19 December 2022).
57. Clinical Center. Convalescent Plasma. Available online: https://clinicalcenter.nih.gov/blooddonor/donationtypes/convalescent_plasma.html (accessed on 22 January 2023).
58. Rojas, M.; Rodríguez, Y.; Monsalve, D.M.; Acosta-Ampudia, Y.; Camacho, B.; Gallo, J.E.; Rojas-Villarraga, A.; Ramírez-Santana, C.; Díaz-Coronado, J.C.; Manrique, R.; et al. Convalescent Plasma in COVID-19: Possible Mechanisms of Action. *Autoimmun. Rev.* **2020**, *19*, 102554. [CrossRef]
59. Franchini, M.; Glingani, C.; Liumbruno, G.M. Potential Mechanisms of Action of Convalescent Plasma in COVID-19. *Diagnosis* **2021**, *8*, 413–420. [CrossRef]
60. Roback, J.D.; Guarner, J. Convalescent Plasma to Treat COVID-19 Possibilities and Challenges. *JAMA* **2020**, *323*, 1561–1562. [CrossRef]
61. U.S. National Library of Medicine. Hyperimmune Plasma for Critical Patients with COVID-19 (COV19-PLASMA). Available online: https://clinicaltrials.gov/ct2/show/NCT04321421 (accessed on 19 December 2022).
62. Perotti, C.; Del Fante, C.; Baldanti, F.; Franchini, M.; Percivalle, E.; Nepita, E.V.; Seminari, E.; De Silvestri, A.; Bruno, R.; Klersy, C. Plasma from Donors Recovered from the New Coronavirus 2019 as Therapy for Critical Patients with COVID-19 (COVID-19 plasma study): A Multicentre Study Protocol. *Intern. Emerg. Med.* **2020**, *15*, 819–824. [CrossRef]
63. Piechotta, V.; Chai, K.L.; Valk, S.J.; Doree, C.; Monsef, I.; Wood, E.M.; Lamikanra, A.; Kimber, C.; McQuilten, Z.; So-Osman, C.; et al. Convalescent Plasma or Hyperimmune Immunoglobulin for People with COVID-19: A Living Systematic Review. *Cochrane Database Syst. Rev.* **2020**, *7*, CD013600. [CrossRef]
64. U.S. National Library of Medicine. Expanded Access to Convalescent Plasma for the Treatment of Patients with COVID-19. Available online: https://clinicaltrials.gov/ct2/show/NCT04338360 (accessed on 19 December 2022).
65. Senefeld, J.W.; Johnson, P.W.; Kunze, K.L.; Bloch, E.M.; van Helmond, N.; Golafshar, M.A.; Klassen, S.A.; Klompas, A.M.; Sexton, M.A.; Soto, J.C.D.; et al. Access to and Safety of COVID-19 Convalescent Plasma in the United States Expanded Access Program: A National Registry Study. *PLoS Med.* **2021**, *18*, e1003872. [CrossRef]
66. U.S. Food & Drug Administration. Recommendations for Investigational COVID-19 Convalescent Plasma. Available online: https://www.fda.gov/vaccines-blood-biologics/investigational-new-drug-applications-inds-cber-regulated-products/recommendations-investigational-covid-19-convalescent-plasma (accessed on 19 December 2022).
67. Drugbank. Sarilumab: Uses, Interactions, Mechanism of Action. Available online: https://go.drugbank.com/drugs/DB11767 (accessed on 16 July 2022).
68. Cooper, S. Sarilumab for the Treatment of Rheumatoid Arthritis. *Immunotherapy* **2016**, *8*, 249–250. [CrossRef]
69. Lamb, Y.N.; Deeks, E.D. Sarilumab: A Review in Moderate to Severe Rheumatoid Arthritis. *Drugs* **2018**, *78*, 929–940. [CrossRef]
70. Scott, L.J. Sarilumab: First Global Approval. *Drugs* **2017**, *77*, 705–712. [CrossRef]

71. June, R.R.; Olsen, N.J. Room for More IL-6 Blockade? Sarilumab for the Treatment of Rheumatoid Arthritis. *Expert Opin. Biol. Ther.* **2016**, *16*, 1303–1309. [CrossRef] [PubMed]
72. U.S. National Library of Medicine. Evaluation of the Efficacy and Safety of Sarilumab in Hospitalized Patients with COVID-19. Available online: https://www.clinicaltrials.gov/ct2/show/NCT04315298 (accessed on 17 December 2022).
73. Regeneron Pharmaceuticals Inc. Regeneron and Sanofi Announce FDA Approval of Kevzara®(Sarilumab) for the Treatment of Moderately to Severely Active Rheumatoid Arthritis in Adult Patients. Available online: https://investor.regeneron.com/news-releases/news-release-details/regeneron-and-sanofi-announce-fda-approval-kevzarar-sarilumab (accessed on 16 July 2022).
74. Regeneron Pharmaceuticals Inc Regeneron and Sanofi Begin Global Kevzara®(Sarilumab) Clinical Trial Program in Patients with Severe COVID-19. Available online: https://newsroom.regeneron.com/news-releases/news-release-details/regeneron-and-sanofi-begin-global-kevzarar-sarilumab-clinical (accessed on 16 July 2022).
75. Sivapalasingam, S.; Lederer, D.J.; Bhore, R.; Hajizadeh, N.; Criner, G.; Hosain, R.; Mahmood, A.; Giannelou, A.; Somersan-Karakaya, S.; O'Brien, M.P.; et al. Efficacy and Safety of Sarilumab in Hospitalized Patients with Coronavirus Disease 2019: A Randomized Clinical Trial. *Clin. Infect. Dis.* **2022**, *75*, e380–e388. [CrossRef] [PubMed]
76. U.S. National Library of Medicine. Sarilumab COVID-19. Available online: https://clinicaltrials.gov/ct2/show/NCT04327388 (accessed on 14 December 2022).
77. Lescure, F.-X.; Honda, H.; Fowler, R.A.; Lazar, J.S.; Shi, G.; Wung, P.; Patel, N.; Hagino, O.; Bazzalo, I.J.; Casas, M.M.; et al. Sarilumab in Patients Admitted to Hospital with Severe or Critical COVID-19: A Randomised, Double-Blind, Placebo-Controlled, Phase 3 Trial. *Lancet Respir. Med.* **2021**, *9*, 522–532. [CrossRef] [PubMed]
78. Drugbank. Tocilizumab: Uses, Interactions, Mechanism of Action. Available online: https://go.drugbank.com/drugs/DB06273 (accessed on 16 July 2022).
79. Sebba, A. Tocilizumab: The First Interleukin-6-Receptor Inhibitor. *Am. J. Health Syst. Pharm.* **2008**, *65*, 1413–1418. [CrossRef]
80. Drugs.com. Tocilizumab. Available online: https://www.drugs.com/ingredient/tocilizumab.html (accessed on 17 July 2022).
81. ACTEMRA®. Starting SC Injections. Available online: https://www.actemra.com/ra/treatment/sc-injections.html (accessed on 22 January 2023).
82. Parsons, L. FDA Approves Roche's Actemra COVID-19 Trial. *PM Live*, 24 March 2020. Available online: https://www.pmlive.com/pharma_news/fda_approves_roches_actemra_covid-19_trial_1329887(accessed on 17 July 2022).
83. U.S. National Library of Medicine. A Study to Evaluate the Safety and Efficacy of Tocilizumab in Patients with Severe COVID-19 Pneumonia (COVACTA). Available online: https://clinicaltrials.gov/ct2/show/NCT04320615 (accessed on 19 December 2022).
84. Rosas, I.O.; Bräu, N.; Waters, M.; Go, R.C.; Malhotra, A.; Hunter, B.D.; Bhagani, S.; Skiest, D.; Savic, S.; Douglas, I.S.; et al. Tocilizumab in Patients Hospitalised with COVID-19 Pneumonia: Efficacy, Safety, Viral Clearance, and Antibody Response from a Randomised Controlled Trial (COVACTA). *Eclinicalmedicine* **2022**, *47*, 101409. [CrossRef]
85. United States Food & Drug Administration. Frequently Asked Questions on the Emergency Use Authorization for Actemra (Tocilizumab) for Treatment of COVID-19. Available online: https://www.fda.gov/media/150345/download (accessed on 22 January 2023).
86. Drugs.com. Avastin FDA Approval History. Available online: https://www.drugs.com/history/avastin.html (accessed on 19 December 2022).
87. Kazazi-Hyseni, F.; Beijnen, J.H.; Schellens, J.H.M. Bevacizumab. *Oncologist* **2010**, *15*, 819–825. [CrossRef]
88. Drugbank. Bevacizumab: Uses, Interactions, Mechanism of Action. Available online: https://go.drugbank.com/drugs/DB00112 (accessed on 17 July 2022).
89. Pang, J.; Xu, F.; Aondio, G.; Li, Y.; Fumagalli, A.; Lu, M.; Valmadre, G.; Wei, J.; Bian, Y.; Canesi, M.; et al. Efficacy and Tolerability of Bevacizumab in Patients with Severe COVID-19. *Nat. Commun.* **2021**, *12*, 814. [CrossRef]
90. U.S. National Library of Medicine. Bevacizumab in Severe or Critical Patients with COVID-19 Pneumonia (BEST-CP). Available online: https://clinicaltrials.gov/ct2/show/NCT04275414 (accessed on 19 December 2022).
91. Haddad, F.; Dokmak, G.; Karaman, R. A Comprehensive Review on the Efficacy of Several Pharmacologic Agents for the Treatment of COVID-19. *Life* **2022**, *12*, 1758. [CrossRef]
92. Zou, R.; Peng, L.; Shu, D.; Zhao, L.; Lan, J.; Tan, G.; Peng, J.; Yang, X.; Liu, M.; Zhang, C.; et al. Antiviral Efficacy and Safety of Molnupiravir against Omicron Variant Infection: A Randomized Controlled Clinical Trial. *Front. Pharmacol.* **2022**, *13*, 939573. [CrossRef]
93. EMC. Paxlovid 150 Mg/100 Mg Film-Coated Tablets. Available online: https://www.medicines.org.uk/emc/product/13145/smpc (accessed on 14 January 2023).
94. Drugbank. Dexamethasone. Available online: https://go.drugbank.com/drugs/DB01234 (accessed on 14 December 2022).
95. Rhen, T.; Cidlowski, J.A. Antiinflammatory Action of Glucocorticoids—New Mechanisms for Old Drugs. *N. Engl. J. Med.* **2005**, *353*, 1711–1723. [CrossRef]
96. U.S. National Library of Medicine. COVID-19-Associated ARDS Treated with Dexamethasone: Alliance COVID-19 Brasil III (CoDEX). Available online: https://clinicaltrials.gov/ct2/show/NCT04327401 (accessed on 19 December 2022).
97. U.S. National Library of Medicine. Efficacy of Dexamethasone Treatment for Patients with ARDS Caused by COVID-19. Available online: https://clinicaltrials.gov/ct2/show/NCT04325061 (accessed on 14 December 2022).
98. World Health Organization. Coronavirus Disease (COVID-19): Dexamethasone. Available online: https://www.who.int/news-room/questions-and-answers/item/coronavirus-disease-Covid-19-dexamethasone (accessed on 23 January 2023).

99. Wongrakpanich, S.; Wongrakpanich, A.; Melhado, K.; Rangaswami, J. A Comprehensive Review of Non-Steroidal Anti-Inflammatory Drug Use in the Elderly. *Aging Dis.* **2018**, *9*, 143–150. [CrossRef]
100. Drugbank. Naproxen: Uses, Interactions, Mechanism of Action. Available online: https://go.drugbank.com/drugs/DB00788 (accessed on 14 July 2022).
101. Giménez, M.; Pujol, J.; Ali, Z.; López-Solà, M.; Contreras-Rodríguez, O.; Deus, J.; Ortiz, H.; Soriano-Mas, C.; Llorente-Onaindia, J.; Monfort, J. Naproxen Effects on Brain Response to Painful Pressure Stimulation in Patients with Knee Osteoarthritis: A Double-Blind, Randomized, Placebo-Controlled, Single-Dose Study. *J. Rheumatol.* **2014**, *41*, 2240–2248. [CrossRef]
102. Terrier, O.; Dilly, S.; Pizzorno, A.; Chalupska, D.; Humpolickova, J.; Bouřa, E.; Berenbaum, F.; Quideau, S.; Lina, B.; Fève, B.; et al. Antiviral Properties of the NSAID Drug Naproxen Targeting the Nucleoprotein of SARS-CoV-2 Coronavirus. *Molecules* **2021**, *26*, 2593. [CrossRef]
103. U.S. National Library of Medicine. Efficacy of Addition of Naproxen in the Treatment of Critically Ill Patients Hospitalized for COVID-19 Infection (ENACOVID). Available online: https://clinicaltrials.gov/ct2/show/NCT04325633 (accessed on 15 December 2022).
104. Drugbank. Remdesivir: Uses, Interactions, Mechanism of Action. Available online: https://go.drugbank.com/drugs/DB14761 (accessed on 15 July 2022).
105. Amirian, E.S.; Levy, J.K. Current Knowledge about the Antivirals Remdesivir (GS-5734) and GS-441524 as Therapeutic Options for Coronaviruses. *One Health* **2020**, *9*, 100128. [CrossRef]
106. Gordon, C.J.; Tchesnokov, E.P.; Woolner, E.; Perry, J.K.; Feng, J.Y.; Porter, D.P.; Götte, M. Remdesivir is a Direct-Acting Antiviral That Inhibits RNA-Dependent RNA Polymerase from Severe Acute Respiratory Syndrome Coronavirus 2 with High Potency. *J. Biol. Chem.* **2020**, *295*, 6785–6797. [CrossRef]
107. Grein, J.; Ohmagari, N.; Shin, D.; Diaz, G.; Asperges, E.; Castagna, A.; Feldt, T.; Green, G.; Green, M.L.; Lescure, F.X.; et al. Compassionate Use of Remdesivir for Patients with Severe COVID-19. *N. Engl. J. Med.* **2020**, *382*, 2327–2336. [CrossRef]
108. Frediansyah, A.; Nainu, F.; Dhama, K.; Mudatsir, M.; Harapan, H. Remdesivir and Its Antiviral Activity against COVID-19: A Systematic Review. *Clin. Epidemiol. Glob. Health* **2021**, *9*, 123–127. [CrossRef]
109. U.S. National Library of Medicine. Study to Evaluate the Safety and Antiviral Activity of Remdesivir (GS-5734TM) in Participants with Severe Coronavirus Disease (COVID-19). Available online: https://clinicaltrials.gov/ct2/show/NCT04292899 (accessed on 19 December 2022).
110. A Olender, S.; Perez, K.K.; Go, A.S.; Balani, B.; Price-Haywood, E.G.; Shah, N.S.; Wang, S.; Walunas, T.L.; Swaminathan, S.; Slim, J.; et al. Remdesivir for Severe Coronavirus Disease 2019 (COVID-19) Versus a Cohort Receiving Standard of Care. *Clin. Infect. Dis.* **2021**, *73*, e4166–e4174. [CrossRef]
111. U.S. National Library of Medicine. Adaptive COVID-19 Treatment Trial (ACTT). Available online: https://clinicaltrials.gov/ct2/show/NCT04280705 (accessed on 4 December 2022).
112. Beigel, J.H.; Tomashek, K.M.; Dodd, L.E.; Mehta, A.K.; Zingman, B.S.; Kalil, A.C.; Hohmann, E.; Chu, H.Y.; Luetkemeyer, A.; Kline, S.; et al. Remdesivir for the Treatment of COVID-19—Final Report. *N. Engl. J. Med.* **2020**, *383*, 1813–1826. [CrossRef]
113. United States Food & Drug Administration. FDA Approves First Treatment for COVID-19. Available online: https://www.fda.gov/news-events/press-announcements/fda-approves-first-treatment-covid-19 (accessed on 23 January 2023).
114. Gilead Sciences. FDA Approves Veklury®(Remdesivir) for the Treatment of Non-Hospitalized Patients at High Risk for COVID-19 Disease Progression. Available online: https://www.gilead.com/news-and-press/press-room/press-releases/2022/1/fda-approves-veklury-remdesivir-for-the-treatment-of-nonhospitalized-patients-at-high-risk-for-covid19-disease-progression (accessed on 23 January 2023).
115. European Medicines Agency. Meeting Highlights from the Committee for Medicinal Products for Human Use (CHMP) 13–16 December 2021. Available online: https://www.ema.europa.eu/en/news/meeting-highlights-committee-medicinal-products-human-use-chmp-13-16-december-2021 (accessed on 23 January 2023).
116. Drugbank. Hydroxychloroquine: Uses, Interactions, Mechanism of Action. Available online: https://go.drugbank.com/drugs/DB01611 (accessed on 15 July 2022).
117. Singh, A.K.; Singh, A.; Singh, R.; Misra, A. Hydroxychloroquine in Patients with COVID-19: A Systematic Review and Meta-Analysis. *Diabetes Metab. Syndr. Clin. Res. Rev.* **2020**, *14*, 589–596. [CrossRef] [PubMed]
118. Pavlova, V.; Hristova, S.; Uzunova, K.; Vekov, T. A Review on the Mechanism of Action of Favipiravir and Hydroxychloroquine in COVID-19. *Res. Rev. Insights* **2021**, *5*, 1–7. [CrossRef]
119. Satarker, S.; Ahuja, T.; Banerjee, M.; E, V.B.; Dogra, S.; Agarwal, T.; Nampoothiri, M. Hydroxychloroquine in COVID-19: Potential Mechanism of Action against SARS-CoV-2. *Curr. Pharmacol. Rep.* **2020**, *6*, 203–211. [CrossRef] [PubMed]
120. U.S. National Library of Medicine. Outcomes Related to COVID-19 Treated with Hydroxychloroquine among In-Patients with Symptomatic Disease (ORCHID). Available online: https://clinicaltrials.gov/ct2/show/NCT04332991 (accessed on 4 August 2022).
121. Casey, J.D.; Johnson, N.J.; Semler, M.W.; Collins, S.P.; Aggarwal, N.R.; Brower, R.G.; Chang, S.Y.; Eppensteiner, J.; Filbin, M.; Gibbs, K.W.; et al. Rationale and Design of ORCHID: A Randomized Placebo-Controlled Clinical Trial of Hydroxychloroquine for Adults Hospitalized with COVID-19. *Ann. Am. Thorac. Soc.* **2020**, *17*, 1144–1153. [CrossRef] [PubMed]
122. Lupkin, S. Malaria Drugs Get FDA "Emergency Use Authorization" For COVID-19. *NPR*. 30 March 2020. Available online: https://www.npr.org/sections/coronavirus-live-updates/2020/03/30/823987540/fda-oks-addition-to-stockpile-of-malaria-drugs-for-covid-19 (accessed on 15 July 2022).

123. Self, W.H.; Semler, M.W.; Leither, L.M.; Casey, J.D.; Angus, D.C.; Brower, R.G.; Chang, S.Y.; Collins, S.P.; Eppensteiner, J.C.; Filbin, M.R.; et al. Effect of Hydroxychloroquine on Clinical Status at 14 Days in Hospitalized Patients with COVID-19: A Randomized Clinical Trial. *JAMA* **2020**, *324*, 2165–2176. [CrossRef]
124. Shippey, E.A.; Wagler, V.D.; Collamer, A.N. Hydroxychloroquine: An Old Drug with New Relevance. *Clevel. Clin. J. Med.* **2018**, *85*, 459–467. [CrossRef]
125. Devaux, C.A.; Rolain, J.-M.; Colson, P.; Raoult, D. New Insights on the Antiviral Effects of Chloroquine against Coronavirus: What to Expect for COVID-19? *Int. J. Antimicrob. Agents* **2020**, *55*, 105938. [CrossRef]
126. Sandler, R. FDA Authorizes Anti-Malarial Drugs Chloroquine and Hydroxychloroquine for Emergency Coronavirus Treatment. Forbes, 30 March 2020. Available online: https://www.forbes.com/sites/rachelsandler/2020/03/30/fda-approves-anti-malarial-drugs-chloroquine-and-hydroxychloroquine-for-emergency-coronavirus-treatment/?sh=2ec6be865e5d (accessed on 15 July 2022).
127. Mundell, E.J. FDA Approves Malaria Drugs to Treat COVID-19, Despite Little Proof They Work. Clinical Collection, 31 March 2020. Available online: https://www.clinicalconnection.com/health-news/news-article/51148/fda-approves-malaria-drugs-to-treat-covid-19-despite-little-proof-they-work (accessed on 19 December 2022).
128. European Medicines Agency. COVID-19: Chloroquine and Hydroxychloroquine Only to Be Used in Clinical Trials or Emergency Use Programmes. Available online: https://www.ema.europa.eu/en/news/covid-19-chloroquine-hydroxychloroquine-only-be-used-clinical-trials-emergency-use-programmes (accessed on 15 July 2022).
129. Peng, S.; Wang, H.; Wang, Z.; Wang, Q. Progression of Antiviral Agents Targeting Viral Polymerases. *Molecules* **2022**, *27*, 7370. [CrossRef]
130. Drugbank. Favipiravir. Available online: https://go.drugbank.com/drugs/DB12466 (accessed on 16 July 2022).
131. Furuta, Y.; Komeno, T.; Nakamura, T. Favipiravir (T-705), a Broad Spectrum Inhibitor of Viral RNA Polymerase. *Proc. Jpn. Acad. Ser. B Phys. Biol. Sci.* **2017**, *93*, 449–463. [CrossRef]
132. U.S. National Library of Medicine. Oral Favipiravir Compared to Standard Supportive Care in Subjects with Mild. Available online: https://www.clinicaltrials.gov/ct2/show/NCT04346628 (accessed on 4 August 2022).
133. Holubar, M.; Subramanian, A.; Purington, N.; Hedlin, H.; Bunning, B.; Walter, K.S.; Bonilla, H.; Boumis, A.; Chen, M.; Clinton, K.; et al. Favipiravir for Treatment of Outpatients with Asymptomatic or Uncomplicated Coronavirus Disease 2019: A Double-Blind, Randomized, Placebo-Controlled, Phase 2 Trial. *Clin. Infect. Dis.* **2022**, *75*, 1883–1892. [CrossRef]
134. U.S. National Library of Medicine. Efficacy and Safety of Favipiravir in Management of COVID-19 (FAV-001). Available online: https://clinicaltrials.gov/ct2/show/NCT04349241 (accessed on 14 December 2022).
135. Hassanipour, S.; Arab-Zozani, M.; Amani, B.; Heidarzad, F.; Fathalipour, M.; Martinez-De-Hoyo, R. The Efficacy and Safety of Favipiravir in Treatment of COVID-19: A Systematic Review and Meta-Analysis of Clinical Trials. *Sci. Rep.* **2021**, *11*, 11022. [CrossRef]
136. Eunethta. Favipiravir for the Treatment of COVID-19. Available online: https://eunethta.eu/doi (accessed on 21 December 2022).
137. Agrawal, U.; Raju, R.; Udwadia, Z.F. Favipiravir: A New and Emerging Antiviral Option in COVID-19. *Med. J. Armed Forces India* **2020**, *76*, 370–376. [CrossRef]
138. Blaising, J.; Polyak, S.J.; Pécheur, E.-I. Arbidol as a Broad-Spectrum Antiviral: An Update. *Antivir. Res.* **2014**, *107*, 84–94. [CrossRef]
139. Huang, L.; Zhang, L.; Liu, Y.; Luo, R.; Zeng, L.; Telegina, I.; Vlassov, V. Arbidol for Preventing and Treating Influenza in Adults and Children. *Cochrane Database Syst. Rev.* **2017**, *2017*, CD011489. [CrossRef]
140. Drugbank. Umifenovir. Available online: https://go.drugbank.com/drugs/DB13609 (accessed on 16 July 2022).
141. U.S. National Library of Medicine. The Efficacy of Lopinavir plus Ritonavir and Arbidol against Novel Coronavirus Infection ELACOI). Available online: https://clinicaltrials.gov/ct2/show/NCT04252885 (accessed on 14 December 2022).
142. Li, Y.; Xie, Z.; Lin, W.; Cai, W.; Wen, C.; Guan, Y.; Mo, X.; Wang, J.; Wang, Y.; Peng, P.; et al. Efficacy and Safety of Lopinavir/Ritonavir or Arbidol in Adult Patients with Mild/Moderate COVID-19: An Exploratory Randomized Controlled Trial. *Med* **2020**, *1*, 105–113.e4. [CrossRef]
143. Yu, B.; Chang, J. The first Chinese Oral Anti-COVID-19 Drug Azvudine Launched. *Innovation* **2022**, *3*, 100321. [CrossRef]
144. Zhang, J.-L.; Li, Y.-H.; Wang, L.-L.; Liu, H.-Q.; Lu, S.-Y.; Liu, Y.; Li, K.; Liu, B.; Li, S.-Y.; Shao, F.-M.; et al. Azvudine is a Thymus-Homing Anti-SARS-CoV-2 Drug Effective in Treating COVID-19 Patients. *Signal Transduct. Target. Ther.* **2021**, *6*, 414. [CrossRef]
145. Granlen News. HIV RT Inhibitor Azvudine Received IND Approval. Available online: http://www.granlen.com/contents/94/9.html (accessed on 14 January 2023).
146. Global Times. China's First Domestic Oral Anti-COVID-19 Drug Is Priced at Less than 300 Yuan per Bottle/35 Tablets. Available online: https://www.globaltimes.cn/page/202208/1272362.shtml (accessed on 14 January 2023).
147. Ren, Z.; Luo, H.; Yu, Z.; Song, J.; Liang, L.; Wang, L.; Wang, H.; Cui, G.; Liu, Y.; Wang, J.; et al. A Randomized, Open-Label, Controlled Clinical Trial of Azvudine Tablets in the Treatment of Mild and Common COVID-19, a Pilot Study. *Adv. Sci.* **2020**, *7*, e2001435. [CrossRef]
148. National Center for Biotechnology Information. Azvudine | C9H11FN6O4—PubChem. Available online: https://pubchem.ncbi.nlm.nih.gov/compound/Azvudine (accessed on 13 January 2023).
149. Yu, B.; Chang, J. Azvudine (FNC): A Promising Clinical Candidate for COVID-19 Treatment. *Signal Transduct. Target. Ther.* **2020**, *5*, 236. [CrossRef]

150. Noreen, S.; Maqbool, I.; Madni, A. Dexamethasone: Therapeutic Potential, Risks, and Future Projection during COVID-19 Pandemic. *Eur. J. Pharmacol.* **2021**, *894*, 173854. [CrossRef]
151. Kivrak, A.; Ulaş, B.; Kivrak, H. A Comparative Analysis for Anti-Viral Drugs: Their Efficiency against SARS-CoV-2. *Int. Immunopharmacol.* **2021**, *90*, 107232. [CrossRef] [PubMed]
152. The Recovery Collaborative Group; Horby, P.; Lim, W.S.; Emberson, J.R.; Mafham, M.; Bell, J.L.; Linsell, L.; Staplin, N.; Brightling, C.; Ustianowski, A.; et al. Dexamethasone in Hospitalized Patients with COVID-19. *N. Engl. J. Med.* **2021**, *384*, 693–704. [CrossRef] [PubMed]
153. Kocks, J.; Kerkhof, M.; Scherpenisse, J.; van de Maat, A.; van Geer-Postmus, I.; le Rütte, T.; Schaart, J.; Gans, R.O.B.; Kerstjens, H.A. A Potential Harmful Effect of Dexamethasone in Non-Severe COVID-19: Results from the COPPER-Pilot Study. *ERJ Open Res.* **2022**, *8*, 00129–2022. [CrossRef] [PubMed]
154. Asadi, M.; Sayar, S.; Radmanesh, E.; Naghshi, S.; Mousaviasl, S.; Jelvay, S.; Ebrahimzadeh, M.; Mohammadi, A.; Abbasi, S.; Mobarak, S.; et al. Efficacy of Naproxen in the Management of Patients Hospitalized with COVID-19 Infection: A Randomized, Double-Blind, Placebo-Controlled, Clinical Trial. *Diabetes Metab. Syndr. Clin. Res. Rev.* **2021**, *15*, 102319. [CrossRef]
155. Nabati, M.; Parsaee, H. Potential Cardiotoxic Effects of Remdesivir on Cardiovascular System: A Literature Review. *Cardiovasc. Toxicol.* **2022**, *22*, 268–272. [CrossRef]
156. Choi, S.W.; Shin, J.S.; Park, S.-J.; Jung, E.; Park, Y.-G.; Lee, J.; Kim, S.J.; Park, H.-J.; Lee, J.-H.; Park, S.-M.; et al. Antiviral Activity and Safety of Remdesivir against SARS-CoV-2 Infection in Human Pluripotent Stem Cell-Derived Cardiomyocytes. *Antivir. Res.* **2020**, *184*, 104955. [CrossRef]
157. Kaur, R.J.; Charan, J.; Dutta, S.; Sharma, P.; Bhardwaj, P.; Sharma, P.; Lugova, H.; Krishnapillai, A.; Islam, S.; Haque, M.; et al. Favipiravir Use in COVID-19: Analysis of Suspected Adverse Drug Events Reported in the WHO Database. *Infect. Drug Resist.* **2020**, *13*, 4427–4438. [CrossRef]
158. Takahashi, H.; Iwasaki, Y.; Watanabe, T.; Ichinose, N.; Okada, Y.; Oiwa, A.; Kobayashi, T.; Moriya, M.; Oda, T. Case Studies of SARS-CoV-2 Treated with Favipiravir among Patients in Critical or Severe Condition. *Int. J. Infect. Dis.* **2020**, *100*, 283–285. [CrossRef]
159. Yamamura, H.; Matsuura, H.; Nakagawa, J.; Fukuoka, H.; Domi, H.; Chujoh, S. Effect of Favipiravir and an Anti-Inflammatory Strategy for COVID-19. *Crit. Care* **2020**, *24*, 413. [CrossRef]
160. Ergür, F.Ö.; Yıldız, M.; Şener, M.U.; Kavurgacı, S.; Ozturk, A. Adverse Effects Associated with Favipiravir in Patients with COVID-19 Pneumonia: A Retrospective Study. *Sao Paulo Med. J.* **2022**, *140*, 372–377. [CrossRef]
161. Alamer, A.; Alrashed, A.A.; Alfaifi, M.; Alosaimi, B.; AlHassar, F.; Almutairi, M.M.; Howaidi, J.; Almutairi, W.; Mohzari, Y.; Sulaiman, T.; et al. Effectiveness and Safety of Favipiravir Compared to Supportive Care in Moderately to Critically Ill COVID-19 Patients: A Retrospective Study with Propensity Score Matching Sensitivity Analysis. *Curr. Med. Res. Opin.* **2021**, *37*, 1085–1097. [CrossRef]
162. Mukhopadhyay, S.; Paul, S.; Samanta, B. Hydroxychloroquine in the Prophylaxis of COVID 19: A Survey of Safety on the Healthcare Workers in India. *Perspect. Clin. Res.* **2021**, *12*, 58–59. [CrossRef]
163. Cortez, M.; Che, C. Malaria Drug Hydroxychloroquine No Better than Regular Coronavirus Care, Chinese Study Says. The Print, 25 March 2020. Available online: https://theprint.in/health/malaria-drug-hydroxychloroquine-no-better-than-regular-coronavirus-care-chinese-study-says/387587/ (accessed on 19 December 2022).
164. Darazam, I.A.; Shokouhi, S.; Mardani, M.; Pourhoseingholi, M.A.; Rabiei, M.M.; Hatami, F.; Shabani, M.; Moradi, O.; Gharehbagh, F.J.; Irvani, S.S.N.; et al. Umifenovir in Hospitalized Moderate to Severe COVID-19 Patients: A Randomized Clinical Trial. *Int. Immunopharmacol.* **2021**, *99*, 107969. [CrossRef]
165. Wang, Z.; Yang, B.; Li, Q.; Wen, L.; Zhang, R. Clinical Features of 69 Cases with Coronavirus Disease 2019 in Wuhan, China. *Clin. Infect. Dis.* **2020**, *71*, 769–777. [CrossRef]
166. Huang, D.; Yu, H.; Wang, T.; Yang, H.; Yao, R.; Liang, Z. Efficacy and Safety of Umifenovir for Coronavirus Disease 2019 (COVID-19): A Systematic Review and Meta-Analysis. *J. Med Virol.* **2021**, *93*, 481–490. [CrossRef]
167. United States Food & Drug Administration. Coronavirus (COVID-19) Update: FDA Authorizes Monoclonal Antibody for Treatment of COVID-19. Available online: https://www.fda.gov/news-events/press-announcements/coronavirus-covid-19-update-fda-authorizes-monoclonal-antibodies-treatment-covid-19 (accessed on 30 September 2022).
168. European Medicines Agency. EMA Receives Application for Marketing Authorisation for Ronapreve (Casirivimab/Imdevimab) for Treatment and Prevention of COVID-19. Available online: https://www.ema.europa.eu/en/news/ema-receives-application-marketing-authorisation-ronapreve-casirivimab-imdevimab-treatment (accessed on 24 January 2023).
169. European Medicines Agency. EMA Issues Advice on Use of REGN-COV2 Antibody Combination (Casirivimab/Imdevimab). Available online: https://www.ema.europa.eu/en/news/ema-issues-advice-use-regn-cov2-antibody-combination-casirivimab-imdevimab (accessed on 24 January 2023).
170. Keam, S.J. Tixagevimab + Cilgavimab: First Approval. *Drugs* **2022**, *82*, 1001–1010. [CrossRef]

MDPI AG
Grosspeteranlage 5
4052 Basel
Switzerland
Tel.: +41 61 683 77 34

Vaccines Editorial Office
E-mail: vaccines@mdpi.com
www.mdpi.com/journal/vaccines

www.ingramcontent.com/pod-product-compliance
Lightning Source LLC
Chambersburg PA
CBHW061547210926
51850CB00009B/2025
* 9 7 8 3 7 2 5 8 6 5 3 4 5 *